# SCIENTIFIC REASONING

# SCIENTIFIC REASONING: THE BAYESIAN APPROACH

**COLIN HOWSON**

**AND**

**PETER URBACH**

*Second Edition*

. . . if this [probability] calculus be condemned, then the whole of the sciences must also be condemned.
—Henri Poincaré

Our assent ought to be regulated by the grounds of probability.
—John Locke

Open Court
Chicago and La Salle, Illinois

**SECOND EDITION**

OPEN COURT and the above logo are registered in the U.S. Patent & Trademark Office.



First printing 1993



Printed and bound in the United States of America.

**Library of Congress Cataloging-in-Publication Data**

Howson, Colin.
Scientific reasoning : the Bayesian approach / Colin Howson and Peter Urbach. — 2nd ed.
p. cm.
Includes bibliographical references and index.
ISBN 0-8126-9234-9. — ISBN 0-8126-9235-7 (pbk.)
1. Science—Philosophy. 2. Reasoning. 3. Bayesian statistical decision theory. I. Urbach, Peter. II. Title.
Q175.H87 1993
501—dc20 93-33935
CIP

# CONTENTS

# PREFACE TO THE SECOND EDITION

The question of how hypotheses should be evaluated is an ancient one. It has been handled in many quite different ways and has always excited lively controversy, not surprisingly, in view of the important practical consequences of adopting one approach rather than another. Our approach, which we set out in this book, is the Bayesian one. It is the idea, accepted by an increasing number of philosophers and scientists, that scientific reasoning is essentially reasoning in accordance with the formal principles of probability.

This edition of our book is much amended and reorganized, corrections have been made, and a good deal of extra material added. We have, in short, attempted to create a comprehensive text on the foundations of scientific reasoning, one, above all, that will be suitable for students. A new chapter (Chapter 6) on Bayesian conditionalization includes discussions of Jeffrey Conditionalization, dynamic Dutch Books, the so-called Reflection Principle, and the Principle of Minimum Information. The methodology of clinical trials receives a more extensive critical treatment in Chapters 11 and 14, as does the theory of sample surveys. The foundations of the standard or 'classical' theory of regression analysis is scrutinized in a new chapter, Chapter 12. And the final chapter defending the Bayesian approach against its critics is expanded to include discussions of the Dempster-Shafer theory of support, calibration, Formal Learning Theory, the so-called Paradox of Ideal Evidence, Popper's putative proofs for the zero probability of universal laws, and the Maximum Entropy Principle. Finally, most chapters have been provided with practice exercises for students.

The text is intended to be self-contained, calling, in the main, on only elementary mathematical and statistical ideas. Nevertheless, some parts are more complex, and some more essential to the overall argument, than others. Accordingly, we would like to suggest that readers who are not already familiar with mathematical probability but who wish to gain an initial understanding of the Bayesian approach and to appreciate its power, adopt the following plan of attack. First, read Chapter 1, which sets the scene, as it were, with a brief historical overview of various approaches to scientific inference. Then look at sections **a, b,** and **c.1** of Chapter 2, which give the simple principles or axioms of the probability calculus, and section **e,** in which are some of the probability theorems that will be found useful in the scientific context: the central theorem here is Bayes's Theorem in its various forms. We then suggest that the reader look at the first few sections of Chapter 7, where Bayes's Theorem is applied to some simple reasoning patterns that are found particularly when deterministic hypotheses are handled; this chapter also compares the Bayesian approach with some others, such as Popper's falsificationist methodology. Chapters 8 through 12 deal with non-deterministic, that is, statistical, hypotheses, giving a critical exposition of the so-called classical, or frequentist, methods that constitute the leading alternative to the Bayesian approach: the main classical ideas can be appreciated from sections **a–d** of Chapter 8 and sections **a–c** of Chapter 9. The final part of the mini-course we are suggesting is to examine Chapter 15, especially sections **a, b, g, h,** and **j,** where some of the main criticisms that have been levelled against the Bayesian approach are answered.

## ACKNOWLEDGEMENTS

In the course of writing this book and preparing this new edition, we have had the advantage of a great deal of helpful advice from several friends and colleagues, and we are pleased to have this opportunity to offer them our warmest thanks. They are Timothy Childers, Allan Franklin, John Howard, Martin Knott, Kevin Korb, Dennis Lindley, Peter Milne, Larry Phillips, and John Worrall. Although all these people would agree with some of what we have written, probably none would agree with it all. Responsibility for the views expressed herein therefore rests entirely with us.

We also express our thanks to Youssef Aliabadi, Helen Brown, Sue Burrett, Alasdair Cameron, Pat Gardner, Rose Gibson, Wenceslao Gonzalez, Kurt Klappholz, Ginny Watkins, and Gay Woolven for friendly advice, research assistance, and help in preparing the manuscript; to the Suntory-Toyota International Centre for Economics and Related Disciplines for financial assistance; and to the staff of the Open Court Publishing Company for their consideration and their painstaking editorial work.

Finally, we thank each other. Although we are separately responsible for particular chapters (CH: 2, 3, 4, 5, 6, 13, 15; PU: 1, 7, 8, 9, 10, 11, 12, 14), we have each benefited from frequent discussions and the reading and rereading of successive drafts of each other's contributions, and this close collaboration, we believe, has produced a unified exposition of the central Bayesian ideas.

C. H.<br>
P. U.<br>
London, June 1993

# PART I

# *Bayesian Principles*

According to the Bayesian view, scientific and indeed much of everyday reasoning is conducted in probabilistic terms. In other words, when evaluating an uncertain claim, one does so by calculating the probability of the claim in the light of given information. Precisely how this is done and why it is reasonable is the topic of this book.

In Part I of the book we shall first introduce the central Bayesian ideas and their intellectual and historical background. This will be Chapter 1. In chapters 2 and 3 we shall present the laws of the probability calculus and some basic facts about distribution and density functions. Chapter 4 describes the older Classical/Logical theory of epistemic probability and why it failed. Chapters 5 and 6 contain the detailed development of the Personalist or Subjective Bayesian theory and the rules that regulate the updating of belief with new evidence. The rest of the book will show in detail why the Bayesian approach is the only one capable of representing faithfully the basic principles of scientific reasoning.

# CHAPTER 1

# *Introduction*

## a THE PROBLEM OF INDUCTION

Hypotheses usually have a general character relative to the empirical observations they are thought to explain. For instance, Mendel's genetic theory apparently concerns all inherited characteristics in all plants and animals, whereas relatively few of these could ever have been observed. If all our information derives from empirical observation, how can we be sure that any particular explanatory theory is the correct one? This is one version of the traditional problem of induction.

It has, however, sometimes been denied that our stock of information is restricted to empirical observations, a number of philosophers having taken the view that we are also capable of cognizing important synthetic principles which enable the gap between observations and scientific theories to be bridged. Immanuel Kant (1783, p. 9), for example, who claimed that his "dogmatic slumber" had been interrupted by the problem of induction, to which he had been alerted by David Hume's brilliant exposition of it, attempted to provide a principle that was both a priori certain and sufficiently rich to guarantee the truth of the theories of physics. His effort was, however, inadequate. The principle he advocated was just that every event has a cause. Much of Kant's endeavour went into showing that this was an a priori truth, and many of his interpreters have worked hard trying to unravel just what his argument was. But whether valid or not, the principle is irrelevant to the issue at hand, which does not concern whether every event has a cause but asks the very different question: how can one be certain, in any particular case, that one has selected the correct cause of an event out of the huge, indeed infinite, number of possible causes?

Another candidate for a bridging principle between empirical observations and scientific theories is the so-called Principle of the Uniformity of Nature, which Hume (1777, section 32) summed up in the phrase “the future will resemble the past”. It is sometimes held that when scientists advocate their theories, they are relying on this principle, at least tacitly.

However, there are two obvious reasons why the theories of science could not be established as definitely true by means of such a principle. First, as it stands, it is empty, for it fails to disclose in what respects the future is supposed to resemble the past. To perform its intended role, the principle would need to be given a specific formulation for application to each case. For example, one such formulation would need to say that, in regard to heated metals, if these have always been observed to expand in the past, then they will do so in the future. It would need a more elaborate formulation to permit the inference that *all* metals would expand if heated, as is usually assumed. But, secondly, as soon as the Uniformity of Nature principle has been made sufficiently specific for it to connect given observations to particular general laws, its inadequacy as a basis for scientific inference becomes manifest, because its own claim to be accepted as true is now just as questionable as the scientific theory it was designed to guarantee.

## b POPPER'S ATTEMPT TO SOLVE THE PROBLEM OF INDUCTION

It would appear then—this is not any longer controversial—that there is no positive solution to the problem of induction, that is to say, no solution by whose means particular explanatory theories could be conclusively shown to be true. However, many philosophers and scientists resist the idea, embraced in recent years with particular vigour by Paul Feyerabend, that all theories are on a par and that, for example, standard scientific claims are no better and no worse than those which would commonly be dismissed as the crackpot ideas of a charlatan. Karl Popper, in particular, was concerned to resist such scepticism and put science on a rational footing. He conceded that since scientific theories are never conclusively verifiable, no positive solution exists to the problem of induction. But Popper maintained that theories may, nevertheless, have some worthwhile epistemic status and in some cases be established as epistemically superior to their rivals, this superiority supposedly being an objective feature, independent of anyone's attitude towards them. To this ambitious purpose he pointed out two facts: first, that while theories cannot be logically proved by empirical observations, they can sometimes be refuted by them, and secondly, that their deductive consequences can sometimes be observationally verified. So, for example, 'All swans are white' is refuted by the sighting of a black swan and is, in Popper's terminology, "corroborated" by the observation of a white one. These facts are, of course, not novel discoveries, but what is original with Popper is the attempt to press them into service to show the rationality of science. The attempt, however, is not a success, nor can it be made into one. It is certainly true that by refuting a theory, one can rule out that particular conjecture as the true one. But this, by itself, is practically no help. For, suppose attention were restricted to unrefuted theories, one would still face an infinite class of alternatives, all of which are equally "corroborated", and the problem would remain: how can one choose, from among these, the theory most reasonably regarded as the true one?

To illustrate the difficulty, consider the task of discovering some general law governing the coloration of swans. If you were interested in the truth of the matter, you would have to consider many theoretical possibilities. Suppose the total number of swans that will ever exist is $n$ and there are just $m$ different colours, then the number of colour-combinations for the class of swans is $m^n$. This, then, represents a lower limit for the number of theories concerning the colours swans might have. If we take account of the further possibilities that swans alter their hues from time to time, and from place to place, and that some of them are multicoloured, then it becomes obvious that the number of mutually exclusive rivals to the simple hypothesis 'All swans are white' is immense, indeed, it is infinite. To be sure, many of these will have been refuted (for example, the conjecture that all swans are red is falsified by the evidence of white swans), but indefinitely many hypotheses would still remain which are not ruled out in this way by our observations. In fact, theories which specify the colour of every swan at any given instant, which have not been refuted by current observations on the colours of these birds, are corroborated by them. The problem of choosing the best hypothesis amongst these then remains and does not seem to have been advanced by Popper's reflections. (Popper's attempt to solve the induction problem is decisively criticized by Lakatos, 1974, and also by Salmon, 1981.)

## c SCIENTIFIC METHOD IN PRACTICE

Popper's view that unrefuted but corroborated hypotheses enjoy some special epistemic advantage, independent of anybody's attitude towards them, led him to recommend that scientists ought to seek out such hypotheses. There was also a descriptive aspect to this recommendation, for Popper assumed that mainstream pure science is in fact conducted more or less as he believed it should be. We shall examine this claim.

Two features of scientific reasoning are immediately reflected in Popper's descriptive account. First, it sometimes happens in scientific work that a theory is refuted by some experimental evidence. When this happens, the scientist usually revises the theory or abandons it altogether. (Of course, by revising a theory, one necessarily abandons it altogether! However, when the new theory bears what one intuitively regards as a family resemblance to the old, it is normally spoken of as merely a revision.) Another typical aspect of scientific thought taken account of in Popper's scheme is this: when investigating a deterministic theory, scientists draw out some of its consequences, check them by means of a suitable experiment, and if they turn out to be true, often conclude that the theory has been confirmed or that its claim on our belief in it is strengthened.

Although accurate in these respects, there are two reasons why Popper's descriptive account can have little further success. One has already been touched on in connection with Popper's—in our opinion—unsuccessful attempt to solve the problem of induction. It is that the principles on which he stands are too weak to narrow down the range of alternative hypotheses sufficiently, and, in practice, scientists do have a system of ranking hypotheses by their value and eligibility for serious consideration. As we shall see in Chapter 7, this fact ensures that many typical modes of scientific reasoning are inaccessible to explanation on Popper's principles.

Another limitation of Popper's account is that it focusses attention just on the logical consequences of a theory, whereas most evidence that scientists consider either for or against theories does not come into this category. This arises in a number of ways. First, many deterministic theories that appear in science have no directly checkable deductive consequences and the predictions by which they are tested and confirmed are necessarily drawn only with the assistance of auxiliary theories. Newton's laws are a case in point. These laws concern the forces that operate between masses in general and the mechanical effects of such forces. Observable consequences about particular masses, such as the planets, can be derived only when the laws are combined with hypotheses about the positions and masses of the planets, the mass-distribution of space, and so on. But although such predictions are not direct logical consequences of Newton's theory, that theory is often regarded as being confirmed by them. (Popper did consider this objection, which was pressed with particular vigour by Lakatos, but as we explain in further detail in Chapter 7, his philosophy seems unable to deal with it.)

Secondly, many scientific theories are explicitly probabilistic and, for this reason, have no logical consequences of a verifiable character. An example is Mendel's theory of inheritance. This states the probabilities with which certain combinations of genes occur during reproduction; but, strictly speaking, the theory does not categorically rule out, nor predict, any particular genetic configuration. Nevertheless, Mendel obtained impressive confirmation from the results of his plant-growing trials, results which his theory did not entail but stated to be relatively probable.

Finally, even deterministic hypotheses are frequently confirmed by evidence that is only assigned some probability, for if it is a quantitative theory, its quantitative consequences may need to be checked with imperfect measuring devices, subject to what is known as experimental error. Take as an example a theory which predicts the position of a planet, this prediction being checked using an appropriate kind of telescope. Because of various unpredictable atmospheric conditions affecting the path of light to the telescope as well as other uncontrollable factors, some connected with the experimenter and some with physical vagaries, the actual reading is acknowledged in experimental work not to be completely reliable. For this reason, if the predicted value of an angle were being ascertained, the result of the measurement would normally be reported in the form of a range of values such as $a \pm b$. Here, $a$ is the reading recorded by the instrument, while the interval $a + b$ to $a - b$ signifies the range in which it is judged the true value probably lies. This calculation is usually based on a theory giving the probability that the instrument reading diverges by different amounts from the true value of the measured quantity. Such theories commonly assume that the experimental reading is normally distributed about the true value with a standard deviation of $b$. (The concepts of a normal distribution and a standard deviation are defined in Chapter 3.) Thus for many deterministic theories, what may appear to be the checking of logical consequences actually involves the examination of experimental effects which are predicted only with a certain probability.

Popper tried to extend the falsificationist idea to the statistical realm, but (as we shall show in Chapter 8) there are insuperable difficulties for any such attempt. The eminent statistician R. A. Fisher was also inspired by the idea that evidence may have a decisive negative impact on a statistical hypothesis, akin to its falsification. He called a statistical hypothesis under test the "null hypothesis" and expressed the view that

> the null hypothesis is never proved or established, but is possibly disproved, in the course of experimentation. Every experiment may be said to exist only in order to give the facts a chance of disproving the null hypothesis. (Fisher, 1947, p. 16)

Fisher's theory of significance tests, which prescribes how statistical hypotheses should be tested, has drawn considerable criticism, and several other theories have been advanced in opposition to it. Most notable amongst these is the modified theory of significance testing due to Jerzy Neyman and Egon Pearson. Though they rejected much of Fisher's methodology, their theory owed a good deal to his work, particularly to his technical results. Above all, they retained the idea of bivalent statistical tests in which evidence determines one of only two possible results, that is, the acceptance or rejection of a hypothesis.

## d PROBABILISTIC INDUCTION: THE BAYESIAN APPROACH

One of the driving forces behind the development of the above-mentioned methodologies was the desire to vanquish, and provide an alternative to, the idea that the theories of science can be and ought to be appraised in terms of their 'probabilities'. In setting themselves against the ideas of probabilistic induction, Popper and the classical statisticians were opposing a well-entrenched tradition in science and philosophy. Although it has long been appreciated that general scientific theories extend beyond any experimental data and hence cannot be verified (in the sense of being logically entailed) by them, there is, as we have mentioned, a strong tendency in the scientific community, and among philosophers and laymen too, to resist a complete scepticism. Their attitude is that while absolute certainty cannot be expected, nevertheless, the explanations thought up by scientists and tested by searching experiments may secure for themselves an epistemic status somewhere between being certainly right and certainly wrong.

This spectrum of degrees of certainty has traditionally been characterised as a spectrum of probabilities. For example, the eminent physicist, mathematician, and philosopher Henri Poincaré reasoned as follows:

> Have we any right, for instance, to enunciate Newton's Law? No doubt numerous observations are in agreement with it, but is not that a simple fact of chance? and how do we know, besides, that this law which has been true for so many generations will not be untrue in the next? To this objection the only answer you can give is: It is very improbable. . . . From this point of view all the sciences would only be unconscious applications of the calculus of probabilities. And if this calculus be condemned, then the whole of the sciences must also be condemned. (Poincaré, 1905, p. 186)

> Thus, in a multitude of circumstances the physicist is often in the same position as the gambler who reckons up his chances. Every time that he reasons by induction, he more or less consciously requires the calculus of probabilities. . . . (Poincaré, 1905, pp. 183–84)

Similarly, the philosopher and economist W. S. Jevons:

> Our inferences . . . always retain more or less of a hypothetical character, and are so far open to doubt. Only in proportion as our induction approximates to the character of perfect induction, does it approximate to certainty. The amount of uncertainty corresponds to the probability that other objects than those examined may exist and falsify our inferences; the amount of probability corresponds to the amount of information yielded by our examination; and the theory of probability will be needed to prevent us from over-estimating or under-estimating the knowledge we possess. (Jevons, 1874, vol. 1, p. 263)

Very many scientists have voiced the same idea, that theories have to be judged in relation to their probability in the light of evidence. In fact, as Jon Dorling (1979, p. 180) has observed, it is rare to find any leading scientist writing in, say, the last three hundred years who did not employ notions of probability when advocating his own ideas or reviewing those of others.

A number of philosophers, from James Bernoulli in the seventeenth century to Rudolf Carnap, Harold Jeffreys, Bruno de Finetti, and Frank Ramsey in this century, have attempted to explicate these intuitive notions of inductive probability. There have been two main strands in this programme. The first regards the probabilities of theories as objective, in the sense of being determined by logic alone, independent of our subjective attitudes towards them. The hope was that one would be able to ascertain the probability that a theory is true and thereby place the comparative evaluation of competing explanations on an objective footing. If this could be done, it would provide some kind of solution to the induction problem and establish what might be regarded as a 'rational' basis for science. Unfortunately, as we shall describe later, this approach foundered upon crippling objections and is saved from inconsistency only by arbitrary and highly questionable stipulations.

The other strand of inductive probability treats the probabilities of theories as a property of our attitude towards them; such probabilities are then interpreted, roughly speaking, as measuring degrees of belief. This is called the *subjectivist or personalist interpretation*. The scientific methodology based on this idea is usually referred to as the methodology of *Bayesianism* because of the prominent role it assigns to a famous result of the probability calculus known as Bayes's Theorem.

Bayesianism has experienced a strong revival amongst statisticians and philosophers in recent years, due, in part, to its intrinsic plausibility and also to the weaknesses which have gradually been exposed in the standard methodologies. In the chapters to come, we shall present a detailed account of the Bayesian methodology, which we shall show may be applied to gain an understanding of the various aspects of scientific reasoning.

## e THE OBJECTIVITY IDEAL

The sharpest and most persistent objection to the Bayesian approach has been that it treats certain subjective factors as relevant to the scientific appraisal of theories. Our reply to this will be that the element of subjectivity admitted in the Bayesian approach is, first of all, minimal and, secondly, exactly right. However, this contradicts an influential school of thought which denies that there should be any subjective element in theory-appraisal at all; such appraisals, according to that school, should be completely objective. Lakatos (1978, vol. 1, p. 1) expressed this objectivist ideal in uncompromising style, thus:

> The *cognitive* value of a theory has nothing to do with its *psychological* influence on people's minds. Belief, commitment, understanding are states of the human mind. But the objective, scientific value of a theory is independent of the human mind which creates it or understands it, its scientific value depends only on what *objective* support these conjectures have in *facts*. [These characteristic italics are in Lakatos's original mimeographed paper, but were removed in the posthumously edited version.]

It was the ambition of Popper, Lakatos, Fisher, Neyman and Pearson, and others of their schools to develop this idea of a standard of scientific merit which is both objective and compelling and yet non-probabilistic. And it is fair to say that their theories, especially those connected with significance testing and estimation, which comprise the bulk of so-called classical methods of statistical inference, have achieved pre-eminence in the field. The procedures they recommended for the design of experiments and the analysis of data have become the standards of correctness with many scientists.

In the ensuing chapters we shall show that these classical methods are really quite unsuccessful, despite their influence amongst philosophers and scientists, and that their pre-eminence is undeserved. Indeed, we shall argue that the ideal of total objectivity is unattainable and that classical methods, which pose as guardians of that ideal, in fact violate it at every turn; virtually none of those methods can be applied without a generous helping of personal judgment and arbitrary assumption.

## f THE PLAN OF THE BOOK

The thesis we shall investigate in this book is that scientific reasoning is reasoning in accordance with the calculus of probabilities. In Chapter 2 we shall introduce that calculus, along with the principal theorems that will later serve in an explanatory role with respect to scientific method. In Chapter 3 we introduce the notion of a probability density and describe some of the main probability distributions and associated theorems needed later in the book. We shall then consider different interpretations of the probability calculus: thus in Chapter 4 we deal with the logical interpretation and show why it fails, and in Chapter 5 we introduce the notion of degrees of belief and explain why they afford a more satisfactory interpretation. Then, in Chapter 6, we consider how, according to the Bayesian approach, opinions should be revised in the light of evidence.

In Part II, which is also Chapter 7, we shall examine the notion of confirmation and look at the ways in which scientists regard deterministic theories as being confirmed by observations. We shall argue that these characteristic patterns of reasoning are best understood as arguments in probability.

In Part III of the book we turn to statistical hypotheses,

where the philosophy of scientific inference has mostly been the preserve of statisticians. Far and away the most influential voice in statistics has been that of the classical statistician, and we shall therefore first give an account of the classical point of view and demonstrate its manifold shortcomings. Thus Chapter 8 deals with Fisher's theory of significance tests, Chapter 9 considers the Neyman-Pearson account of significance tests, and Chapter 10, the classical theory of estimation.

In Part IV we consider two areas where statistical inference is routinely used and where disagreements over the principles of inference have a substantial practical importance. Thus, in Chapter 11, we consider methods for testing causal hypotheses, particularly hypotheses relating to the efficacy of medical treatments, and in the following chapter we examine ways in which relations between quantitative variables are studied using the techniques of regression analysis.

In Part V we present an outline of the Bayesian approach to statistical inference. This will involve first considering the nature of the statistical probabilities that statistical hypotheses purport to be about (are they real or imaginary?). This is done in Chapter 13. Then, in Chapter 14, we show that the difficulties afflicting classical statistical inference are absent from the Bayesian approach, which, we argue, is perfectly satisfactory. Finally, in Chapter 15, which constitutes Part VI of the book, we answer the main criticisms that have been levelled against Bayesianism.

## EXERCISES

1. Suppose that a certain number of metals have on a certain number of occasions been reliably observed to expand shortly after being heated. If we wished to combine this information with a Uniformity of Nature principle, in order to derive the conclusion that *all* metals expand *whenever* they are heated, what form would that principle have to take?

2. Suppose you are investigating the relationship between the volume of a certain gas and its simultaneous pressure, and you have $n$ different, joint volume-pressure observations. Show that your data are compatible with infinitely many,

possible, mutually incompatible relationships. How does this fact undermine Popper's attempt to solve the problem of induction?

3. According to Popper "[w]e should prefer as a basis for action the best-tested theory", though he admitted that there can be no "absolute reliance". Comment on the following argument Popper puts for his position.

> . . . since we *have* to choose, it will be 'rational' to choose the best-tested theory. This will be 'rational' in the most obvious sense of the word known to me: the best-tested theory is the one which, in the light of our *critical discussion,* appears to be the best so far, and I do not know of anything more 'rational' than a well-conducted critical discussion.
>
> Of course, in choosing the best-tested theory as a basis for action, we 'rely' on it, in some sense of the word. It may therefore even be described as the *most* 'reliable' theory available, in some sense of this term. Yet this does not say it is 'reliable'. It is not 'reliable' at least in the sense that we shall always do well even in practical action, to foresee the possibility that something may go wrong with our expectations. (Popper, 1972, p. 22)

4. Bertrand Russell describes what he called the "Principle of Induction" (Russell, 1912, p. 37) in terms of the following four propositions:

> **(a)** When a thing of a certain sort A has been found to be associated with a thing of a certain other sort B, and has never been found dissociated from a thing of the sort B, the greater the number of cases in which A and B have been associated, the greater is the probability that they will be associated in a fresh case in which one of them is known to be present.
>
> **(b)** Under the same circumstances, a sufficient number of cases of association will make the probability of a fresh association nearly a certainty, and will make it approach certainty without limit.
>
> **(a′)** The greater the number of cases in which a thing of the sort A has been found associated with a thing of the sort B, the more probable it is (if no cases of failure of association are known) that A is always associated with B.

> **(b′)** Under the same circumstances, a sufficient number of cases of the association of A with B will make it nearly certain that A is always associated with B, and will make this general law approach certainty without limit.

Not all of these propositions are generally true. Explain why.

# CHAPTER 2

# The Probability Calculus

## a INTRODUCTION

Everything in this book rests upon some property or other of the *probability calculus*. In view of this it will be best to give a brief account of the calculus at the outset and to derive those properties which will be central to later developments. In this way we can avoid breaking up the argument time and again in order to demonstrate that some equation or inequality does indeed follow from basic principles. All the derivations are extremely simple and most require nothing more than elementary algebra.

In this chapter we shall develop the characteristic laws of the calculus purely formally, and then introduce the principal applications of those laws as the book proceeds. However, a glance at the literature on probability will reveal that there are currently two, apparently distinct, formal developments of the probability calculus: in one, probabilities are defined on sentences; and in the other, on subsets of some given set (the latter set often being called the space of elementary possibilities). The second, set-theoretical development, was introduced only in this century, by the Soviet mathematician A. N. Kolmogorov, though the mathematical theory of probability goes back to the seventeenth century. That earlier theory ascribed probabilities directly to hypotheses (to types of sentences or statements), but Kolmogorov's treatment (1933) has now become the standard way of presenting the calculus among statisticians and mathematicians.

The sentence-based presentation tends to be favoured by people who are primarily interested in the application of probability theory to the problems of inductive inference, because there the primary bearers of probabilities are explicitly linguistic entities, namely, hypotheses of one form or other. Since these problems are our primary concern also, we shall adopt that approach too. However, the difference between the two types of development is really just a formal one: the sets in the domain of Kolmogorov's probability function are actually disguised propositions, as we shall show presently.

To the uninitiated all this may sound very cryptic. We shall provide a much fuller discussion of these issues later in this chapter, in the course of which we shall also develop the principal consequences of the probability calculus in terms of assignments of probabilities directly to sentences. We shall then introduce, in as informal a way as possible, the notion of a random variable and discuss probability distributions and probability-density distributions over the values of random variables. We shall not be aiming at any great mathematical rigour or purity; there are many texts available on the mathematics of probability, and where readers think they will benefit by more information on any of these topics, they are invited to consult any or all of these texts. We want only to introduce as much technical material as is necessary to understand what is going on, and so there will be few proofs, and many if not all mathematical corners will be cheerfully cut.

## ■ b SOME LOGICAL PRELIMINARIES

We shall be employing some, but not many, of the notions of elementary logic—principally just the so-called *logical connectives* 'and', 'or', and 'not'. We shall assume that the sentences we shall be referring to in what follows are drawn from some otherwise unspecified descriptive 'language' *L*. We have in mind something like ordinary English as it is used to formulate factual hypotheses, make predictions, record events, and so on, except that we shall use a standard logical notation instead of the vernacular to discuss the logical structure of English sentences. Accordingly, we shall symbolise 'and' by '&', 'or' by '**v**', and 'not' by '~'. The connectives are sentence-building operations: by means of them new sentences can be constructed from existing ones in such a way that the truth-value ('true', 'false') of the new sentence depends only on the truth-values of the sentences from which it was so constructed (for this reason the connectives are also called *truth-functional* operations). For example, *a* & *b*, called the *conjunction* of the sentences $a$ and $b$, will be taken to be true when $a$ and $b$ are both true, and false in all other cases. This information is conveniently summed up in the *truth table* for &:

| $a$ | $b$ | $a \& b$ |
|---|---|---|
| T | T | T |
| T | F | F |
| F | T | F |
| F | F | F |

$T$ and $F$ stand for true and false respectively, and the truth table exhibits the truth-value of $a \& b$ corresponding to each of the four possible distributions of truth values to $a$ and $b$. $a \vee b$ is called the *disjunction* of $a$ and $b$ and is understood inclusively: this means that $a \vee b$ will be regarded as being false only when $a$ and $b$ are both false, and true otherwise. In other words, it is defined by the following truth table:

| $a$ | $b$ | $a \vee b$ |
|---|---|---|
| T | T | T |
| T | F | T |
| F | T | T |
| F | F | F |

$\sim a$, the *negation* of $a$, will, of course, be false when $a$ is true and true when $a$ is false:

| $a$ | $\sim a$ |
|---|---|
| T | F |
| F | T |

Occasionally we shall make use of the biconditional '$\leftrightarrow$': $a \leftrightarrow b$ will be true when $a$ and $b$ have the same truth-value, and will be false otherwise:

| $a$ | $b$ | $a \leftrightarrow b$ |
|---|---|---|
| T | T | T |
| T | F | F |
| F | T | F |
| F | F | T |

We shall use the notation $a \vdash b$ to signify that *a deductively entails,* or implies, *b*. To say that *a* deductively entails *b* is to say that, whatever the actual state of the world might be, the logical structure of *a* and *b* ensures that *a* can never be true and *b* false, subject to a uniform interpretation of their denoting terms. A trivial example is $a \vdash a$. Some more examples are provided in the exercises below.

The notation $a \Leftrightarrow b$ will signify that *a* is logically equivalent to *b;* this means simply that whatever the actual state of the world might be, the logical structure of *a* and *b* ensures that *a* and *b* must have the same truth-value. $a \Leftrightarrow a$, $a \Leftrightarrow {\sim}{\sim}a$, $a \ \&\ b \Leftrightarrow b \ \&\ a$, $a \vee b \Leftrightarrow b \vee a$, are some simple logical equivalences, as the reader can easily check with truth tables.

We have talked about *logical* equivalence and *deductive* entailment; despite this, our notions of entailment and logical equivalence are a bit stronger than the purely logical ones, since we shall, as is usual, understand them as incorporating all of contemporary mathematics. Thus, for example, if $x$ is any individual, and $A$ and $B$ any two sets, then $x \in A \ \&\ x \in B$ will be regarded as deductively entailing $x \in A \cap B$. $\in$ as usual signifies the membership relation, and $A \cap B$ the intersection of the two sets $A$ and $B$, that is to say, the set whose members are common to $A$ and $B$. The union $A \cup B$ of two sets $A$ and $B$ is the set whose members are in $A$ or in $B$ or in both $A$ and $B$. The complement $A - B$ of a set $A$ with respect to $B$ is the set whose members are the members $B$ excluding all those which are also members of $A$. The empty set is, as usual, denoted by the symbol $\emptyset$.

## c THE PROBABILITY CALCULUS

### c.1 The Axioms

We can now proceed to state the rules governing the assignment of probabilities to sentences. These are the rules, or axioms, of the so-called probability calculus. It is customary among writers of probability texts to start by stipulating that probabilities are values of a function $P$ whose domain (i.e., the things it assigns values to) is a set $C$ of statements, which is closed under the operations of forming negations, conjunctions, and disjunctions. Closure under these operations means simply that $C$ contains $a$ & $b$, $a \vee b$, $\sim a$, and $\sim b$, whenever it contains $a$ and $b$. We are not going to follow this practice, because in one of the interpretations of the probability calculus, the subjective interpretation, the closure condition is usually very unrealistic. We shall not assume even that there is a well-defined domain for $P$, but merely that $P$ assigns values to some arbitrary collection of sentences which can at any time be augmented ($P$ is not in that case strictly speaking a function, since a function has a fixed domain; we shall ignore this nicety and nevertheless refer to $P$ as a function, but one whose domain may be indeterminate).

The probability axioms themselves are four in number. The first states that probabilities are non-negative real numbers. The second says that every tautology is assigned the value 1; and the third says that the sum of the probabilities of two mutually inconsistent sentences is equal to the probability of their disjunction. In symbols:

**(1)** $P(a) \geq 0$ for all $a$ in the domain of $P$.

**(2)** $P(t) = 1$ if $t$ is a tautology.

**(3)** $P(a \vee b) = P(a) + P(b)$ if $a$ and $b$ and $a \vee b$ are all in the domain of $P$, and $a$ and $b$ are mutually exclusive; i.e., the truth of one entails the falsity of the other.

It may be that the domain $D$ of $P$ includes no mutually exclusive statements, and though (1)–(3) are all satisfied, (3) is satisfied only vacuously. In such circumstances it is not always possible to extend $D$ in such a way that (1)–(3) remain satisfied. A consequence of (1)–(3), which we shall prove later, is that if $a \vdash b$ and some extension $D'$ of $D$ contains certain truth-functions of $a$ and $b$, then $P(a) \leq P(b)$. Hence if $D$ were to contain only $a$ and $b$ and $P(a) > P(b)$, then $P$ could not be extended to $D$ in such a way that (1)–(3) are satisfied (Williams, 1976). To avoid this situation arising we therefore have to add a postulate to the effect that whatever sentences $D$ contains at any given time, $P$ is such that it can be extended to any domain containing arbitrary truth-functional combinations of statements in $D$.

Axioms 1–3 suffice to generate that part of the probability calculus dealing with so-called *absolute*, or *unconditional, probabilities*. (3) is often called the *Additivity Principle*, since it states that $P$ adds over disjunctions of pairs of mutually inconsistent statements. As we shall show shortly, (3) together with the other axioms implies that $P$ adds over all finite disjunctions of mutually exclusive statements.

A good deal of what follows will be concerned with probability functions of two variables, unlike $P$ above, which is a function of one only. These two-place probability functions are called *conditional probabilities*, and the conditional probability of $a$ given $b$ is written $P(a \mid b)$. There is a systematic connection between conditional and unconditional probabilities, however, and it is expressed in our fourth axiom:

$$\textbf{(4)}\ P(a \mid b) = \frac{P(a \ \&\ b)}{P(b)},$$

where $a$, $b$, and $a$ & $b$ are in the domain of $P$, and where $P(b) \neq 0$. Many authors take $P(a \mid b)$ actually to be defined by this condition. We prefer to regard (4) as a postulate on a par with (1)–(3). The reason for this is that in some interpretations of the calculus, independent meanings are given conditional and unconditional probabilities, and equation (4) becomes a synthetic, not an analytic, truth.

### c.2 Two Different Interpretations of the Axioms

We have presented the axioms of the probability calculus in a rather unmotivated and abstract way because, as has been remarked since the beginning of the nineteenth century, there are at least two quite distinct notions of probability, both of which appear to satisfy the formal conditions (1)–(4) above. According to one of these, the probability calculus expresses the fundamental laws regulating the assignment of objective physical probabilities to events defined in the outcome spaces of stochastic experiments (a classical example of a stochastic trial, and, because of its simplicity, one we shall make much use of subsequently, is that of tossing a coin and noting which face falls uppermost).

The other notion of probability is epistemic. This type of probability is, to use Laplace's famous words, "relative in part to [our] ignorance, in part to [our] knowledge" (1820, p.6): it expresses numerically degrees of uncertainty in the light of data. We shall be discussing these two notions in considerable detail in the following chapters; we mention them here not only because they involve distinct interpretations of the probability-values themselves, but also because the statements to which they assign probabilities are of quite distinct types. In the latter, epistemic, interpretation, the statements to which the probabilities are assigned are specific hypotheses, like 'the Labour Party will not win the next General Election in the UK'. As we shall see, however, there is more than one epistemic interpretation, an ostensibly person-independent one, and a frankly subjective one.

There is also more than one objectivist interpretation of the probability function, and in at least one of these, the statements describe *generic* events which can arise as possible outcomes of a stochastic trial or experiment. But here we are faced with an apparent difficulty: 'the coin lands heads' is true or false relative to specific tosses of specific coins. How can a sentence describe the generic event of landing heads?

The answer is, in brief, that it does so by leaving the referents of the appropriate singular terms in the sentence unspecified within the type, or class, from which they come. In a natural language such as English, we are not accustomed to the notion of a syntactically well-formed but partially uninterpreted sentence. Within the notation of formal logic, however, the notion is easily characterised. Thus, $B(a)$, where $a$ is an individual name, or constant, and $B$ a predicate symbol, describes a *specific* individual event when $a$ and $B$ are both fixed ($a$ might, for example, be made to refer to the next toss of this coin, and $B$ be the predicate, *lands heads*).

The same sentence $B(a)$ will be said to describe the *generic* event of this coin's landing heads when $a$ is not specified as any one of the tosses of this coin, but $B$ remains fixed as the predicate *lands heads*, referred to the class of tosses of this coin. Whether a sentence is to make specific or generic reference will be made clear in context. The reader should note that even a sentence like 'This coin lands heads on the $i$th toss' is as ambiguous between specific and generic reference as 'this coin lands heads'. The term 'the $i$th toss' refers implicitly to a finite or infinite sequence of tosses of this coin, but again, we may choose to make that reference generic or specific.

The motive, speaking for objective probability theorists of a certain stripe, for attaching probabilities to generic events, or rather to the sentences characterising those generic events, is, as we shall see in Chapter 9, that the associated probability numbers are not intended to describe features of the outcome of any particular performance of the experiment, but, on the contrary, to express the frequency with which the event in question occurs in long sequences of performances of that experiment. But some people have also tried to construct theories of objective probability in which these probabilities are attached to predictions about the outcome of a specific performance of some stochastic trial. We shall defer all discussion of these attempts to the appropriate chapter, however, and proceed now to derive the familiar 'laws' of the probability calculus from the axioms (1)–(4).

### c.3 Useful Theorems of the Calculus

The first result states the well-known fact that the probability of a sentence and that of its negation sum to 1:

**(5)** $P(\sim a) = 1 - P(a)$

Proof.

$a \vdash \sim\sim a$. Hence by (3) $P(a \vee \sim a) = P(a) + P(\sim a)$. But by (2) $P(a \vee \sim a) = 1$, whence (5).

Next, it is simple to show that contradictions have zero probability:

**(6)** $P(f) = 0$, where $f$ is any contradiction.

Proof.

$\sim f$ is a tautology. Hence $P(\sim f) = 1$ and by (5) $P(f) = 0$.

Our next result states that equivalent sentences have the same probability:

**(7)** If $a \Leftrightarrow b$ then $P(a) = P(b)$.

Proof.

First, note that $a \vee \sim b$ is a tautology if $a \Leftrightarrow b$. Assume that $a \Leftrightarrow b$. Then $P(a \vee \sim b) = 1$. Also if $a \Leftrightarrow b$ then $a \vdash \sim\sim b$; so $P(a \vee \sim b) = P(a) + P(\sim b)$. But by (5) $P(\sim b) = 1 - P(b)$, whence $P(a) = P(b)$.

We can now prove the important property of probability functions that they respect the entailment relation; to be precise, the probability of any consequence of $a$ is at least as great as that of $a$ itself:

**(8)** If $a \vdash b$ then $P(a) \leq P(b)$.

Proof.

If $a \vdash b$ then $[a \vee (b \,\&\, {\sim}a)] <=> b$. Hence by (7) $P(b) = P[a \vee (b \,\&\, {\sim}a)]$. But $a \vdash {\sim}(b \,\&\, {\sim}a)$ and so $P[a \vee (b \,\&\, {\sim}a)] = P(a) + P(b \,\&\, {\sim}a)$. Hence $P(b) = P(a) + P(b \,\&\, {\sim}a)$. But by (1) $P(b \,\&\, {\sim}a) \geq 0$, and so $P(a) \leq P(b)$.

From (8) it follows that probabilities are numbers between 0 and 1 inclusive:

**(9)** $0 \leq P(a) \leq 1$, for all $a$ in $S$.

Proof.

By axiom 1, $P(a) \geq 0$, and since $a \vdash t$, where $t$ is any tautology, we have by (8) that $P(a) \leq P(t) = 1$.

We shall now demonstrate the general (finite) additivity condition:

**(10)** Suppose $a_i \vdash {\sim}a_j$, where $1 \leq i < j \leq n$. Then $P(a_1 \vee \ldots \vee a_n) = P(a_1) + \ldots + P(a_n)$.

Proof.

$P(a_1 \vee \ldots \vee a_n) = P[(a_1 \vee \ldots \vee a_{n-1}) \vee a_n]$, assuming that $n > 1$; if not the result is obviously trivial. But since $a_i \vdash {\sim}a_j$, for all $i \neq j$, it follows that $(a_1 \vee \ldots \vee a_{n-1}) \vdash {\sim}a_n$, and hence $P(a_1 \vee \ldots \vee a_n) = P(a_1 \vee \ldots \vee a_{n-1}) + P(a_n)$. Now simply repeat this for the remaining $a_1, \ldots, a_{n-1}$ and we have (10). (This is essentially a proof by mathematical induction.)

*Corollary*. If $a_1 \vee \ldots \vee a_n$ is a tautology, and $a_i \vdash {\sim}a_j$ for $i \neq j$, then $1 = P(a_1) + \ldots + P(a_n)$.

Our next result is often called the 'theorem of total probability'.

**(11)** If $P(a_1 \vee \ldots \vee a_n) = 1$, and $a_i \vdash {\sim}a_j$ for $i \neq j$, then $P(b) = P(b \,\&\, a_1) + \ldots + P(b \,\&\, a_n)$, for any sentence $b$.

Proof.

$b \Leftrightarrow (b \,\&\, a_1) \vee \ldots \vee (b \,\&\, a_n) \vee [b \,\&\, \sim(a_1 \vee \ldots \vee a_n)]$. Furthermore, all the disjuncts on the right-hand side are mutually exclusive. Let $a = a_1 \vee \ldots \vee a_n$. Hence by (10) we have that $P(b) = P(b \,\&\, a_1) + \ldots + P(b \,\&\, a_n) + P(b \,\&\, \sim a)$. But $P(b \,\&\, \sim a) \leq P(\sim a)$, by (8), and $P(\sim a) = 1 - P(a) = 1 - 1 = 0$. Hence $P(b \,\&\, \sim a) = 0$ and (11) follows.

*Corollary 1.* If $a_1 \vee \ldots \vee a_n$ is a tautology, and $a_i \vdash \sim a_j$ for $i \neq j$, then $P(b) = \Sigma P(b \,\&\, a_i)$.

*Corollary 2.* $P(b) = P(b \mid c)\, P(c) + P(b \mid \sim c)\, P(\sim c)$, for any $c$ such that $P(c) > 0$.

Another useful consequence of (11) is the following:

**(12)** If $P(a_1 \vee \ldots \vee a_n) = 1$ and $a_i \vdash \sim a_j$ for $i \neq j$, and $P(a_i) > 0$, then for any $b$, $P(b) = P(b \mid a_1)\, P(a_1) + \ldots + P(b \mid a_n)\, P(a_n)$.

Proof.

A direct application of (4) to (11).

(12) itself can be generalised to

If $P(a_1 \vee \ldots \vee a_n) = 1$ and $P(a_i \,\&\, a_j) = 0$ for all $i = j$, and $P(a_i) > 0$, then for any $b$, $P(b) = P(b \mid a_1)\, P(a_1) + \ldots + P(b \mid a_n) P(a_n)$.

We shall now develop some of the important properties of the function $P(a \mid b)$. We start by letting $b$ be some fixed sentence such that $P(b) > 0$ and defining the function $Q(a)$ of one variable to be equal to $P(a \mid b)$, for all $a$.

Now define '$a$ is a tautology modulo $b$' simply to mean '$b \vdash a$' (for then $b \vdash (t \leftrightarrow a)$, where $t$ is a tautology, so that relative to $b$, $a$ and $t$ are equivalent), and '$a$ and $c$ are exclusive modulo $b$' to mean '$b \,\&\, a \vdash \sim c$'; then

**(13)** $Q(a) = 1$ if $a$ is a tautology modulo $b$; and the corollary

**(14)** $Q(b) = 1$;

**(15)** $Q(a \vee c) = Q(a) + Q(c)$, if $a$ and $c$ are exclusive modulo $b$.

Now let $Q'(a) = P(a \mid c)$, where $P(c) > 0$; in other words, $Q'$ is obtained from $P$ by fixing $c$ as the conditioning statement, just as $Q$ was obtained by fixing $b$. Since $Q$ and $Q'$ are probability functions on $S$, we shall assume that axiom 4 also holds for them: i.e., $Q(a \mid d) = \frac{Q(a \,\&\, d)}{Q(d)}$, where $Q(d) > 0$, and similarly for $Q'$. We can now state an interesting and important invariance result:

**(16)** $Q(a \mid c) = Q'(a \mid b)$.

Proof.

$$Q(a \mid c) = \frac{Q(a \,\&\, c)}{Q(c)} = \frac{P(a \,\&\, c \mid b)}{P(c \mid b)} = \frac{P(a \,\&\, b \,\&\, c)}{P(b \,\&\, c)} = \frac{P(a \,\&\, b \mid c)}{P(b \mid c)} = \frac{Q'(a \,\&\, b)}{Q'(b)} = Q(a \mid b).$$

*Corollary*. $Q(a \mid c) = P(a \mid b \,\&\, c) = Q'(a \mid b)$.

(16) and its corollary say that successively conditioning $P$ on $b$ and then on $c$ gives the same result as if $P$ were conditioned first on $c$ and then on $b$, and the same result as if $P$ were simultaneously conditioned on $b \,\&\, c$. This property of conditional probabilities will come into prominence when we discuss so-called probability kinematics in Chapter 4.

**(17)** If $h \vdash e$ and $P(h) > 0$ and $P(e) < 1$, then $P(h \mid e) > P(h)$.

This is a very easy result to prove (we leave it as an exercise), but it is of fundamental importance to the interpretation of the probability calculus as the logic of inductive inference. It is for this reason that we employ the letters $h$ and $e$; in the inductive interpretation of probability $h$ will be some hypothesis and $e$ some evidence. *(17) then states if h predicts e then the occurrence of e will, if the conditions of (17) are satisfied, raise the probability of h.*

(17) is just one of the results that exhibit the truly inductive nature of probabilistic reasoning. It is not the only one, and more celebrated are those that go under the name of *Bayes's Theorems*. These theorems are named after the eighteenth-century English clergyman Thomas Bayes. Although Bayes, in a posthumously published and justly celebrated

Memoir to the Royal Society of London (1763), derived the first form of the theorem named after him, the second is due to the great French mathematician Laplace.

***Bayes's Theorem (first form)***

**(18)** $P(h \mid e) = \dfrac{P(e \mid h)\, P(h)}{P(e)}$, where $P(h), P(e) > 0$.

Proof.

$$P(h \mid e) = \frac{P(h \,\&\, e)}{P(e)} = \frac{P(e \mid h)\, P(h)}{P(e)}$$

Again we use the letters $h$ and $e$, standing for hypothesis and evidence. This form of Bayes's Theorem states that the probability of the hypothesis conditional on the evidence (or the *posterior probability* of the hypothesis) is equal to the probability of the data conditional on the hypothesis (or the *likelihood* of the hypothesis) times the probability (the so-called *prior probability*) of the hypothesis, all divided by the probability of the data.

***Bayes's Theorem (second form)***

**(19)** If $P(h_1 \vee \ldots \vee h_n) = 1$ and $h_i \vdash \sim h_j$ *for* $i \neq j$, and $P(h_i), P(e) > 0$ then

$$P(h_k \mid e) = \frac{P(e \mid h_k)\, P(h_k)}{\Sigma P(e \mid h_i)\, P(h_i}$$

*Corollary*. If $h_1 \vee \ldots \vee h_n$ is a tautology, then if $P(e)$, $P(h_i) > 0$ and $h_i \vdash \sim h_j$ for $i \neq j$, then

$$P(h_k \mid e) = \frac{P(e \mid h_k)\, P(h_k)}{\Sigma P(e \mid h_i)\, P(h_i)}$$

***Bayes's Theorem (third form)***

**(20)** $$P(h \mid e) = \frac{P(h)}{P(h) + \dfrac{P(e \mid \sim h) P(\sim h)}{P(e \mid h)}}$$

From the point of view of inductive inference, this is one of the most important forms of Bayes's Theorem. For, since $P(\sim h) = 1 - P(h)$, it says that $P(h \mid e) = \mathrm{f}\left(P(h), \dfrac{P(e \mid \sim h)}{P(e \mid h)}\right)$ where $f$ is an increasing function of the prior probability $P(h)$ of $h$ and a decreasing function of the *likelihood ratio* $\frac{P(e \mid \sim h)}{P(e \mid h)}$. In other words, for a given value of the likelihood ratio, the posterior probability of $h$ increases with its prior, while for a given value of the prior, the posterior probability of $h$ is the greater, the less probable $e$ is relative to $\sim h$ than to $h$.

## d RANDOM VARIABLES

Bayes's Theorem or Theorems can be expressed not only as results about probabilities but also as results about *probability densities*. In order to explain what probability densities are, we must first introduce the notion of a *random variable*. A random variable is a quantity $X$ capable, depending on the actual state of affairs, of taking on different numerical values and having a definite probability of being found in any open or closed interval of real numbers—that is to say, since we are assigning probabilities to statements, the statement that $X$ lies in any such interval is assigned a definite probability. In textbooks of mathematical probability, random variables are given a rather more elaborate and formal definition, but the simple and informal account will suffice for now.

Since it is a quantity which may take different values relative to different possible states of affairs, a random variable is formally a *function* whose domain is the set of those distinct possibilities $w$ and whose values $X(w)$ are real numbers. Suppose, for example, we are interested in talking about the probability that a randomly selected person from a population $S$ has a height lying in some specified range. The class of possible outcomes of the random selection procedure is therefore $S$ itself, and for each person $w$, we can set $X(w)$ to be the height of $w$ in feet and inches, or whatever units are chosen.

The outcome space of an experiment is often most naturally represented as a set of numbers. For example, we might be measuring the number of decay particles emitted per unit time by some radioactive element, or the number of heads obtained in $n$ tosses of some coin, and in such cases we should record the outcomes simply by means of these numbers. In our height example, we did not have to make $S$ the actual population of people; we could instead have taken it to be the set of possible height measurements themselves. The random variables in examples like these, in which the outcome space $S$ is itself a set of numbers, are then formally just the identity functions on those various sets, although in practice we usually call the numbers in $S$ themselves the random variables. This deliberate confusion of random variables, which are really functions, with numbers, which are really scalars, is harmless in most cases, but there are nonetheless occasions where the confusion can lead to paradox (*see* Chapter 11, section **e**).

Random variables are a very convenient method of describing experimental outcomes, even when those outcomes appear not to be at all numerical in nature. In particular, random variables taking just the values 0 and 1 are often used for the description of those possible outcomes of an observation in which merely the presence or absence of some particular character $C$ is all one is interested in. In such cases we might set $X(w) = 1$ if $w$ has $C$, and $X(w) = 0$ if not; in this way the set of possible outcomes of the experiment is collapsed into just two ($X$ is called the *indicator* of $C$). Clearly, by means of appropriately chosen random variables defined on a set $S$ of outcomes of some experiment, we can partition $S$ in any way we wish. Conversely, any partition of a space $S$ can be regarded as induced by the values of an appropriate random variable.

## ■ e KOLMOGOROV'S AXIOMS

Probabilities are also assigned to statements of the form '$X$ is in $I$', where $X$ is a random variable and $I$ is what is called a Borel set, in the set-theory-based approach deriving from Kolmogorov's classic work (1933). This may sound paradoxical, because as we remarked earlier, in that approach probabilities are assigned to sets, not statements. But there is really no antithesis here, since the sets involved are the *extensions* of statements like '$X$ is in $I$'—that is to say, the sets are those possible states of affairs within the specified universe of possibilities which make those statements true. Before proceeding further, let us state exactly what Kolmogorov's axioms say.

These first of all require that the probability function is defined on a *field* $\boldsymbol{F}$ of subsets of some set $S$. $\boldsymbol{F}$ is a field of subsets of $S$ if $S$ is in $\boldsymbol{F}$, if $\boldsymbol{F}$ contains all complements of sets in $\boldsymbol{F}$ relative to $S$, and also if $\boldsymbol{F}$ contains all unions and intersections of sets in $\boldsymbol{F}$. $\boldsymbol{F}$ is often the set of all subsets of $S$, which is obviously a field, but it does not have to be, and sometimes, for mathematical reasons, it cannot be. It is then stipulated that

**(1′)** $P(A) \geq 0$, for all $A$ in $\boldsymbol{F}$

**(2′)** $P(S) = 1$

**(3′)** $P(A \cup B) = P(A) + P(B)$, if $A \cap B = \emptyset$

**(4′)** $P(A \mid B) = \dfrac{P(A \cap B)}{P(B)}$, where $P(B) > 0$

(Kolmogorov included a further axiom, that of *continuity,* and a further condition on the field $\boldsymbol{F}$, both of which we shall discuss shortly.) The similarity to our axioms (1)–(4) is obvious, but not surprising when we point out that (1)–(4) were chosen deliberately with (1′)–(4′) in mind!

$S$ is intended to represent the finest partition into possible states of affairs which can be made using some contextually specified conceptual structure, or *language* (often, but not necessarily, specifying the outcome space of an experiment). The members of $\boldsymbol{F}$ are the extensions of a more general class of descriptions, *guaranteed incidentally to include all those of the form 'X is in I',* and which will be closed under the truth-functional operations of conjunction, disjunction, and negation. To give a simple example, suppose $S$ is to represent the outcome space of throwing a die and recording the number uppermost. The elementary outcomes are of the form 'The outcome is the number $i$', $1 \leq i \leq 6$, and their extensions, comprised in $S$, are naturally represented by letting $S$ be simply $\{1,2,3,4,5,6\}$. The singleton set $\{2,4,6\}$ is the extension of the sentence 'The outcome is an even number', and also of '$X = 1$', where $X$ is the indicator of the characteristic of being an even outcome.

## f PROPOSITIONS

Some people have been misled into believing that the Kolmogorov formalism represents a new and distinct *interpretation* of the probability calculus (Popper calls it the "measure-theoretic interpretation", for example). In fact, of course, this is not so, and Kolmogorov's differs from the formalism we have given only in its assigning probabilities to the extensions of sentences rather than to the sentences themselves. To that extent it is simpler, since there can be more than one sentence with the same extension. Indeed, we have just seen an example of one in the sentences 'The outcome is an even number' and '$X = 1$' where $X$ is the indicator for evenness.

Philosophers call the common extension of a class of equi-extensional statements the *proposition* expressed by them. *Thus the field* **F** *in Kolmogorov's formalism can formally be identified as a set of propositions generated by some universe of discourse.* It is not difficult to see that the logical connectives &, **v**, and ~ on sentences are replaced by the set-theoretical operations of intersection, union, and complementation on the corresponding propositions. If we represent the proposition expressed by $a$ as $A$ and $M$ the mapping that associates $A$ with $a$, we have

$$M(a \vee b) = A \cup B = M(a) \cup M(b)$$

$$M(a \,\&\, b) = A \cap B = M(a) \cap M(b)$$

$$M(\sim a) = V - A = V - M(a),$$

where $V$ is the universal set (universal relative to the language) of possibilities, or the proposition corresponding to a tautology. These equalities are summed up by saying that the structure of sentences and that of propositions are *homomorphic*. The condition that $a$ and $b$ are mutually exclusive in axiom 3 of the probability calculus translates into the set-theoretic condition $M(a) \cap M(b) = \emptyset$.

It should now be clear that under the mapping $M$ Kolmogorov's axioms (1′)–(4′) translate into our axioms (1)–(4). The only reason for opting for one rather than the other presentation of the axioms is one of convenience. Kolmogorov's formalism, as we have said, is simpler, but it also possesses another characteristic which makes it appeal to mathematicians: it is language-free. Because of this, as we shall see in the next section, it allows for probabilities to be assigned to propositions which in many cases exceed what can be stated by sentences of any natural language.

## g INFINITARY OPERATIONS

For example, there are the denumerably infinite unions and intersections of sets in ***F*** ($\{A_i\}$ is a denumerably *infinite* family of sets if the index *i* ranges over all the positive integers). Kolmogorov stipulated that ***F*** should be closed not only under the operations corresponding to 'and', 'or', and 'not', but under these infinite operations as well. Any field of sets obeying this further closure condition is called a *sigma field.*

These infinite unions and intersections would correspond, under the mapping *M,* to infinite disjunctions and conjunctions were such available in natural languages: $\cup A_i$ would correspond to the 'disjunction' $a_1 \vee a_2 \vee a_3 \vee, \ldots,$ and $\cap A_i$ to the 'conjunction' $a_1$ & $a_2$ & $a_3$ &. . . . While infinite disjunction and conjunction operators do not exist in natural languages, those languages can in some cases express the same propositions by other means. Thus the true statement 'Somc positive whole number is even and prime' clearly represents the infinite disjunction 'One is even and prime or two is even and prime or . . .', and 'Every prime greater than two is odd' says the same as the infinite conjunction 'Three is odd and five is odd and seven is odd and . . .'.

These infinite operations are always expressible, however, in the language of mathematics, set theory, so a smooth formal development becomes much easier if one adopts set theory as one's lingua franca at the outset, which is the reason why mathematicians largely choose to work within that framework. We shall continue to work within a standard logical language, however, since, as we said at the outset, it is more natural to formulate inductive reasoning in such a framework. We shall be eclectic even there and freely add to our descriptive apparatus, where we feel that they help, infinite conjunction and disjunction operators in the ad hoc but serviceable form of writing 'and so on' or ' . . . ' after the appropriate finite conjunctions and disjunctions. Adding corresponding explicit infinitary operations to a standard first-order language results, incidentally, in a system (called $L_{\omega_1,\omega}$) which shares many features of first-order ones: its class of valid sentences is axiomatisable, for example.

## h COUNTABLE ADDITIVITY

Having stipulated that the domain of $P'$, that is $\boldsymbol{F}$, be a sigma field, Kolmogorov adopted as an additional axiom over and above (1′)–(4′) his so-called Axiom of Continuity. This is equivalent to requiring that where $UA_i$ is any denumerable union of mutually disjoint sets in $\boldsymbol{F}$, then $P'(UA_i)$ is equal to $\Sigma P'(A_i)$. (There is no need to stipulate that the infinite sum exists: this follows from the fact that the sequence of partial, finite sums is a bounded monotone sequence.)

This infinite additivity condition, often called the *Principle of Countable Additivity,* is very much stronger than that of finite additivity, as can be seen in the context of the following simple example. Suppose that we are considering the 'experiment' of selecting a positive integer in some way. If we adopt the Principle of Countable Additivity, then it is impossible for each integer to have the same probability of being selected, for the probability of selecting some integer is one, which is equal to the probability of selecting one or two or three or . . . , which is equal, given the infinite additivity condition, to the probability of selecting one plus the probability of selecting two plus the probability of selecting three, plus . . . , and so on. But it is obvious that if these summands were equal and positive, the sum would diverge, while if they were equal and zero, their sum would still be zero in the limit. If the probabilities are only finitely additive, on the other hand, it is not difficult to see that it is consistent to make the probability of each integer being selected the same for all integers, namely 0.

We shall not list the Principle of Countable Additivity among the fundamental axioms of the probability calculus, because while it is a legitimate constraint in one of the two principal interpretations of the calculus we shall be looking at, it is certainly not so in the other. Chapter 3 and Chapter 9 will discuss these two interpretations in detail.

## EXERCISES

**1.** The connective '$\rightarrow$' is defined by the truth table

| $a$ | $b$ | $a \to b$ |
|---|---|---|
| $T$ | $T$ | $T$ |
| $T$ | $F$ | $F$ |
| $F$ | $T$ | $T$ |
| $F$ | $F$ | $T$ |

'$\to$' is variously called the arrow, or conditional, or material implication. Show using truth tables that

a. $a \to b \Leftrightarrow \sim a \vee b \Leftrightarrow \sim (a \,\&\, \sim b)$
b. $a \vdash b$ if $a \to b$ is a tautology

**2.** Show that

a. if $f$ is a contradiction and $t$ a tautology and $a$ is any statement, $f \vdash a \vdash t$
b. $a \Leftrightarrow b$ if $a \leftrightarrow b$ is a tautology

**3.** Show that $a \Leftrightarrow b$ iff $a \vdash b$ and $b \vdash a$.

**4.** Show that $P(a \vee b) = P(a) + P(b) - P(a \,\&\, b)$. (Hint: $a \vee b \Leftrightarrow (a \,\&\, b) \vee (a \,\&\, \sim b) \vee (\sim a \,\&\, b)$, $a \Leftrightarrow (a \,\&\, b) \vee (a \,\&\, \sim b)$, $b \Leftrightarrow (a \,\&\, b) \vee (\sim a \,\&\, b)$, and the disjuncts on the right-hand sides of each equivalence are mutually exclusive. Use (10).)

**5.** Hence show that $P(a) = P(b) = 1 \Leftrightarrow P(a \,\&\, b) = 1$, and that $P(a) = P(b) = 0 \Leftrightarrow P(a \vee b) = 0$.

**6.** Prove consequence (12) of the probability axioms.

**7.** Show that $Q(a)$ (p. 26) is a probability function.

**8.** Prove (13)–(15) above.

**9.** Show that if $P(a), P(b) > 0$, then $P(a \mid b) = 1$ if and only if $P(\sim b \mid \sim a) = 1$.

**10.** Show that if $P(b), P(c) > 0$ then if $P(a \mid b) = 1$ and $P(b \mid c) = 1$ then $P(a \mid c) = 1$.

**11.** Show that if $P(b), P(c) > 0$ then if $P(a \mid b) = 1$ and $P(a \mid c) = 1$ then $P(a \mid b \,\&\, c) = 1$.

**12.** Prove (18)–(20) above.

# CHAPTER 3

# Distributions and Densities

## a DISTRIBUTIONS

Statements of the form '$X < x$', '$X \leq x$', play a fundamental role in mathematical statistics. Clearly, the probability of any such statement will vary with the choice of the real number $x$; it follows that this probability is a function $F(x)$, the so-called *distribution function*, of the random variable $X$. Thus, where $P$ is the probability measure concerned, the value of $F(x)$ is defined to be equal, for all $x$ to $P(X < x)$ (although $F$ depends therefore also on $X$ and $P$, these are normally apparent from the context and $F$ is usually written as a function of $x$ only). Some immediate consequences of the definition of $F(x)$ are that

**(i)** $F(-\infty) = 0$, and $F(+\infty) = 1$,
**(ii)** if $x_1 < x_2$ then $F(x_1) \leq F(x_2)$, and
**(iii)** $P(x_1 \leq X < x_2) = F(x_2) - F(x_1)$.

Distribution functions are not necessarily functions of one variable only. For example, we might wish to describe a possible eventuality in terms of the values taken by a number of random variables. For example, consider the 'experiment' which consists in noting the heights ($X$, say) and weights ($Y$) jointly of members of some human population. It is usually accepted as a fact that there is a joint (objective) probability distribution for the vector variable $(X,Y)$, meaning that there is a probability distribution function $F(x,y) = P(X < x \ \& \ Y < y)$. Mathematically this situation is straightforwardly generalised to distribution functions of $R$ variables.

## b PROBABILITY DENSITIES

It follows from (iii) that if $F(x)$ is differentiable at the point $x$, then the *probability density* at the point $x$ is defined and is equal to $f(x) = \frac{dF(x)}{dx}$; in other words, if you divide the probability that $X$ is in a given interval $(x,x + h)$ by the length $h$ of that interval and let $h$ tend to 0, then if $F$ is differentiable, there is a probability density at the point $x$, which is equal to $f(x)$. If the density exists at every point in an interval, then the associated probability distribution of the random variable is said to be continuous in that interval. The simplest continuous distribution, and one which we shall refer to many times in the following pages, is the so-called *uniform distribution*. A random variable $X$ is uniformly distributed in a closed interval $I$ if it has a constant positive probability density at every point in $I$ and zero density outside that interval.

Probability densities are of great importance in mathematical statistics—indeed, for many years the principal subject of research in that field was finding the forms of density functions of random variables obtained by transformations of other random variables. They are so important because many of the probability distributions in physics, demography, biology, and similar fields are continuous, or at any rate approximate continuous distributions. Few people believe, however, in the real—as opposed to the mathematical—existence of continuous distributions, regarding them as only idealisations of what in fact are discrete distributions.

Many of the famous distribution functions in statistics are identifiable only by means of their associated density functions; more precisely, those cumulative distribution functions have no representation other than as integrals of their associated density functions. Thus the famous *normal distributions* (these distributions, of fundamental importance in statistics, are uniquely determined by the values of two parameters, their mean and standard deviation, which we shall discuss shortly) have distribution functions characterised as the integrals of the density functions.

Some terminology. Suppose $X$ and $Y$ are jointly distributed random variables with a continuous distribution function $F(X,Y)$ and density function $f(x,y)$. Then $F(X) = \int f(x,y)dy$ is called the *marginal distribution* of $X$. The operation of obtaining marginal distributions by integration in this way is the continuous analogue of using the theorem of total probability to obtain the probability $P(a)$ of $a$ by taking the sum $\Sigma P(a \,\&\, b_i)$.

Indeed, if $X$ and $Y$ are discrete, then the marginal distribution for $X$ is just the sum

$$P(X = x_i) = \Sigma P(X = x_i \ \& \ Y = y_j).$$

The definitions are straightforwardly generalised to joint distributions of $n$ variables.

Finally, a word about notation. For the sake of simplicity, we shall in the subsequent chapters often use $P(x)$, instead of $F(x)$ and $f(x)$, to refer to both distribution and density functions. We trust that no confusion will be caused thereby.

## ■ c EXPECTED VALUES

The *expected value* of a function $g(X)$ of $X$ is defined to be (where it exists) the probability-weighted average of the values of $g$. To take a simple example, suppose that $g$ takes only finitely many values $g_1, \ldots, g_n$ with probabilities $a_1, \ldots, a_n$. Then the expected value $E(g)$ of $g$ always exists and is equal to $\Sigma g_i a_i$. If $X$ has a probability density function $f(x)$ and $g$ is integrable, then $E(g) = \int_{-\infty}^{\infty} g(x)f(x)dx$ where the integral exists.

In most cases, functions of random variables are themselves random variables. For example, the sum of any $n$ random variables is a random variable. This brings us to an important property of expectations: they are so-called *linear functionals*. In other words, if $X_1, \ldots, X_n$ are $n$ random variables, then if the expectations exist for all the $X_i$, then, because expectations are either sums or limits of sums, so does the expected value of the sum $X = X_1 + \ldots + X_n$ and $E(X) = E(X_1) + \ldots + E(X_n)$.

## ■ d THE MEAN AND STANDARD DEVIATION

Two quantities which crop up all the time in statistics are the mean and standard deviation of a random variable $X$. The *mean value* of $X$ is the expected value $E(X)$ of $X$ itself, where that expectation exists; it follows that the mean of $X$ is simply the probability-weighted average of the values of $X$. The *variance* of $X$ is the expected value of the function $(X - m)^2$, where that expectation exists. The *standard deviation* of $X$ is the square root of the variance. The square root is taken because the standard deviation is intended as a characteristic measure of the spread of $X$ away from the mean and so should be expressed in units of $X$. Thus, if we write s.d. $(X)$ for the standard deviation of $X$, s.d.$(X) = \sqrt{E[(X - m)^2]}$, where the expectation exists. The qualification 'where the expectation exists' is important, for these expected values do not always exist, even for some well-known distributions. For example, if $X$ has the Cauchy density $\frac{a}{\pi(a^2 + x^2)}$, then it has neither mean nor variance.

We have already mentioned the family of *normal distributions* and its fundamental importance in statistics. This importance derives from the facts that many of the variables encountered in 'nature' are normally distributed and also that the sampling distributions of a great number of statistics tend to the normal as the size of the sample tends to infinity (a statistic is a numerical function of the observations, and hence a random variable). For the moment we shall confine the discussion to normal distributions of one variable. Each member of this family of distributions is completely determined by two parameters, its mean $\mu$ and standard deviation $\sigma$. The normal distribution function itself is given by the integral over the values of the real variable $t$ from $-\infty$ to $x$ of the density we mentioned above, that is, by

$$f(t) = \frac{1}{\sigma\sqrt{2\pi}} e^{-\frac{1}{2}\left(\frac{t - \mu}{\sigma}\right)^2}$$

It is easily verified from the analytic expression for $F(x)$ that the parameters $\mu$ and $\sigma$ are indeed the mean and standard deviation of $X$. The curve of the normal density is the familiar bell-shaped curve symmetrical about $x = \mu$ with the points $x = \mu \pm \sigma$ corresponding to the points of maximum slope of the curve (figure 1 ). For these distributions the mean coincides with the *median,* the value of $x$ such that the probability of the set $\{X < x\}$ is one half (these two points do not coincide for all other types of distribution, however). A fact we shall draw on later is that the interval on the $x$-axis determined by the distance of 1.96 standard deviations centred on the mean supports 95% of the area under the curve, and hence receives 95% of the total probability.

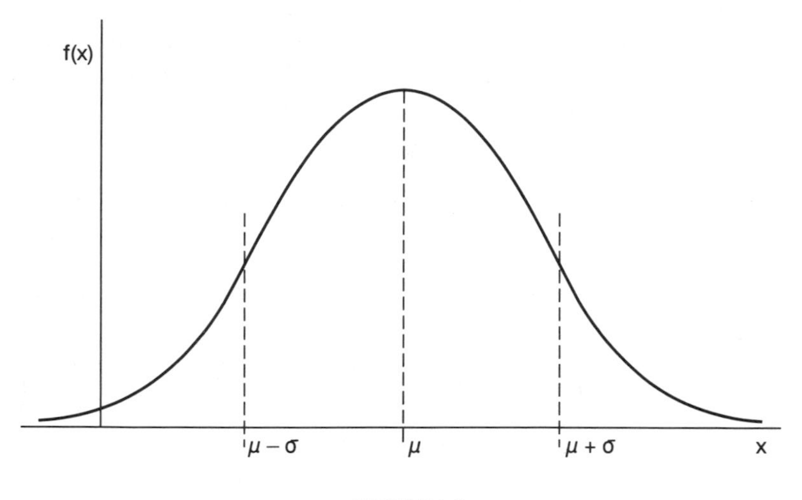


**FIGURE 1**

## e PROBABILISTIC INDEPENDENCE

Two sentences $h_1$ and $h_2$ are said to be *probabilistically independent* (relative to some given probability measure *P*) if and only if $P(h_1 \& h_2) = P(h_1)P(h_2)$. It follows immediately that, where $P(h_1)$ and $P(h_2)$ are both greater than zero, so that the conditional probabilities are defined, $P(h_1 \mid h_2) = P(h_1)$ and $P(h_2 \mid h_1) = P(h_2)$, just in case $h_1$ and $h_2$ are probabilistically independent.

Let us consider a simple example, which is also instructive in that it displays an interesting relationship between probabilistic independence and the so-called Classical Definition of probability. A repeatable experiment is determined by the conditions that a given coin is to be tossed twice and the resulting uppermost faces are to be noted in the sequence in which they occur. Suppose each of the four possible types of outcome—two heads, two tails, a head at the first throw and a tail at the second, a tail at the first throw and a head at the second—has the same probability, which of course must be one quarter. A convenient way of describing these outcomes is in terms of the values taken by two random variables $X_1$ and $X_2$, where $X_1$ is equal to 1 if the first toss yields a head and 0 if it is a tail, and $X_2$ is equal to 1 if the second toss yields a head and 0 if a tail.

According to the Classical Definition, or, as we shall call it, the Classical Theory of Probability, which we look at in the next chapter (and which should not be confused with the Classical Theory of Statistical Inference, which we shall also discuss), the probability of the sentence '$X_1 = 1$' is equal to the ratio of the number of those possible outcomes of the experiment which satisfy that sentence, divided by the total number, namely four, of possible outcomes. Thus, the probability of the sentence '$X_1 = 1$' is equal to $\frac{1}{2}$, as is also, it is easy to check, the probability of each of the four sentences of the form '$X_i = x_i$', $i = 1$ or 2, $x_i = 0$ or 1. By the same Classical criterion, the probability of each of the four sentences '$X_1 = x_1$ & $X_2 = x_2$' is $\frac{1}{4}$. Hence

$$P(X_1 = x_1 \ \& \ X_2 = x_2) = P(X_1 = x_1)\, P(X_2 = x_2)$$

and consequently the pairs of sentences '$X_1 = x_1$', '$X_2 = x_2$' are probabilistically independent (we have avoided answering, or trying to answer here, the question of what criteria justify the application of the Classical 'definition'; in the next chapter we shall discuss that definition in more detail).

The notion of probabilistic independence is generalised to $n$ sentences as follows: $h_1, \ldots, h_n$ are said to be probabilistically independent if and only if for every subset $h_{i1}, \ldots, h_{ik}$ of $h_1, \ldots, h_n$,

$$P(h_{i1} \ \& \ldots \& \ h_{ik}) = P(h_{i1}) \times \ldots \times P(h_{ik}).$$

It is easy to see, just as in the case of the pairs, that if any set of sentences is probabilistically independent, then the probability of any one of them conditional on any of the others, where the conditional probabilities are defined, is the same as its unconditional probability. It is also not difficult to show (and it is, as we shall see shortly, important in the derivation of the binomial distribution) that if $h_1, \ldots, h_n$ are independent, then so are all the $2^n$ sets $\pm h_1, \ldots, \pm h_n$, where $+h$ is $h$ and $-h$ is $\sim h$.

Any $n$ random variables $X_1, \ldots, X_n$ are said to be independent if for all sets of intervals $I_1, \ldots, I_n$ of values of $X_1, \ldots, X_n$ respectively, the sentences $X_1 \in I_1, \ldots, X_n \in I_n$ are probabilistically independent. We have, in effect, already seen that the two random variables $X_1$ and $X_2$ in the example above are probabilistically independent. If we generalise that example to that of the coin's being tossed $n$ times, and define the random variables $X_1, \ldots, X_n$ just as we defined $X_1$ and $X_2$, then again a consequence of applying the Classical 'definition' to this case is that $X_1, \ldots, X_n$ are probabilistically independent. It is also not difficult to show that a necessary and sufficient condition for any $n$ random variables $X_1, \ldots, X_n$ to be independent is that

$$F(x_1, \ldots, x_n) = F(x_1 \times \ldots \times F(x_n)$$

where $F(x_1, \ldots, x_n)$ is the joint distribution function of the variables $X_1, \ldots, X_n$ and $F(x_i)$ is the marginal distribution of $X_i$ (for a proof see Cramér, 1946, pp. 159–60). Similarly, if it exists, the joint density $f(x_1, \ldots, x_n)$ factors into the product of marginal densities $f(x_1)\, X \ldots X\, f(x_n)$ if the $X_i$ are independent.

## f CONDITIONAL DISTRIBUTIONS

According to the conditional probability axiom, axiom 4,

$$\textbf{(1)}\quad P(X < x \mid y \leq Y < y + \delta y) = \frac{P(X < x \,\&\, y \leq Y < y + \delta y)}{P(y \leq Y < y + \delta y)}.$$

The left-hand side is an ordinary conditional probability. Note that if $F(x)$ has a density $f(x)$ at the point $x$, then $P(X = x) = 0$ at that point. We noted in the discussion of (4) that $P(a \mid b)$ is in general only defined if $P(b) > 0$. However, it is in certain cases possible for $b$ to be such that $P(b) = 0$ and for $P(a \mid b)$ to take some definite value. Such cases are afforded where $b$ is a sentence of the form $Y = y$ and there is a probability density $f(y)$ at that point. For then, if the joint density $f(x,y)$ also exists, then multiplying top and bottom in (1) by $\delta y$, we can see that as $\delta y$ tends to 0, the right-hand side of that equation tends to the quantity

$$\int_{-\infty}^{x} \frac{f(u,y)du}{f(y)},$$

where $f(y)$ is the marginal density of $y$, which determines a distribution function for $X$, called the *conditional distribution function of* $X$ with respect to the event $Y = y$. Thus in such cases there is a perfectly well-defined conditional probability

$$P(x_1 \leq X < x_2 \mid Y = y),$$

even though $P(Y = y) = 0$.

The quantity $\frac{f(x,y)}{f(y)}$ is the density function at the point $X = x$ of this conditional distribution (the point $Y = y$ being regarded now as a parameter), and is accordingly called the *conditional probability density of X at x*, relative to the event $Y = y$. It is of great importance in mathematical statistics and it is customarily denoted by the symbol $f(x \mid y)$. Analogues of (18) and (19), the two forms of Bayes's Theorem, are now easily obtained for densities: where the appropriate densities exist

$$f(x \mid y) = \frac{f(y \mid x)f(x)}{f(y)}$$

and

$$f(x \mid y) = \frac{f(y \mid x)f(x)}{\int_{-\infty}^{\infty} f(y \mid x)f(x)dx}.$$

## g THE BIVARIATE NORMAL

We can illustrate some of the abstract formal notions we have discussed above in the context of a very important multivariate distribution, the *bivariate normal distribution*. This distribution is, as its name implies, a distribution over two random variables, and it is determined by five parameters. The marginal distributions of the two variables $X$ and $Y$ are both themselves normal, with means $\mu_X, \mu_Y$ and standard deviations $\sigma_X, \sigma_Y$. One more parameter, the correlation coefficient $\rho$, completely specifies the distribution. The bivariate density is given by

$$f(x,y) = \frac{e^{-\frac{1}{2(1-\rho^2)}\left[\left(\frac{x-\mu_X}{\sigma_X}\right)^2 - 2\rho\left(\frac{x-\mu_X}{\sigma_X}\right)\left(\frac{y-\mu_Y}{\sigma_Y}\right) + \left(\frac{y-\mu_Y}{\sigma_Y}\right)^2\right]}}{2\pi\sigma_X\sigma_Y\sqrt{1-\rho^2}}.$$

This has the form of a more-or-less pointed, more-or-less elongated hump over the $x,y$ plane, whose contours are ellipses with eccentricity (i.e., departure from circularity) determined by $\rho$. $\rho$ lies between $-1$ and $+1$ inclusive. When $\rho = 0$, $X$ and $Y$ are *uncorrelated,* and the contour ellipses are circles. When $\rho$ is either $+1$ or $-1$ the ellipses degenerate into straight lines. In this case all the probability is carried by a set of points of the form $y = ax + b$, for specified $a$ and $b$, which will depend on the means and standard deviations of the marginal distributions. It follows that the conditional probability $P(X = x \mid Y = y)$ is 1 if $y = ax + b$, and 0 if not.

The conditional distributions obtained from bivariate (and more generally multivariate) normal distributions have great importance in the area of statistics known as *regression analysis.* It is not difficult to show that the mean $\mu(X \mid y) = \int xf(x \mid y)dx$ (or the sum where the conditional distribution is discrete) has the equation $\mu(X \mid y) = \mu_x + \rho\frac{\sigma_X}{\sigma_Y}(y - \mu_y)$. In other words, the dependence of the mean on $y$ is linear, with gradient proportional to $\rho$, and this relationship defines what is called the regression of $X$ on $Y$. The linear equation above implies the well-known phenomenon of *regression to the mean.* Suppose $\sigma_X = \sigma_Y$ and $\mu_X = \mu_Y = m$. Then $\mu(X \mid y) = m + \rho(y - m)$, which is the point located a proportion $\rho$ of the distance between $y$ and $m$. For example, suppose that people's heights are normally distributed and that $Y$ is the average of the two parents' height and $X$ is the offspring's height. Suppose also that the means and standard deviations of these two variables are the same and that $\rho = \frac{1}{2}$. Then the mean value of the offspring's height is halfway between the common population mean and the two parents' average height. It is often said that results like this explain what we actually observe, but explaining exactly how parameters of probability distributions are linked to what we can observe turns out to be a hotly disputed subject, and it is one which will occupy a sustantial part of the remainder of this book.

Let us leave that topic in abeyance, then, and end this brief outline of that part of the mathematical theory of probability which we shall have occasion to use, with the derivation and some discussion of the limiting properties of the first non-trivial random-variable distribution to be investigated thoroughly, and which has no less a fundamental place in statistics than the normal distribution, to which it is intimately related.

## h THE BINOMIAL DISTRIBUTION

This was the binomial distribution. It was through examining the properties of this distribution that the first great steps on the road to modern mathematical statistics were taken, by James Bernoulli, who proved (in *Ars Conjectandi,* published posthumously in 1713) the first of the limit theorems for sequences of independent random variables, the so-called *Weak Law of Large Numbers,* and Abraham de Moivre, an eighteenth-century Huguenot mathematician settled in England, who proved that, in a sense we shall make clear shortly, the binomial distribution tends for large $n$ to the normal. Although Bernoulli demonstrated his result algebraically, it follows, as we shall see, from de Moivre's limit theorem.

Suppose (i) $X_i$, $i = 1, \ldots, n$, are random variables which take two values only, which we shall label 0 and 1, and that the probability that each takes the value 1 is the same for all $i$, and equals $p$:

$$P(X_i = 1) = P(X_j = 1) = p.$$

Suppose also (ii) that the $X_i$ are independent; i.e.,

$$P(X_1 = x_1 \,\&\, \ldots \,\&\, X_n = x_n) = P(X_1 = x_1) \times \ldots \times P(X_n = x_n),$$

where $x_i = 1$ or 0. In other words, the $X_i$ are independent, identically distributed random variables. Let $Y = X_1 + \ldots + X_n$. Then for any $r$, $0 \leq r \leq n$,

$$\textbf{(2)}\quad P(Y = r) = {}^nC_r p^r (1 - p)^{n-r}$$

since using the additivity property, the value of $P$ is obtained by summing the probabilities of all conjunctions

$$X_1 = x_1 \,\&\, \ldots \,\&\, X_n = x_n,$$

where $r$ of the $x_i$ are ones and the remainder are zeros. There are ${}^nC_r$ of these, where ${}^nC_r$ is the number of ways of selecting $r$ objects out of $n$, and is equal to $\frac{n!}{(n-r)!r!}$, where $n!$ is equal to $n(n - 1)(n - 2) \ldots 2 \cdot 1$, and 0! is set equal to 1. By the independence and constant probability assumptions, the probability of each conjunct in the sum is $p^r(1 - p)^{n-r}$, since $P(X_i = 0) = 1 - p$.

$Y$ is said to possess the *binomial distribution.* The mean of $Y$ is $np$, as can be easily seen from the facts that

$$E(X_1 + \ldots + X_n) = E(X_1) + \ldots + E(X_n)$$

and that

$$E(X_i) = p \cdot 1 + (1-p) \cdot 0 = p.$$

The squared standard deviation, or variance of $Y$, is

$$\begin{aligned} E(Y - np)^2 &= E(Y^2) + E(np)^2 - E(2Ynp) \\ &= E(Y^2) + (np)^2 - 2npE(Y) \\ &= E(Y^2) - (np)^2. \end{aligned}$$

Now

$$\begin{aligned} E(Y^2) &= \sum E(X^2_i) + \sum_{i \neq j} E(X_i X_j) \\ &= np + n(n-1)p^2. \end{aligned}$$

Hence

$$\text{s.d.}(Y) = \sqrt{-np^2 + np} = \sqrt{np(1-p)}.$$

## THE WEAK LAW OF LARGE NUMBERS

The significance of these expressions is apparent when $n$ becomes very large. De Moivre showed that for large $n$, $Y$ is approximately normally distributed with mean $np$ and standard deviation $\sqrt{np(1-p)}$ (the approximation is very close for quite moderate values of $n$). This implies that the so-called *standardised variable* $Z = \dfrac{(Y - np)}{\sqrt{np(1-p)}}$ is approximately normally distributed for large $n$, with mean 0 and standard deviation 1. Hence

$$P(-k < Z < k) \approx \Phi(k) - \Phi(-k),$$

where $\Phi$ is the normal distribution function with zero mean and unit standard deviation. Hence

$$P\left(p - k\sqrt{\frac{pq}{n}} < \frac{Y}{n} < p + k\sqrt{\frac{pq}{n}}\right) \approx \Phi(k) - \Phi(-k),$$

where $q = 1 - p$. So, setting $\epsilon = k\sqrt{\dfrac{pq}{n}}$,

$$P(p - \epsilon < \frac{Y}{n} < p + \epsilon) \approx \Phi\left(\epsilon\sqrt{\frac{n}{pq}}\right) - \Phi\left(-\epsilon\sqrt{\frac{n}{pq}}\right).$$

Clearly, the right-hand side of this equation tends to 1, and we have obtained the *Weak Law of Large Numbers:*

$$P(\left|\frac{Y}{n} - p\right| < \epsilon) \to 1, \text{ for all } \epsilon > 0.$$

This is one of the most famous theorems in the history of mathematics. James Bernoulli proved it originally by purely combinatorial methods. It took him twenty years to prove, and he called it his "golden theorem". It is the first great result of the discipline now known as mathematical statistics and the forerunner of a host of other limit theorems of probability. Its extramathematical significance lies in the fact that sequences of independent binomial random variables with constant probability, or Bernoulli sequences as they are called, are thought to model many types of sequence of repeated stochastic trials (the most familiar being tossing a coin $n$ times and registering the sequence of heads and tails produced). What the theorem says is that for such sequences of trials the relative frequency of the particular character concerned, like heads in the example we have just mentioned, is with arbitrarily great probability going to be situated arbitrarily close to the parameter $p$.

The Weak Law, as stated above, is only one way of appreciating the significance of what happens as $n$ increases. As we saw, it was obtained from the approximation

$$P(p - k\sqrt{\frac{pq}{n}} < \frac{Y}{n} < p + k\sqrt{\frac{pq}{n}}) \approx \Phi(k) - \Phi(-k),$$

where $q = 1 - p$, by replacing the variable bounds (depending on $n$) $\pm k\sqrt{\frac{pq}{n}}$ by $\epsilon$, and replacing $k$ on the right-hand side by $\epsilon\sqrt{\frac{n}{pq}}$. The resulting equation is equivalent to the first. In other words, the Weak Law can be seen either as the statement that if we select some fixed interval of length $2\epsilon$ centred on $p$, then in the limit as $n$ increases, all the distribution will lie within that interval, or as the statement that if we first select any value between 0 and 1 and consider the interval centred on $p$ which carries that value of the probability, then the end-points of the interval move towards $p$ as $n$ increases, and in the limit coincide with $p$.

Throughout the eighteenth and nineteenth centuries people took these facts to justify inferring, from the observed relative frequency of some given character in long sequences of apparently *causally* independent trials, the approximate value of the postulated binomial probability. While such a practice may seem *suggested* by Bernoulli's theorem, it is not clear that it is in any way justified. While doubts were regularly voiced over the validity of this 'inversion', as it was called, of the theorem, the temptation to see in it a licence to infer to the value of $p$ from 'large' samples persists, as we shall see in Chapter 13, where we shall discuss the issue in more detail.

The Weak Law of Large Numbers is only one of the theorems of mathematical probability theory which go under the name of 'laws' of large numbers. Another, equally famous for pointing to a connection between probabilities and frequencies in sequences of identically distributed, independent binomial random variables, is the Strong Law, which is usually stated as a result about actually infinite sequences of such variables: it asserts that with probability equal to 1, the limit of $y/n$ exists (that is to say, the relative frequency of ones converges to some finite value) and is equal to $p$. So stated, the Strong Law requires for its proof the axiom of countable additivity, though there is a version for finite sequences of random variables which needs only finite additivity (*see*, for example, Feller, 1950, p. 203). We shall leave the discussion of the mathematics of probability there and now turn our attention to the foundations of the Bayesian methodology and to a discussion of an epistemic interpretation of probabilities.

## EXERCISES

1. Show that if $x_1 < x_2$ then $F(x_1) \leq F(x_2)$, where $F$ is the distribution function for $X$.

2. Show that $F(x_2) - F(x_1) = P(x_1 \leq X < x_2)$.

**3.** Where $X_1$ and $X_2$ each take only finitely many values, show that $E(X_1 + X_2) = E(X_1) + E(X_2)$.

**4.** Show that the uniform density $f(x)$ in $[a,b]$ has the form $f(x) = |b-a|^{-1}$ for $a \leq x \leq b$, and $f(x) = 0$ outside that interval. Show that

a. its distribution function has the form $F(x) = |b-a|^{-1}\,|x-a|$

b. its mean value is $\frac{(a+b)}{2}$

**5.** Show that if $h$ and $h'$ are probabilistically independent, then so are each of the pairs $\{h,\sim h'\}$, $\{\sim h,h'\}$, $\{\sim h,\sim h'\}$.

**6.** Where $Y = X_1 + \ldots + X_n$, and the $X_i$ are $n$ independent random variables taking the values 1,0 with probabilities $p,1-p$ respectively, graph the three binomial distributions corresponding to the values (i) $p = \frac{1}{2}$, $n = 3$, (ii) $p = \frac{1}{2}$, $n = 5$, and (iii) $p = \frac{1}{2}$, $n = 7$ (i.e., plot $P(Y = r)$ against $r$ in each of the three cases).

**7.** Suppose $Y$ is as in the previous question and $n = 10$. What are the values of the random variable $V_n = \frac{Y}{n}$? Sketch what happens to the graphs of $P(V_n = v)$ as $n$ increases to infinity.

# CHAPTER 4

# *The Classical and Logical Theories*

## a INTRODUCTION

The purpose of this book is to explain characteristic features of scientific inference in terms of the probabilities, relative to the available data, of the various explanatory hypotheses under consideration. In this chapter we shall consider the nature of these probabilities. We shall first discuss, and reject, the thesis, associated in this century with the names of Keynes and Carnap, that they are logical quantities, determined in a quite objective way by the logical structure of the hypotheses and the data. We shall argue that this idea is untenable and that they should instead be understood as subjective assessments of credibility, regulated by the requirement that they be overall consistent; and we will show that a necessary and sufficient condition for consistency is agreement with the rules of the probability calculus.

The first developed theory of probability was what we now know as the Classical Theory but which is, as we shall show presently, better regarded as the forerunner of Keynes's and Carnap's explicitly logical theories. The Classical Theory sought to provide a foundation both for the gambler's calculations of his expected gains and losses and for the philosopher's and scientist's belief in the validity of inductive inference; let us see how.

## b THE CLASSICAL THEORY

The theory was first expounded in the late seventeenth century (it appears, for example, in James Bernoulli's *Ars Conjectandi,* 1713), but the account which became classic, and one of the most widely read of all works ever written on probability, is

Laplace's *Philosophical Essay on Probabilities,* 1820, published originally as the preface to his monumental *Analytical Theory of Probabilities.* The basic principle of the theory was that for suitable hypotheses $h$ and data $e$, there is a measurable, numerical degree of certainty ("degree of certainty" is actually Bernoulli's term) which should be entertained in the truth of $h$ in the light of $e$. These degrees of certainty are located in the closed unit interval, where they are given uniquely as the values of a probability function, with 1 corresponding to the certainty that $h$ is true and 0 to the certainty that $h$ is false. Where the probability in any given case could be calculated, its value, call it $p$, was also regarded as providing, in the ratio $p{:}1-p$, the fair odds on $h$ in the light of $e$, or the odds relative to which the rational expectation of gain for each side of a bet at those odds is zero (we shall discuss this notion of fair odds later in this chapter).

The pioneers did not use modern notation, nor did they distinguish explicitly between conditional and unconditional probabilities; only in the work of Keynes and later authors, like Jeffreys (1961) and more recently Carnap (1950, 1952, and 1971), do we find such probabilities represented explicitly as conditional probabilities. Nevertheless, it seems clear that after the discovery of Bayes's Theorem and the use to which it was put by Laplace, the probability function which he himself took to measure the rational degree of certainty of $h$ in the light of $e$ was a conditional probability. We shall in what follows represent these probabilities by the modern notation $P(h \mid e)$.

### b.1 The Principle of Indifference

Implicit in the Classical Theory is the following principle:

> if there are $n$ mutually exclusive possibilities $h_1, \ldots, h_n$, and $e$ gives no more reason to believe any one of these more likely to be true than any other, then $P(h_i \mid e)$ is the same for all $i$.

The principle was later called by von Kries (1886) the *Principle of Insufficient Reason* and by Keynes (1921) the *Principle of Indifference* (we shall follow Keynes's terminology, merely because it is more compact). If the $h_i$ are exhaustive as well as being exclusive, a probability distribution which assigns them all the same probability is called *uniform.* Thus the principle prescribes uniform distributions over sets of mutually exclusive and exhaustive possibilities (such sets are called *partitions*) where there is no reason to believe one more likely to be realised than any other.

The principle is of fundamental importance within the Classical Theory, because the judgment that $e$ is indifferent, as it were, or epistemically neutral, between a set of exclusive alternatives seems to be one which can be made easily in certain cases. If, for example, $e$ were to say merely that a number between 1 and $n$ will be selected, and nothing else, then it seems plausible to say that $e$ is indifferent between the hypotheses of the form '$i$ is selected', for each integer $i$ between 1 and $n$.

Let us see what follows from supposing that there are sets of alternatives $h_1, \ldots, h_n$ between which such judgments of equality are valid. In particular, let us suppose that the $h_i$ are also an exhaustive set and that $h$ is a hypothesis equivalent, given $e$, to a disjunction of $r$ of them. In Laplace's now famous terminology, the $h_i$ represent "cases equally possible" in the light of $e$, and the $r$ hypotheses whose disjunction is equivalent to $h$ are the 'cases favourable to $h$'. It follows from (14)–(16), Chapter 2, that $P(h_i \mid e) = \frac{1}{n}$, for all $i$, and that $P(h \mid e) = \frac{r}{n}$. Let us quote Laplace:

> The theory of chance consists in reducing all the events of the same kind to a certain number of cases equally possible, that is to say, to such as we may be equally undecided about in regard to their existence, and in determining the number of cases favourable to the event whose probability is sought. The ratio of this number to that of all the cases possible is the measure of this probability, which is thus simply a fraction whose numerator is the number of favourable cases and whose denominator is the number of all the cases possible. (Laplace, 1820, pp. 6–7)

The earliest applications of the Classical Theory were to games of chance, like rolling dice or picking tickets from an urn or tossing coins, because their classes of elementary outcomes are finite and also, as a consequence of the carefully randomized structure of those games, we seem very much to be in a prior state of "equal indecision", to use Laplace's terminology, as to the occurrence of each of those outcomes. Later, in the pioneering work of Bayes (1763), the scope of the Classical Theory was extended to outcome spaces which could be represented by a bounded one- or many-dimensional interval of real numbers: if all the points in the interval $[a,b]$ are the possible values of a real parameter $T$, then the probability that $T$ will lie in a subinterval $[c,d]$ is, by analogy with discrete sets of alternatives, plausibly equal to $\frac{|d - c|}{|b - a|}$.

It might be objected that the randomizing procedures used in games of chance, like shuffling decks of cards or making tossed coins as nearly perfectly balanced as physically possible, are regarded as legitimate grounds for distributing equal degrees of certainty over the elementary outcomes of the apparatus involved only after we have observed that such procedures are as a matter of fact associated with roughly equal frequencies in those outcomes. We are therefore, so the objection continues, appealing not to a criterion of 'epistemic neutrality' at all, but covertly to an unjustified equation of degree of certainty with long-run observed frequency.

The Classical theorists after Laplace had an answer to this, which was that the objection puts the cart before the horse. Laplace, building on earlier work of Bayes, seemed in a famous demonstration to have shown that the identity between observed frequencies and degrees of certainty is not an assumption, covert or otherwise, which needs to be made but, on the contrary, appears to be a consequence of assuming that at the outset one knows *nothing whatever* about the propensity of the apparatus to generate the different possible frequencies of outcomes. If one assumes literally nothing about such propensities, the Principle of Indifference then seems to imply that the various frequencies are of equal epistemic standing relative to that null data set.

It was on the basis of such an application of the Principle of Indifference that Laplace was able to prove his result, the famous, or infamous, Rule of Succession. In so doing he seemed also to have answered Hume's famous thesis that one cannot without circularity prove that the observation of past events gives any guidance to what will occur in the future. Because Laplace's result appears to justify a very strong type of enumerative induction, one moreover which in this century has been accorded the status of a purely logical principle, and because it introduces in an informal way the central ideas of Carnap's later, explicitly logical theory, we shall give an outline of Laplace's argument.

### b.2 The Rule of Succession

Laplace's result was called the Rule of Succession by John Venn (1866), and the name stuck. It states that if an event $A$ is one of the possible outcomes of instantiating some set of repeatable conditions, and if in $s$ repetitions $A$ is observed to occur $r$ times, then the probability that it will occur at the $(s + 1)$th repetition, conditional upon the data, is $\frac{r+1}{s+2}$. Since as $s$ grows large $\frac{r+1}{s+2}$ tends to $\frac{r}{s}$, the Rule of Succession appears to provide a justification for estimating epistemic probabilities by sample frequencies; hence, when $r = s$, that is to say, when we have observed only $A$'s in the sample, $\frac{r+1}{s+2}$ tends quickly to 1, the value of certainty.

Laplace obtained the result by considering a random sample from an 'infinite urn' containing black and white tickets in an unknown 'proportion'. A proportion in an infinite set is obviously a fiction unless understood in some limit sense, and Laplace's result is in fact a mathematically, if not philosophically, quite respectable limiting result obtained by letting the number of tickets in a finite urn grow very large. We shall reconstruct his reasoning, as later Classical theorists did, in terms of a large finite urn, letting its size tend to infinity. The reason for considering very large numbers of tickets at all is that Laplace wanted, as we shall see, to use urn-sampling as a model of an experiment with outcomes $A$ and not-$A$, which is capable of being repeated an arbitrary number of times.

Suppose the urn contains $n$ tickets, an unknown number of which are white; $s$ tickets are withdrawn and not replaced, $r$ of which are white. What is the probability, relative to this information, that the next ticket to be withdrawn will be white? The way this probability is obtained is exactly the way Carnap was later to compute it in the 1950s, and it is an exercise in permutations and combinations. Suppose we regard the tickets as numbered 1 to $n$ in the order in which they are withdrawn. It clearly does not matter how we *name* the tickets; this method merely makes the discussion simpler.

Let $e$ be the sentence 'There are $r$ white tickets among the $s$ drawn', and let $b(q)$ be the sentence 'There are $q$ tickets initially in the urn'. We do not of course know the value of $q$. Finally, let $h$ be the sentence 'The $s + 1$th ticket to be drawn is white'. We want to compute $P(h \mid e)$, and Classical theorists found an ingenious method for doing this. By the probability calculus

$$\textbf{(1)}\quad P(h \mid e) = \frac{P(h\&e)}{P(e)} = \frac{\Sigma P[h \,\&\, e \mid b(q)]\, P[b(q)]}{\Sigma P[e \mid b(q)]\, P[b(q)]}$$

where the sums are over the possible values of $q$. The Principle of Indifference seems to enable us to compute all the terms on the right-hand side of (1). Consider $P[e \mid b(q)]$. First, what is $P[b(q)]$? Well, there are ${}^nC_q$ ways in which $q$ white tickets may be distributed among the numbers 1 to $n$. We have no reason to suppose any one of these more likely than any other, so by the Principle of Indifference each has the same probability; call it $p$. Hence by the additivity principle $P[b(q)] = p \cdot {}^nC_q$. We shall see that the value of $p$ is unimportant to the calculation.

To find $P[e \,\&\, b(q)]$ and hence $P[e \mid b(q)]$ we need to know how many of the ${}^nC_q$ distributions of the $q$ white tickets in the urn have $r$ white tickets distributed among the tickets numbered 1 to $s$. This is not difficult to compute: there are ${}^sC_r$ ways in which there might be $r$ white tickets among the $s$ drawn, and for each of these possibilities there are ${}^{n-s}C_{q-r}$ ways in which the remaining $n-s$ tickets can have $q-r$ white tickets distributed among them. Hence there are altogether ${}^sC_r{}^{n-s}C_{q-r}$ ways in which the statement 'There are $r$ white tickets among the $s$ drawn' can be satisfied relative to any distribution of $q$ white tickets initially in the urn. Since each has the probability $p$, $P[e \mid b(q)] = p\,{}^sC_r{}^{n-s}C_{q-r}$. It follows that the required conditional probability is equal to $\dfrac{{}^sC_r{}^{n-s}C_{q-r}}{{}^nC_q}$.

$p$ does not appear in this expression, but in order to obtain $P(e)$ we need to multiply the expression by $P[b(q)]$ and sum over the possible values of $q$, as (1) makes clear; and $P[b(q)]$, as we have seen, does depend on $p$. Now for the miracle: enter the Principle of Indifference again, as deus ex machina. For since by assumption we know nothing about the value of $q$, we can simply appeal to the Principle to make each of the $n+1$ possible values, $0,1,2, \ldots n$, equiprobable, i.e., to give them each the probability $(n+1)^{-1}$. Again, we do not need to know $p$, though we are implicitly making it satisfy the equation $p = \frac{1}{(n+1)\,{}^{n}C_{q}}$. Note the dependence of $p$ on $q$. We shall find when we come to the work of Carnap that what he does is very familiar, for it is just this equation in $p$, $n$, and $q$ which defines $c^*$, Carnap's most favoured confirmation function (in Carnap, 1950).

To proceed. We now have

$$(2)\ P(e) = \Sigma P[e \mid b(q)]P[b(q)] = \frac{\Sigma\,{}^{s}C_{r}\,{}^{n-s}C_{q-r}}{(n+1)\,{}^{n}C_{q}}.$$

Now for $P(h\&e)$. It is not difficult to see, by reasoning similar to the above, that

$$(3)\ P(h\ \&\ e) = \frac{1}{n+1}\,\frac{\Sigma\,{}^{s}C_{r}\,{}^{n-s-1}C_{q-r-1}}{{}^{n}C_{q}}.$$

$P(h \mid e)$ is obtained by dividing (3) by (2), and a little algebraic manipulation eventually leads to a value of the quotient equal to $\frac{r+1}{s+2}$, i.e., to the Rule of Succession. This expression does not depend on $n$, the number of tickets in the urn, and so holds in the limit as $n$ tends to infinity, yielding Laplace's result.

The same result can also be obtained by noting that $\frac{{}^{s}C_{r}\,{}^{n-s}C_{q-r}}{{}^{n}C_{q}}$ is approximately equal, for large $n$, to ${}^{s}C_{r}x^{r}(1-x)^{s-r}$, where $x = \frac{q}{n}$. The weights $(n+1)^{-1}$ on these factors in the right-hand side of (2) approach the uniform density $f(x) = x$, and the sum is replaced by the integral ${}^{s}C_{r} \int_{0}^{1} x^{r}(1-x)^{s-r}\,dx$. Similarly, (3) approaches the integral ${}^{s}C_{r} \int_{0}^{1} x^{r+1}(1-x)^{s-r}\,dx$. Integrating by parts we find that $\int_{0}^{1} x^{r}(1-x)^{s-r}\,dx$ is equal to $\frac{r!(s-r)}{(s+1)!}$ and $\int_{0}^{1} x^{r+1}(1-x)^{s-r}\,dx$ is equal to $\frac{(r+1)!(s-r)!}{(s+2)!}$. Dividing the second of these expressions by the first we again obtain $\frac{r+1}{s+2}$.

Exactly the same integral formulas had been discovered earlier by Bayes (they appear, though disguised in the geometrical form favoured in England at the time, in Bayes's famous posthumous *Memoir,* 1763). Bayes himself used as a model for the occurrence/non-occurrence of an arbitrary event $A$ not an urn, as Laplace did, but a billiard table of length $l$ on which are rolled two billiard balls. $A$ occurs when the second ball exceeds the distance $d$ travelled by the first, and Bayes took the proportional distance $\frac{d}{l}$ to represent an unknown probability of $A$'s occurring. Now let $e$ state that $A$ was observed to occur $r$ times out of $s$. Adopting a tactic that was later to become the hallmark of Bayesian statistics, Bayes took the postulated unknown probability of $A$ to be a random variable, set $P(e \mid X = x) = {}^sC_r\, x^r (1 - x)^{s-r}$, and assigned $X$ a uniform prior probability distribution, with density $f(x) = 1$ in the unit interval. By the theorem of total probability for the continuous case he then obtained $P(e) = {}^sC_r \int_0^1 x^r (1 - x)^{s-r}\, dx$.

Bayes was interested in obtaining the posterior probability distribution for $X$ conditional on having observed $e$ (the distribution he obtained is what is called a beta distribution, and we shall discuss this and its properties in Chapter 14); he did not himself take the immediate next step and derive the Rule of Succession. That was left to Laplace, who also pointed out the immediate corollary of the Rule, that if $A$ has occurred at every observation $(r = s)$, then the probability that it will occur at the next tends quite quickly to 1. Using this corollary, Laplace, in a much-quoted passage, proceeded to compute the exact odds on the sun's rising the next day ("It is a bet of 1826214 to one that it will rise again tomorrow" Laplace, 1820, p. 19).

The problem of induction, which Hume had declared, with the most telling arguments, to be insoluble, seemed, half a century later, to be solved. Hume had argued that all "probable arguments" from past to future are necessarily circular, presupposing what they set out to prove. Yet the Rule of Succession appears to presuppose nothing except a prior epistemic neutrality between possibilities. As we shall see, however, Hume is not so easily refuted, and nobody today sees in the Rule of Succession a solution of that most intractable of problems, the problem of induction.

**b.3 The Principle of Indifference and the Paradoxes**

As a matter of fact, by the end of the first half of this century the Classical Theory and the Rule of Succession were largely discredited. One reason for this state of affairs was the feeling that the Rule was just too powerful to be plausible. Keynes, in a memorable passage, wrote that

> no other formula in the alchemy of logic has exerted more astonishing powers. For it has established the existence of God from total ignorance, and it has measured with numerical precision the probability that the sun will rise tomorrow. (Keynes, 1921, p. 89)

The second, probably more important, reason why the Classical Theory became discredited was that the Principle of Indifference was discovered to yield inconsistencies with alarming ease. Consider an urn again. Suppose that you draw a ball from an urn which you are told contains white and coloured balls in some unknown proportion, and that the coloured balls are either red or blue. What sort of ball will you draw? Your data seem to be neutral, in the first instance, between its being white or coloured. Hence according to the Principle of Indifference, the probability that it will be white is one half. But if it is coloured, the ball is red or blue, and the data are surely neutral between the ball's being either white or blue or red. Hence, according to the Principle of Indifference, the probability of the ball's being white is one third.

'Paradoxes' (as they came to be called) of this type abound. Consider another, which occasioned a good deal of discussion. In the example of Laplace's urn it seemed a straightforward application of the Principle of Indifference to ascribe equal probabilities to the $n + 1$ possible numbers of white tickets initially in the urn. It was by this strategy that we eliminated the nuisance parameter $p$ from the computation. But was that strategy really justified? Recall that we know nothing about the distribution of white tickets initially in the urn, and so we have no information concerning the whiteness or non-whiteness of each of those $n$ tickets. Call a *constitution* of the urn a particular specification for each of the $n$ tickets of whether it is white or not. Since ticket number 1 might be white or not white, and similarly for tickets 2, 3, and so on up to $n$, there are $2 \cdot 2 \cdot \ldots \cdot 2$ ($n$ times) = $2^n$ possible constitutions of the urn. Our information does not distinguish any one of these as being more likely than any other, and so by the Principle of Indifference each should have the same probability, named $2^{-n}$. But that means that the probability of $b(q)$, the statement that says that $q$ of the tickets initially in the urn are white, is ${}^nC_q 2^n$, which varies with $q$, and not $(n + 1)^{-1}$, as we took it to be above, by another application of the Principle of Indifference. But the two applications of the Principle cannot both be correct, since they give clearly inconsistent results. Yet it is difficult to see why one should be any more correct than the other.

Bertrand's well-known paradoxes of geometrical probability showed that the Principle of Indifference generated inconsistencies also in the continuous domain (a very clear presentation is given by Neyman, 1952, pp. 15–17). A simple example of the type of inconsistency which can be generated by continuous variables is afforded by considering a parameter $T$ about which nothing is known except that its value lies in some specified bounded interval $[a,b]$, $0 < a$. Suppose that we take our essentially null information to justify a prior density for $T$ equal to $(b - a)^{-1}$. If all we know about $T$ are the limits of its possible values, however, then all we know about the parameter $T^2$ are the limits of *its* possible values. $T^2$ should therefore possess, by the same token, a uniform prior density equal to $(b^2 - a^2)^{-1}$. We now have a contradiction; for it follows from the assumption that the density of $T$ is uniform in $[a,b]$, that the density of $T^2$ is not uniform but is equal to $\frac{1}{2x(b - a)}$ at the point $T = x$.

The idea of a probabilistic logic of inductive inference based on some form of the Principle of Indifference nevertheless retained a powerful appeal. Keynes (1921) continued in the programme inaugurated by Leibniz and Bernoulli, but elaborated and sharpened their thesis that the probability calculus is a more general logic than the logic of deduction with the claim that between any two propositions $a$ and $b$ there is a relation of partial entailment, such that the degree to which $a$ entails $b$ is measured by a unique conditional probability $P(b \mid a)$, and this degree of *partial entailment* is equal to the degree of belief which it would be rational to repose in $b$ were your data restricted to $a$ (Keynes, 1921, p. 16; Keynes also thought, however, that $P(b \mid a)$ did not always take numerical values, and indeed that probabilities in general are only partially ordered).

When it came to evaluating these ostensibly logical probabilities for those pairs of statements for which $P$ did take numerical values, Keynes recommended a form of the Principle of Indifference which he hoped retained what he regarded as its valid core without, however, generating paradoxes and inconsistencies. His modified version of the Principle is as follows. Suppose (a) that the data $e$ determine a finite set of $n$ exclusive possibilities $h_i$, (b) that $e$ has 'the same form' relative to each of the $h_i$, and (c) that none of those hypotheses is further 'divisible' or decomposable into subalternatives of the same form as some other hypothesis in the original set. Then and only then should the $h_i$ be assigned equal probabilities relative to $e$ (Keynes, 1921, pp. 60–66). This modification seems to meet the difficulty generated by cases such as the urn containing white and coloured balls. For it is now no longer legitimate to infer that the prior probabilities of the drawn ball's being either white or coloured are equal, since the latter possibility decomposes into a disjunction of two narrower possibilities, of the same simple colour kind as the former, being red or being blue.

But it is not clear that this modification of the Principle of Indifference saves it from its problems with continuous transformations of random variables whose possible values are an interval on the line. Keynes asserted that a legitimate application is to divide up the interval (if it is bounded) into a finite number $m$ of subintervals of equal length and to assign equal probabilities to each subinterval, letting $m$ then tend to infinity. But he conceded that the resulting uniform density distribution is not maintained under a large class of transformations, a fact he attempted to dismiss by saying that it is an example of the phenomenon, well known to mathematical physicists, where often the form of a density distribution depends on the passage to the limit (1921, p. 69). This is true, but does not save him the arbitrary choice of how in any given case to proceed to the limit (that is, which particular variable to take as determining the uniform distribution).

Also, it is far from clear that the colour categories in the example above *are* of the same kind: white light, for example, is a superposition of light of other wavelengths. 'Being white' is a colour category of the same form as 'being blue' only relative to a particular linguistic system, which does not make distinctions which might be made by a more refined one. So there seems to be implicit in Keynes's analysis reference to some appropriate language, sufficiently well defined syntactically to permit judgments of sameness or difference of form of its component predicates.

Keynes, writing in England in the second decade of this century, was geographically and temporally just too distant from the Continental work, still in its pioneering stage, which would revolutionise the study of formal logic and give a precise sense to the sorts of syntactic criteria he attempted to enunciate. Thirty years later Carnap was able to remedy the deficiency, and in so doing find reasons, which seemed to him conclusive, against adopting the Keynesian form of the Principle of Indifference. What Carnap was not able to do, however, was eliminate the arbitrariness inherent in the language-relativity of the probability distributions he found more congenial.

## ■ c CARNAP'S LOGICAL PROBABILITY MEASURES

### c.1 Carnap's $c^{\dagger}$ and $c^{*}$

Carnap, in his first detailed (and monumental) work on probabilistic inductive logic (1950), attempted to exploit the advances in logical notation and theory made in the previous decades to create a purely logical theory of induction, in which the value of an appropriate conditional probability function, which he wrote $c(h,e)$, measures the degree of rational credibility of $h$ relative to data or evidence $e$. While in the work cited he deliberately avoided the term 'degree of rational belief', because of what he called its "psychologism", and regarded $c(h,e)$ as a formal explication of, among other things, the degree of confirmation of $h$ by $e$, there is no doubt that his work is a continuation of the Bernoulli-Laplace-Keynes tradition. For $c(h,e)$ is also regarded by Carnap as determining, in the quantity $c(h,e):1 - c(h,e)$, the fair odds on $h$ in the light of $e$; and indeed, the very term 'degree of rational belief' makes an appearance in his later work (1971, p. 9, for example).

Carnap followed Keynes in regarding $c(h,e)$ also as a measure of the extent to which $h$ is entailed by $e$ (though as a matter of fact the idea goes much farther back, originating in Bolzano, 1837). One of the fruits of the development of formal logic between the date of Keynes's *Treatise* and Carnap's *Logical Foundations* was that Carnap was able to demonstrate that any conditional probability function $P(b \mid a)$ defined on the sentences of some language can be understood as determining a measure of the extent to which $a$ entails $b$. It is easy to see this. According to the standard definition of logical consequence, $a$ is a consequence of $b$ if all the possible structures interpreting a predicate language $L$ (or $L$-possible worlds) which make $b$ true also make $a$ true. Using the notation of Chapter 2, section **f,** let $M(a)$ be the set of all such structures making $a$ true. As we saw there, every probability $P$ on the sentences $a$ of $L$ determines a real-valued, non-negative, additive function $P'$ on the sets $M(a)$. Such a function is technically a normalised measure. Now any finite measure on the same field obeys the same formal laws as the ordinary counting measure which, if a set $A$ is finite, simply tells us how many things there are in $A$. Now

$$P(b \mid a) = \frac{P(b \,\&\, a)}{P(b)} = \frac{kP'[M(a) \cap M(b)]}{kP'[M(b)]}$$

for all nonzero $k$ and all $b$ such that $P(b) > 0$. But $kP'$ is a finite measure on the sets $M(a)$, and so the ratio on the right-hand side, and therefore $P(b \mid a)$, can be regarded as defining a generalised 'proportion' of those structures making $b$ true which also make $a$ true, and hence determining a measure of the 'degree' to which $b$ entails $a$.

But this is not very interesting, since any conditional probability function $P$ has this property, independently of what its actual values are. What Carnap attempted to do was to impose constraints of a methodological/epistemological kind on $c(h,e)$ which would eventually determine the function uniquely as the foundation for a quantitative system of inductive logic. This he never succeeded in doing, though he did consider two such functions, called by Carnap $c^{\dagger}$ and $c^{*}$, as candidates for that role. We shall now show, for the simplest type of Carnapian language, how Keynes's modified Principle of Indifference leads directly to one of these functions, $c^{\dagger}$, which fails conspicuously to meet the methodological criteria

Carnap regarded as appropriate to an inductive probability measure. This fact may explain why Carnap never explicitly in his 1950 work or thereafter invoked that principle, though $c^*$, the other of the two functions and the one favoured by Carnap, is, as we shall see, none other than Laplace's measure, generating the Rule of Succession.

Consider a very simple predicate language $L(A,n)$ which enables us to discuss the possession or non-possession of an attribute $A$ by any or all of $n$ distinct individuals $a_i$. We can list all the completely specified possible states of affairs in such a simple universe by means of the conjunctions

$$\pm A(a_1) \ \& \ \pm A(a_2) \ \& \ \ldots \ \& \pm A(a_n),$$

where $+A(a_i)$ is $A(a_i)$ and $-A(a_i)$ is $\sim A(a_i)$. There are clearly $2^n$ such sentences, called the *state-descriptions* of $L(A,n)$ by Carnap. The sentences of $L(A,n)$ describing the possible numbers of $A$'s in the universe are equivalent to those disjunctions of all state-descriptions containing the same number of unnegated $A$'s; these disjunctions Carnap called the *structure-descriptions* of $L(A,n)$. (This terminology is misleading since the state-descriptions more truly determine the types of structure describable within $L(A,n)$.) It should be clear that one possible interpretation of these state-descriptions is as the different constitutions of Laplace's urn in section **b.2** above, by letting $a_i$ denote the ticket numbered $i$ and $A$ the property of being white. The structure-descriptions correspond to the statements $b(q)$ in that earlier section, for each $q = 1,2, \ldots ,n$.

We shall return to the discussion of the urn shortly. For the moment we shall remain at the purely abstract level and note that the state-descriptions are complete specifications of the sorts of structure describable within $L(A,n)$: they constitute an ultimate partition relative to the descriptive resources available within $L(A,n)$, and so are not further decomposable within $L(A,n)$. We may therefore apply the Keynesian form of the Principle of Indifference to infer that the probabilities of each of the state-descriptions relative to null data are equal to $2^{-n}$. Such probabilities are unconditional probabilities (though they can of course always be represented as conditional probabilities relative to a tautology: $P(h)$ is always equal to $P(h \mid t)$ where $t$ is a tautology). Carnap represented the unconditional probability of a sentence $h$ by $m(h)$, calling $m$ the *measure function* associated with the conditional probability function $c$. Therefore $c(h,e) = \frac{m(h \,\&\, e)}{m(e)}$. The measure which assigns the state descriptions the equal weights of $2^{-n}$ he wrote as $m^{\dagger}$; and the conditional probability function $c$ based on $m^{\dagger}$ is $c^{\dagger}$.

Any sentence $h$ of $L(A,n)$ can be represented as a disjunction of $z$ of these state-descriptions, where $z$ is a number between 0 and $2^n$ inclusive. Hence $m^{\dagger}(h) = z2^{-n}$. In $L(A,n)$ there is, in particular, a sentence $e(r,s)$ which states that $r$ of the individuals $a_1, \ldots, a_s$, $s \leq n$, possess $A$. In the Laplace urn interpretation of $L(A,n)$, $e(r,s)$ corresponds to the statement $e$ in section **b.2** that $r$ of the $s$ withdrawn tickets are white. It is not difficult to work out (and the reader might wish to attempt it as an exercise) that $m^{\dagger}[e(r,s)] = {}^sC_r 2^{n-s}\, 2^{-n}$. Clearly, $A(a_{s+1})$ is the statement $h$ in **b.2** that the $s + 1$th ticket to be withdrawn will be white. Thus $m^{\dagger}\ [A(a_{s+1}) \ \&\ e(r,s)] = {}^sC_r 2^{n-s-1}\, 2^{-n}$. It follows that $c^{\dagger}\ [A(a_{s+1}),e(r,s)] = \frac{1}{2}$, *independent of both* s *and* r. In other words, no sample parameters convey any information about the probability of the next individual to be observed being an $A$. Where has the Rule of Succession gone?

We shall answer that question in a moment. It follows by the reasoning above that if all the *constitutions* of the Laplace urn, i.e., all possible distributions of the property of being white among $n$ numbered tickets, are all equally probable, then all the statements of the form 'the $i$th ticket is white (not white)', $i = 1, \ldots, n$, are probabilistically independent with probability $\frac{1}{2}$. In particular, the probability of $A(a_{s+1})$ conditional on $e(r,s)$ is equal to the unconditional probability of $A(a_{s+1})$. Laplace's use of the Principle of Indifference to justify the equiprobability of the possible *numbers* of white tickets is now seen to be crucial to generating non-trivial inductive inferences on the basis of sample evidence. But we have shown that if the Principle is to apply to the elementary possibilities relative to a language with names for all the individuals, then such inferences are impossible: probabilistic independence is the order of the day. Keynes himself was quite clear that the Principle should apply *only* to the state-descriptions: he wrote that "the equiprobability of each 'constitution' [i.e., the measure $m^{\dagger}$] is alone legitimate, and the equiprobability of each numerical ratio erroneous" (1921, p. 61).

Carnap not surprisingly regarded $c^{\dagger}$ as a wholly unsuitable foundation for a system of quantitative inductive logic, and hence in effect abandoned the Keynesian form of the Principle of Indifference. He opted rather to follow Laplace (though he did not say this explicitly) in making equiprobable the possible numbers of $A$'s in the universe of $L(A,n)$, and then, for each $q$ between 0 and $n$ inclusive, making equiprobable all the ${}^{n}C_{q}$ possible ways in which $q$ of the $n$ individuals may be $A$'s. In other words, he opted for a measure function, called $m^{*}$, which made all the $n + 1$ structure-descriptions of $L(A,n)$ equiprobable also, and all the state-descriptions of which that structure-description are disjunctively composed equiprobable also. Thus the $m^{*}$ probability of each structure-description is $(n + 1)^{-1}$, and the $m^{*}$ probability of each state-description in the structure-description saying that there are $q$ $A$'s in the universe is $\frac{1}{(n + 1){}^{n}C_{q}}$. This is, it will be recalled, just the value of the probability $p$ we arrived at in **b.2**, when we 'derived' the Rule of Succession.

### c.2 The Dependence on A Priori Assumptions

$c^{*}$ is simply one among many other conditional probability functions whose values are sensitive to sample data. Carnap was aware of this and conceded that any choice of one among these was bound to be arbitrary to a greater or lesser extent: $c^{\dagger}$ and $c^{*}$ were merely the two functions "which are most simple and suggest themselves as the most natural ones". Indeed, all he could find to say positively in favour of $c^{*}$ was that it "is the only one which is not entirely inadequate" (1950, p. 565). In his 1952 book he considered a continuum of other 'inductive methods', or conditional probability functions defined on simple languages like $L(A,n)$, each method corresponding to a value of a real-valued non-negative parameter $\lambda$; $c^{*}$ corresponds to setting $\lambda$ equal to 2 for $L(A,n)$, while the measure $c^{\dagger}$, assigning equal probabilities to all the state-descriptions, corresponds to setting $\lambda$ equal to infinity.

$c^{*}$ is moderately sensitive to sample data, as we have seen, and $c^{\dagger}$ is totally insensitive to sample data. It is in fact easy to show that $\lambda$ determines how willing you are or should be to accept the past as a guide to the future, for all the $c$-measures in the $\lambda$-continuum satisfy the following condition:

$$\textbf{(4)}\quad c[A(a_{i+1}),e(r,s)] = f(\lambda,r,s) = \frac{r + \lambda/2}{s + \lambda}$$

(This is for sentences of $L(A,n)$; in the general case 2 in the right-hand expression is replaced by a parameter depending on the precision of the predicate $A$ relative to $L$.) It is a simple exercise to show that if $r = s$ then $f$ is a strictly decreasing function of $\lambda$, with values between $\frac{1}{2}$ and 1. Setting $\lambda = 0$ we obtain the so-called 'straight rule': you simply extrapolate the observed frequency $\frac{r}{s}$ as the probability of observing an $A$ next time. So if $r = s$, $f = 1$. Setting $\lambda = \infty$, on the other hand, determines the measure $c^{\dagger}$: $f = \frac{1}{2}$, independent of the sample parameters $r$ and $s$.

Carnap later in his life (Carnap and Jeffrey, 1971) considered an even more extensive class of what he called "credibility" measures, defined relative to very much more expressive, set-theoretical languages, while Hintikka (1965, 1966, 1968, and elsewhere) and his followers have experimented with languages in which no bound is placed upon the size of the universe of discourse and where consequently the classes of $L$-elementary possibilities cannot be characterised in terms of state-descriptions. In particular, Hintikka was able to exhibit a systematic method of assigning a priori probabilities (the unconditional probabilities corresponding to Carnap's measure functions $m$) to these possibilities in such a way that universal hypotheses can have positive probabilities; in Carnap's $\lambda$-continuum they are all assigned zero probability by all positive values of $\lambda$, and Hintikka and many others regarded this as unacceptable in a theory which proposes to justify inductive procedures—for we certainly do regard universal laws as supported by appropriate experimental evidence.

Whereas Carnap's Continuum of Inductive Methods is parametrised just by one real parameter, $\lambda$, Hintikka introduced more. The resulting systems are rather complex (though for an excellent simplifying discussion, *see* Kuipers 1978 and 1980), and we shall only sketch their features salient to this discussion. In these systems the elementary possibilities are the so-called *constituents* of the language. A constituent is a sentence of the language that says which predicates, including those definable within the language in terms of the primitive predicates, identity, and any individual names there might be, are instantiated and which not. There are usually infinitely many $L$-definable predicates, where $L$ is the (usually first-order) language, each of which can be characterised by the minimum degree of complexity of its defining formula. A given degree of complexity also determines, therefore, a particular set of constituents, and a result due to Hintikka himself is that every sentence in a first-order language is equivalent to a disjunction of finitely many constituents of a given complexity. Since the constituents are mutually exclusive, any set of a priori probabilities assigned to the constituents in $L$ then determine the probabilities, and the conditional probabilities, of all other sentences in $L$.

The advantage of distributing probabilities via constituents rather than state-descriptions is that state-descriptions are easily defined only for simple monadic languages, whereas the former method imposes no such restrictive constraints. Also, it is easy to see, using the constituent method, how to distribute probabilities in such a way as to endow universal sentences with positive a priori probabilities. Consider the language possessing only one predicate symbol, $B$, say, and no individual names or identity. There are three constituents only in this language, namely the statement that all the individuals are $B$; the statement that some individuals, but not all, are $B$; and the statement that no individuals are $B$. One a priori probability-weighting is that which assigns equal probabilities of one third to each constituent, giving the two law statements in this language a priori probabilities each of one-third independently of the size of the domain within which they are to be interpreted.

Hintikka and his followers have experimented with various types of a priori distribution over constituents, in languages often much more sophisticated than this one. The details do not concern us, for we are concerned here with a certain basic strategy common to Carnap and Hintikka, and one which is implicit in the entire programme of constructing a probabilistic epistemology on the lines laid down by Bernoulli and Laplace. For any conditional probability distribution over the sentences of a language $L$ necessarily involves the assignment of unconditional probabilities to any partition of the space of possibilities representable within $L$. It does so because $P(a|t)$, where $t$ is a tautology and $a$ any sentence, is equal to $P(a)$. But what considerations can possibly justify any such a priori distribution?

One possible answer, and the one implicit in the Keynesian form of the Principle of Indifference, is that the distribution of prior probabilities is legitimate if it is neutral, or, as it is sometimes put, 'informationless', with respect to the class of elementary possibilities determinable within that language: in other words, all the finest-characterised possibilities admitted by that language should receive equal a priori probabilities.

This answer will not do, however. In the first place, such a priori assignments necessarily exhibit a more-or-less strong bias against certain types of world, which we have no right, in advance of all empirical information, to indulge in. This a priori bias can take extreme forms. For example, consider the classification of worlds according to whether they admit the truth of at least one universal law or whether they admit the truth of none. The measure assigning equal probabilities to the state-descriptions of the language $L(A,n)$ is as far as it is possible to be from being neutral between these two categories of world, telling us that in the limit as $n$ increases, it is with probability one the latter sort of world we live in, for there are only two distinct universal hypotheses in $L(A,n)$: the state-description in which every individual possesses $A$ and the state-description in which no individual possesses $A$. Their disjunction is therefore equivalent to the statement 'Some universal law is true'. But each of them has zero probability with respect to that measure, and hence so does their disjunction. Indeed, however neutral is any proposed a priori measure as between the members of one partition of possibilities, one can always find another with respect to which that measure is as biased as one likes. All attempts to achieve neutrality in this way will in fact be very partial towards and against types of possible world; uniform neutrality is impossible, just as attempts to iron out a wrinkle in a badly made suit of clothes will not remove it but simply send it elsewhere.

Secondly, elementary possibilities are elementary only relative to some language, and language is a human artifact whose ultimate categories stand on a footing of equality only as a result, therefore, of a collective decision that they should do so, a decision which may consequently be revoked. Often the decision to refine or otherwise amend these categories (and even, possibly, abandon them) is taken as a result of the development of a particular scientific theory and the belief that the latter provides the most adequate representation of some empirical domain to date. But the systems of inductive logic we have been discussing are supposed to adjudicate decisions of empirical adequacy.

That these systems themselves may depend on some prior inductive judgment is not a novel suggestion, however. Carnap has himself suggested (1952, p. 55) that the parameter $\lambda$ may be evaluated empirically, by a process which today we call *calibration* and which consists in comparing the class of predictions assigned $x$ percent probability with the frequency with which those predictions were true. If there is a significant discrepancy between the probability and the truth-frequency (if the inductive method is not calibrated, in other words), then Carnap recommends adjusting the value of $\lambda$ appropriately.

But this suggestion calls into question the fundamental role assigned his systems of inductive logic by Carnap. If their adequacy is itself to be decided empirically, then the validity of whatever criterion we use to assess that adequacy is in need of justification, not something to be accepted uncritically. As if aware of this objection, Carnap subsequently reverted to his earlier position that "in principle it is never necessary to refer to experience in order to judge the rationality of a $c$-function" (Carnap, 1968, p. 264; *see also* Carnap, 1971, p. 25). But the position this retreat represents is hardly, if we are correct, any more tenable, for it is that same a priorism we have seen cause to reject.

The $\lambda$-continuum is, we believe, a striking illustration of Hume's thesis, which we have already touched on in the discussion of the Rule of Succession, that 'probable arguments' from past to future presuppose implicitly what they set out to prove. $c$ tells you how inductive you should be, but in order to determine $c$ you need to set a value for $\lambda$, which itself is determined by deciding how inductive you should be. It has been argued, on the other hand (Good, 1983, p. 96), that we can construct a *metainductive* argument to determine the most probable value of 'first order' parameters like $\lambda$. In other words, we should—so the suggestion goes—treat $\lambda$ itself as a random variable, with a prior distribution, and derive a posterior distribution for it in the light of the data. We can also then regard (4) as giving the probability of $A(a_{i+1})$ conditional on both $e(r,s)$ and $\lambda$, and then 'integrate out' $\lambda$ from the equation by means of the theorem of total probability for a continuous variable.

This is in the spirit of Carnap's suggestion that empirical considerations might determine $\lambda$, by means of a higher-order induction; indeed, it seems to offer a practical method for so doing. But this is illusory. $\lambda$ does not actually *measure* any quantity. Carnap (1952) lists a set of adequacy constraints on $c$-functions and shows that they determine a class of $c$-functions up to a parameter $\lambda$ satisfying (4). So $\lambda$ is merely an undetermined parameter which characterises a particular inductive method; it makes no real sense to assign it a probability distribution. Even if $\lambda$ could legitimately be interpreted as a measure of, say, how regular the world is, treating it as a random variable doesn't banish a priorism: it merely shifts it up a level, to the prior distribution for $\lambda$. You might, if you wish, make that distribution also a random variable, in which you get a priorism at the third level; and so on. Clearly, this procedure can be repeated ad infinitum, but equally clearly (we hope), it is a formal trick which does not evade the arbitrariness endemic in this approach.

To sum up: 'logical' probability measures, whether based on the Principle of Indifference or on some other method of distributing probabilities a priori, do not, we believe, possess a genuinely logical status. For such systems are ultimately quite arbitrary, and we take logic to be essentially noncommittal on substantive matters. (Popper has repeatedly made the same general point about Carnap, but this did not stop him advancing his own a priori distribution—*see* Chapter 15, section **c.**) Carnap's claim in the preface to his *Logical Foundations of Probability* that "all principles and theorems of inductive logic are analytic . . . hence the validity of inductive reasoning is not dependent upon any synthetic presuppositions" is not true—not, at any rate, of the probabilistic inductive logic he went on to expound.

Nor do we feel that any of the measures we have discussed qualify as rationality constraints either. How can it be rational to accept a probability distribution as a guide to belief and action which is constructed independently of all empirical evidence? But if we accept this, are we not left with the conclusion that there are no valid objective criteria on which to ground assessments of the probabilities of hypotheses?—a conclusion which has led many people to reason further that the probability calculus is incapable of providing a foundation for an objective theory of scientific inference. Indeed, this is what Fisher declared in one of his (very influential) criticisms of the Principle of Indifference, which, he claimed,

> leads to apparent mathematical contradictions. In explaining these contradictions away, advocates of inverse probability [this was the traditional name for the use of Bayes's Theorem to generate posterior probabilities] seem forced to regard mathematical probability . . . as measuring merely psychological tendencies, theorems respecting which are useless for scientific purposes. (Fisher, 1947, pp. 6–7)

It is the burden of the rest of this book that Fisher's inference is incorrect and that theorems respecting such psychological tendencies are so far from being useless for scientific purposes that they form the everyday logic—and a genuine logic at that—of scientific inference. This statement might seem to stand in direct contradiction to our previous remarks. It does not. A first step on the way to showing exactly why it does not is to establish what Fisher himself took for granted, that the "psychological tendencies" of which he spoke —and which we shall characterise explicitly as individuals' degrees of belief—can be numerically represented as mathematical probabilities. To this task we now turn.

## EXERCISES

**1.** Suppose that the probabilities of each of the 36 joint outcomes *(X,Y),* where *X* and *Y* may take any of the values 1,2, . . . ,6, of simultaneous throws of two dice, are judged equal. What are the probabilities of the following outcomes

a. $X = 3$, $Y = 4$
b. *exactly* one of *X,Y* is a 4
c. *at least* one of *X,Y* is a 4
d. $X = 4$ conditional on *X* being even
e. $X = 4$ conditional on $Y = 3$
f. $X = 4$ conditional on $X + Y = 8$
g. $X = 4$ conditional on $X + Y$ being odd

**2.** Carefully derive equation 1 in section **b.2.**

**3.** Suppose that the range of the random variable *T* is the interval $[a,b]$, $0 \leq a \leq b$, in which it is uniformly distributed. Show that if $x$ is any point in that interval, then the probability density of $T^2$ is equal to $\frac{1}{2n(b - a)}$ at the point $T = x$.

**4.** The notation $\forall x Q(x)$ means that every individual in the domain under discussion has the property *Q*. Show that in *L(A,n)*

a. $m^\dagger[\forall x A(x)] = 2^{-n}$
b. $m^*[\forall x(x)] = (n+1)^{-1}$
c. $c^\dagger[\forall x A(x), A(a_1)\& \ldots \&A(a_m)] = 2^{m-n}$
d. $c^*[\forall x A(x), A(a_1)\& \ldots \&A(a_m)] = \frac{(m + 1)}{(n + 1)}$
e. Hence show that all probabilities in **a.–d.** tend to *0* as *n* tends to infinity.

**5.** Where *e(r,s)* is defined as in section **c.1,** show that $m^\dagger[e(r,s)] = {}^{\mathrm{B}}C_r 2^{n-\mathrm{B}} 2^{-n}$.

**6.** A bag contains blue, green, and yellow balls. Three balls are drawn: one green, one blue, and one yellow. According to the Rule of Succession, the probability that the next ball to be drawn will be green is $\frac{2}{5}$; and similarly, the probability is $\frac{2}{5}$ that it will be blue, and $\frac{2}{5}$ that it will be yellow. Hence the probability is $\frac{6}{5}$ that it will have one of the three colours.

What has gone wrong?

7. The gambler's fallacy is to regard it as less probable that there will be an $A$ at the next trial if there has just been a long run of $A$'s. Is this a fallacy? Is regarding another $A$ as *more* probable equally fallacious?

# CHAPTER 5
# Subjective Probability

## a DEGREES OF BELIEF AND THE PROBABILITY CALCULUS

### a.1 Betting Quotients and Degrees of Belief

Our point of departure is the theory of betting odds. In actual betting practice, odds are non-zero numbers $k$ which are offered by one party (the bookmaker) to be accepted or not by another (the punter). The odds are offered usually against the occurrence of some event $E$, and the punter nominates a sum $Q$ such that he or she will contract to receive from the bookmaker the sum $Qk$ if $E$ occurs and forfeit $Q$ if it does not.

In what follows we shall talk of the truth and falsity of hypotheses rather than the occurrence and non-occurrence of events. Our particular interest is going to be in those odds on a hypothesis $h$ which you believe confer no positive advantage or disadvantage to either side of a bet on $h$ at those odds, in the ideal world in which the bet is immediately and veraciously settled after the bet. We shall also suppose that these advantage-equilibrating odds are unique: values above or below would, you believe, confer advantage to one or other side. This is a strong idealising assumption; we shall consider what happens, in section **b,** when it is relaxed. For the time being suppose it holds. Such odds, if you can determine them, we shall call your *subjectively fair* odds on $h$.

This definition of subjectively fair odds does not presuppose that any odds are fair *in fact*. We shall discuss later the question of whether any odds are actually fair. We assume only that people do, rightly or wrongly, *think* that some odds are fair, and we believe this assumption to be borne out in the fact that people frequently bet. This is not, of course, to say that the odds they bet at are the ones they find fair. Usually this will not be the case, for most people bet only when they think the odds advantageous to them. But this does mean that they have a notion of advantageous and disadvantageous odds, and indeed in certain cases are capable of narrowing down the band between the odds they deem advantageous and those they think disadvantageous to a number correct to so many places of decimals. One way this quantity can be elicited is by asking people which they would prefer: a reward if the event in question occurs or that same reward if another event with agreed odds occurs, where the latter can be manipulated at will (we give an example, due to Lindley, in section **c.3** below).

We are less concerned with elicitation, however, than with the fact that there are subjectively fair odds there to be elicited, for it is this fact we shall exploit to provide a convenient measure of people's degrees of belief. For the odds you take to be fair on $h$ will clearly reflect the extent to which you believe it likely that $h$ will turn out to be true. Indeed, we would make your assessment of the fair odds on $h$ the measure of your belief in $h$ but for the inconvenient fact that on the odds scale, length of interval will not measure the *difference* between degrees of belief. The odds scale goes from 0 to plus infinity, with 1 as the point of indifference; hence the difference between being cognitively indifferent between $h$ and $\sim h$ and being certain that $\sim h$ is true is 1, whereas the difference between being certain that $h$ is true and being cognitively indifferent between $h$ and $\sim h$ is infinite. The standard solution to the problem is to transform the semi-infinite odds scale, with $\infty$ appended, into the closed unit interval by means of the one-to-one mapping $p = \frac{k}{(1 + k)}$. Odds of $\frac{1}{1}$, that is to say, even money odds, go to $\frac{1}{2}$ under this mapping; 0 goes to 0; and $\infty$ goes to 1, giving the desired symmetry about the point of indifference between $h$ and $\sim h$.

The quantity $p = \frac{k}{(1 + k)}$, where $k$ are the odds on $h$ you believe fair, *will therefore be taken as the numerical measure of your degree of belief in h. p* is called the *betting-quotient* associated with the odds $k$. Odds can be recovered uniquely from betting-quotients by means of the reverse transformation $k = \frac{p}{(1 - p)}$.

Characterising degrees of belief in terms of characteristic odds or betting-quotients commenced with Ramsey (1931), and most authors have since followed him in making a willingness actually to bet in suitable circumstances the criterion of strength of belief (Ramsey, 1931, p. 79). This leads, as we shall see in section **c.1** below, to severe if not intractable problems when the question is posed why these behaviourally elicited quantities should satisfy the probability calculus.

We emphasise that we are *not* assuming that the intellectual judgment that odds are fair commits the judge to any behavioural display whatever. To believe odds fair is tantamount to believing the price of a gamble a fair one. But you can certainly believe a price fair without buying the good in question, and only in special circumstances would you actually do so. This may sound an obvious point, but it has been traditional in the literature to measure the strength of belief in terms of a willingness to bet at all odds up to some maximum value. Kyburg, for example, writes (1983, p. 64) that "The time-honored way of finding out how seriously someone believes what he says he believes is to invite him to put his money where his mouth is". But even equipped with enough capital to withstand betting losses, backing up judgments with financial commitment is not to everyone's taste, and declining to do so is no necessary indicator of belief.

Attempts to measure the values of options in terms of utilities are traditionally the way people have sought to forge a link between belief and action, and much contemporary Bayesian literature takes this as its starting point. We do not want to deny that beliefs have behavioural consequences in appropriate conditions, they clearly do, but stating what those conditions are with any precision is a task fraught with difficulty, if not impossible. Our view is that the fewer special —and questionable—assumptions that have to be made, the better, and the more secure the conclusions that one draws. Fortunately, we can derive our desired conclusion without assuming, or presuming, anything at all about the nature of the link between belief and action. For the conclusion we want to derive, that beliefs infringing a certain condition are inconsistent, can be drawn merely by looking at the consequences of what *would* happen if anyone *were* to bet in the manner and in the conditions specified.

### a.2 Why Should Degrees of Belief Obey the Probability Calculus?

Following de Finetti (1937), we are going to assume a canonical form for bets between two individuals $A$ and $B$ as a contract whereby $A$ pays the sum $pS$ (dollars, pounds, or whatever) to $B$ in exchange for the payment of the sum $S$ if the hypothesis bet on is true, and 0 if it is not (we shall assume that $S$ is arbitrarily finely divisible). The payoff conditions therefore look like this

| $h$ | Payoff to $A$ |
|---|---|
| $T$ | $S - pS$ |
| $F$ | $-pS$ |

where $T$ stands for 'true' and $F$ stands for 'false'. $A$ is clearly betting *on* $h$ at odds $pS{:}S - pS = p{:}1 - p$, and $B$ is betting *against* $h$ at the reciprocal odds $1 - p{:}p$; $p$ can therefore be identified as the betting-quotient on $h$. In future when we refer to *a bet on h with betting-quotient p* we shall mean a contract of the above form. $S$ is often called the *stake*. We can also speak of $A$ buying from $B$ a bet on a paying $S$ for the price $pS$. Clearly, $B$ strictly speaking needs no separate name; he or she is merely the other side of the bet.

Such bets can be brought into the traditional form described at the beginning of the chapter, given by the payoff table

| $h$ | Payoff to $A$ |
|---|---|
| $T$ | $Q k$ |
| $F$ | $-Q$ |

where $k = \frac{(1-p)}{p}$, by writing $Q = pS$. We use the de Finetti $(S,p)$ representation for bets rather than the $(Q,k)$ one since our focus of interest is $p$ rather than $k$, and the constraints to be imposed on $p$ emerge more simply in that formalism.

Now define a *betting strategy* with respect to a set of hypotheses $\{h_1,h_2 \ldots\}$ to be a set of instructions of the form 'bet on (against) $h_i$', for each $i$. Suppose that $p_1,p_2, \ldots$ is a set of betting-quotients on the $h_i$. A celebrated theorem, proved independently by F. P. Ramsey and B. de Finetti, shows that

> if the $p_i$ do not satisfy the probability axioms, then there is a betting strategy and a set $S_i$ of stakes such that whoever follows this betting strategy will lose a finite sum whatever the truth-values of the hypotheses turn out to be.

The Ramsey-de Finetti theorem is often also called the *Dutch Book Theorem,* because a Dutch Book is a system of stakes which ensures a net loss.

The significance of the theorem lies in its corollary that betting-quotients which do not satisfy the probability axioms cannot consistently be regarded as fair. For (i) fair odds have been characterised as odds which offer zero advantage to either side of a bet; (ii) the sum of finitely (or even denumerably) many zeros is zero; hence the net advantage of a *set* of bets at fair odds is zero; and, finally, (iii) if a particular betting strategy is *assured* of a positive net gain or loss for whoever adopts it, then the net advantage in betting at the odds involved cannot be zero. We conclude that the assurance of a net gain or loss from finitely many simultaneous bets implies that they cannot all be fair. It follows immediately that *if your degrees of belief are measured by the betting-quotients you think fair, then consistency demands that they satisfy the probability axioms.* Thus agreement with the probability axioms is a necessary condition of consistency; in section **a.6** below we shall show that it is also sufficient.

### a.3 The Ramsey–de Finetti Theorem

Ramsey's and de Finetti's theorem involves only elementary algebra and is very simple to prove, as we shall now show (the proof we give here owes much to Skyrms [1977, Ch. VI]). For each axiom of the probability calculus we shall show how its infraction entails the existence of a betting strategy leading to a necessary loss for one of the bettors.

**(i) Axiom 1.** Suppose that $p < 0$ and that you buy a bet on a proposition $a$ paying one dollar, for the price $p$. Clearly, you will make a sure gain of $1 + |p|$ if $a$ is true, and $|p|$ if $a$ is false. Hence your fair betting-quotient on $a$ must be non-negative.

**(ii) Axiom 2.** Suppose that you buy a bet on a tautology $t$ paying one dollar for a price $p$. If $p < 1$, then you will make a certain gain of $1 - p$; if $p > 1$, then you will make a certain loss of $p - 1$. So the only fair betting-quotient on $t$ is 1.

**(iii) Axiom 3.** Suppose that you buy bets on two mutually exclusive propositions $a$ and $b$, each bet paying one dollar, for the prices $p$ and $q$ respectively. Then your net gain is as below (remember that $a$ and $b$ cannot both be true):

| a | b | net gain |
|---|---|---|
| T | F | $1 - p - q = 1 - (p + q)$ |
| F | T | $-p + 1 - q = 1 - (p + q)$ |
| F | F | $-p - q = -(p + q)$ |

This diagram is clearly equivalent to the following:

| $a \vee b$ | net gain |
|---|---|
| T | $1 - (p + q)$ |
| F | $-(p + q)$ |

Thus your separate bets on $a$ and $b$ determine a bet on the disjunction $a \vee b$ paying one dollar and with betting-quotient $p + q$. Were you now also to bet against that disjunction with a betting-quotient $r$ not to equal $p + q$, where the stake is also one dollar, then you will have a net gain of $r - (p + q)$ (positive or negative) *whatever the truth-values of* a *and* b. For if the first two bets are labelled (i) and (ii), and the bet against the disjunction is (iii), then the net gain from (i) + (ii) + (iii) is as below:

| $a \vee b$ | (i)+(ii) | + (iii) |
|---|---|---|
| T | $1-(p + q) - (1 - r)$ | $= r - (p + q)$ |
| F | $-p + q + r$ | $= r - (p + q)$ |

Hence if your fair betting-quotient on $a$ is $p$ and on $b$ is $q$, your fair betting-quotient on the disjunction can only be $p+q$, and we have proved the additivity axiom.

Before we turn to the remaining axiom, that of conditional probability, we shall show that the same type of argument requires not merely finite but also countable additivity. Consider a class of mutually exclusive hypotheses $h_i$, $i = 1,2,3,\ldots$ Suppose that a unit stake is placed on each of the 'even-number' hypotheses $h_{2i}$ and that you bet on all these hypotheses simultaneously, with betting-quotients $p_2$,$p_4$, etc. If the infinite 'disjunction' of those hypotheses is true, then exactly one of them, $h_{2j}$ say, is true, and the net gain is $-p_2 -p_4 \ldots +(1-p_{2j})- \ldots = 1-(p_2+ \ldots +p_{2n}+ \ldots)$, which is independent of $j$. Hence if $h$ is true, the net gain from all these bets is $1 - (p_2+ \ldots +p_{2n}+ \ldots)$. If $h$ is false, then you lose the quantity $p_2+ \ldots +p_{2n}+\ldots$. So a set of simultaneous bets on all the $h_{2i}$ with the same stake on each is equivalent to a bet on $h$ with betting-quotient $(p_2+p_4+ \ldots)$. The fair betting-quotient on $h$ must equal $(p_2 + p_4+ \ldots)$. QED.

There are, however, vigorous critics of the thesis that subjective probabilities are countably additive. De Finetti, for example, has produced many counter-arguments. To reassure the reader that we are not dismissing out of hand these objections from someone whose authority is certainly not to be considered lightly, let us consider briefly one of the most seductive of these counter-arguments.

This considers the example of a positive integer chosen 'at random'. It might seem natural in these circumstances to require a uniform, zero, degree of belief in each integer being selected. This is quite consistent with finite additivity, but not countable additivity, as we saw in Chapter 2, section **h.** But, as Spielman (1977) points out, it is not at all clear what selecting an integer at random could possibly amount to: any actual process would inevitably be biased toward the 'front end' of the sequence of positive integers, and so there is in reality little force in de Finetti's counter-example. Let us now move on to consider the remaining probability axiom, axiom 4.

### a.4 Conditional Betting-Quotients

Axiom 4 we shall take to impose a condition on so-called *conditional betting-quotients*. A conditional betting-quotient is a betting quotient for a conditional bet, where a conditional bet on $a$ given $b$ is a bet on $a$ which is to proceed in the event of $b$'s turning out true and is called off if $b$ is false. We imagine a scenario in which the truth-value of $b$ is announced as soon as the contract has been made. The payoff conditions for the bettor-on in such a bet, with conditional betting quotient $p$ and stake $S$, are therefore:

| $a$ | $b$ | payoff |
|---|---|---|
| T | T | $S(1 - p)$ |
| F | T | $-pS$ |
|  | F | 0 |

We shall define your conditional degree of belief in $a$ given $b$ to be the betting rate you think fair in a conditional bet of this type. We can interpret this in a possibly more illuminating way as follows. Your degree of belief in a proposition $c$ is what you believe the fair betting-quotient on $c$ to be. This is less a personal statement about yourself than a claim about which betting-quotient you believe to be fair relative to the information which you happen to possess. So we can gloss your conditional degree in $a$ given $b$ to be what you believe the fair betting-rate on *a would be* relative to the same information stock augmented by the additional information consisting of the statement that $b$ is true. Note that this is not the same as saying that your conditional degree of belief in $a$ given $b$ is what you now believe the fair betting-quotient on $a$ would be were you to come to know $b$ in addition to what you already know, and no more (as we erroneously stated in the first edition of this book).

It is tempting to think of a conditional degree of belief in $a$ given $b$ as a degree of belief in a conditional 'proposition' $a \mid b$. The temptation should be resisted. We shall show in this section that consistent conditional degrees of belief, as we have defined them, are formally conditional probabilities, and David Lewis (1976) has shown that the usual rules of the probability calculus will not permit an interpretation of a conditional probability as the probability of a conditional sentence, even a non-truth-functional one.

We shall now proceed to show that axiom 4 of the probability calculus is a consistency condition for conditional degrees of belief as defined. In particular, we shall show that if axiom 4 is not satisfied, then there is a betting strategy involving conditional bets which will lead to an inevitable loss for one party.

The proof proceeds by showing that bets on a suitable combination of hypotheses determine some other bet, in this case a conditional bet. To be precise, we shall show that by setting appropriate stakes on $b$ and $a$ & $b$, simultaneous bets on those two statements are equivalent to a bet on $a$ conditional on $b$, and that any odds placed on $b$ and $a$ & $b$ can therefore be made to determine the odds for a bet on $a$ conditional on $b$.

Suppose your fair betting-quotients on $a$ & $b$ and $b$ are $q$ and $r$ respectively, where $r>0$. Suppose you were to bet at these rates on $a$ & $b$ with stake $r$ and against $b$ with stake $q$. Your net payoff is as follows:

| $a$ & $b$ | $b$ | net payoff |
|---|---|---|
| T | T | $r(1-q) - q(1-r) = r(1 - \frac{q}{r})$ |
| F | T | $-rq - q(1-r) = -q = -r(\frac{q}{r})$ |
| | F | $-rq + qr = 0$ |

*But this is clearly the payoff matrix of a bet on a conditional on b, with stake r and conditional betting-quotient* $\frac{q}{r}$*, i.e., the ratio of the betting quotients q on a & b and r on b. As with two mutually exclusive hypotheses, therefore, simultaneous bets with appropriate stakes also determine a further bet—in this case, a conditional one. Hence, if you were to state a fair conditional betting-quotient which differed from* $\frac{q}{r}$*, you would implicitly be assigning different conditional betting-quotients to the same hypothesis.*

It does not follow, however, that you would necessarily make a positive net loss by buying a bet-on at your dearer price and selling one at your cheaper, with the same stake. For $b$ may turn out to be false, whereupon the net gain from all the bets would be zero; the net gain is only non-zero if $b$ is true. Nevertheless, anyone who believes that the betting-quotients $q,r$, and the conditional betting-quotient $p \neq \frac{q}{r}$, are all fair is no less inconsistent in that belief; indeed, it is quite easy to show that by suitably extending $A$'s bets, a non-zero (positive or negative) net gain is assured whether $b$ turns out to be true or false.

For suppose you were to bet (i) on $a$ & $b$ with stake $r$, (ii) against $b$ with stake $q$, (iii) conditionally against $a$ given $b$, with stake $r$, and finally (iv) on $b$ with stake $q - pr$. As before, suppose your fair betting-quotient on $a$ & $b$ is $q$, on $b$ is $r$, and on $a$ given $b$ is $p$. Bets (i) and (ii) above determine, as we saw, a conditional bet on $a$ given $b$ with stake $r$ and betting-quotient $\frac{q}{r}$. Taking on bet (iii) simultaneously with (i) and (ii) guarantees, as we also saw, a net gain of $pr - q$ if $b$ is true, with zero gain if not. It is straightforward to work out that making bet (iv) simultaneously with all the others guarantees an overall net gain (positive or negative) equal to $r(pr - q)$ whatever the truth-values of $a$ and $b$. Given $r > 0$, this will be zero if and only if $p = \frac{q}{r}$, i.e., if and only if $P(a \mid b) = \frac{P(a \ \& \ b)}{P(b)}$. (Note, if you have not already done so, that a positive net gain can be turned into a positive net loss of the same magnitude by reversing the direction of all the bets.)

This completes the proof that if a set of betting-quotients does not satisfy the probability calculus, then they cannot all be fair (in **a.6** below we shall prove a form of converse to this). As we pointed out earlier, this result is independent of any formal characterisation of fairness of odds beyond the stipulation that they confer no advantage to either side of a bet at those odds. To round off the discussion, we shall now consider a particular method, used since the eighteenth century, of *computing* the advantage to taking a particular side in a bet.

### a.5 Fair Odds and Zero Expectations

Laplace (1820, p.20) defined the *advantage* to taking a given side in a wager to be the expected value of the bet. Thus advantage, so defined, is calculated in the same units as the stake $S$, and so can be subjected to straightforward arithmetical operations, like taking sums of separate advantages. Carnap, Laplace's twentieth-century successor, calls that same expected value the *"estimated gain"* (1950, p. 170), and a bet *fair* just when the "estimated gain" is zero, where the expectation is computed relative to an appropriate Carnapian $c$-function.

Assuming your fair betting-quotients are consistent, a bet on $a$ with stake $S$ is formally a random variable $X_a$ which takes the value $S(1 - p)$ if $a$ is true and $-pS$ if not, where $p$ is your fair betting-quotient on $a$. A bet against $a$ with the same stake is $-X_a$. Simultaneously making bets on or against $n$ hypotheses is the arithmetical sum of the corresponding random variables. Thus if we explicitly define the *advantage* of the bet represented by $X_a$ to be its *expected value,* relative to your subjective probability distribution, then we deduce as theorems (i) that the advantage, *as you see it,* of betting at odds determined by your degree of belief is zero, and (ii) that the advantage attached to a betting strategy, as we defined it in the previous section, is the sum of the advantages of each of the bets separately which comprise that strategy (because the expectation of a sum of random variables is equal to the sum of their expected values). (i) is very easily seen, since $E(X_a) = S(1 - p)p - pS(1 - p) = 0$.

We have, in other words, found a mathematical representation of the informal notion of advantage which yields as a consequence the results that degrees of belief are subjectively fair betting-quotients and that the net advantage to placing $n$ bets is the sum of each separately. These results do not of course prove anything substantially new. They merely show that the informal notion of subjective fairness can, to use Carnapian terminology, be given a formal explication which preserves all the desired consequences.

### a.6 Fairness and Consistency

We have laid a foundation for a theory of consistent degrees of belief, characterised as subjectively fair odds, whose methodological consequences we shall explore in the subsequent chapters. A natural question to arise at this point is whether there are any odds other than those on tautologies and contradictions which are in some clear and objective sense fair. One candidate for a criterion of objective fairness was, as we have seen, having zero expectation relative to a 'logical' probability distribution of the type Laplace, Keynes, and Carnap tried to define. We have seen that their attempts foundered on the rock of pure arbitrariness. However, there is famously an alternative criterion: *odds are fair when they are determined by the real physical probabilities of the events concerned, where those probabilities exist*. We believe that, with certain qualifications, this claim is true, and indeed we shall base our theory of statistical inference on it. But any argument for that thesis must await a discussion of the notion of physical probability itself, a notion which, as we shall see, is fraught with difficulties. We shall take up that discussion again in Chapter 13.

Ramsey (1931) used the term "consistent" to characterise degrees of belief having the formal structure of probabilities. We have shown in sections **a.2** and **a.3** that your system of beliefs is inconsistent if the betting quotients you believe fair do not satisfy the probability axioms. But what about the converse—are we justified in claiming that your belief system is consistent if the betting-quotients you believe fair do satisfy the probability axioms? This would amount to the claim that if $P$ is a set of betting-quotients over a set $H$ of $n$ hypotheses, and if $P$ satisfies the probability axioms 1–4, then for any betting strategy and any system of stakes, it is not the case that for every truth-value distribution over the members of $H$ the net gain is uniformly negative (or positive: remember that a negative net gain can be transformed into a positive one by reversing the directions of the bets). For consider any set of bets with arbitrary stakes on or against each of the hypotheses in $H$. These, as we know from the previous section, are random variables $X_1, \ldots, X_n$ and if the value of their sum $Y$ were always negative, say, then the expected value of $Y$ would clearly be negative also. But as we also know, $E(Y) = \Sigma E(X_i)$, and $E(X_i) = 0$ for each $i$, by (i) of the previous section. Hence, if any set of betting quotients satisfies the probability calculus, then no betting strategy can generate a positive or negative gain come what may.

So, consistency for partial beliefs is equivalent to their being formally probabilities. Today it is usual, following de Finetti, to use the adjective 'coherent' to mean that partial beliefs satisfy the probability axioms. This seems to us to direct the attention away from the all-important logical fact that the probability calculus is a complete axiomatisation of consistent partial belief. The probability axioms, as Ramsey emphasised, do have therefore a purely logical interpretation; not, as Keynes and Carnap believed, as a calculus of partial entailment, but as the logic of consistent partial belief.

## ■ b UPPER AND LOWER PROBABILITIES

We promised that we would return to discuss those hypotheses and data sets where it might seem to be unrealistic to suppose that one would have point-valued degrees of belief. To borrow an example from Suppes (1981, p. 41): if we consider the question of whether it will rain at some specified time in Fiji, we can certainly suggest a value $k_1$ such that odds less than $k_1$ on that hypothesis are, in our opinion, unrealistically low, and we can also suggest odds $k_2$, such that odds greater than $k_2$ are unrealistically high. But we might also say that there is an intermediate interval of odds between which we feel quite unable to discriminate. The typical indefiniteness of one's knowledge would, it seems, be more faithfully reflected by an interval-valued function which only in certain cases takes degenerate intervals, or points, as values.

We believe that this suggestion reveals a confusion as to what subjective probabilities actually are. The whole point of introducing the apparatus of subjective probability is precisely because one's knowledge is typically indefinite: subjective probabilities *express* that indefiniteness by taking non-extreme values. Nevertheless, we have defined your subjective probability of $h$ as the betting-quotient on $h$ you believe to be fair in the present circumstances, and this does leave open the possibility that you may feel unable to specify an exact value. Indeed, the occasions on which you feel that you can specify a unique number with confidence may well turn out to be exceptions rather than the rule.

It turns out that very little is lost in conceding that what we have supposed to be point-valued degrees of belief are actually interval-valued, so long as the intervals are small. Suppose that $P_*(a)$ is the least upper bound (supremum) of all the betting quotients on $a$ at which you definitely think a bet on $a$ advantageous to the bettor-on, and $P^*(a)$ is the greatest lower bound (infimum) of betting-quotients at which you definitely think a bet on $a$ would be advantageous to the bettor-against. For all intermediate values you have no opinion at all about the relative advantages of either side of the bet. $P_*(a)$ is called your *lower probability* of $a$ and $P^*(a)$ is your *upper probability* of $a$.

We can define consistency for upper and lower probabilities analogously to consistency for point-probabilities. We then find that consistent upper and lower probabilities turn out to have some simple formal properties:

**(i)** $0 \leq P_* \leq P^* \leq 1$.

**(ii)** $P_*(a) = 1 - P^*(\sim a)$.

**(iii)** If $a_1 \vdash \sim a_2$, then $P_*(a_1) + P_*(a_2) \leq P_*(a_1 \vee a_2) \leq P^*(a_1 \vee a_2) \leq P^*(a_1) + P^*(a_2)$.

**(iv)** Where $P_*(a) = P^*(a)$, then that quantity is simply called *the probability of h*. Probabilities so defined obey the usual rules of the probability calculus.

**(v)** There is a set **P** of probability functions such that $P^*(h)$ is the greatest lower bound of $P(h)$ for every $P$ in **P**, and $P_*(h)$ is the least upper bound of $P(h)$ for every $P$ in **P.**

The formal theory of upper and lower probabilities has been developed from a variety of starting points by Good (1962), Smith (1961), Dempster (1968), Williams (1976), and, most recently and most comprehensively, by Walley (1991). The theory is an obvious generalisation of classical probability theory, with an appealing interpretation, in terms of the indefiniteness of one's opinions, for the Bayesian. It also corroborates Keynes's well-known thesis that degree-of-belief probabilities are only partially ordered, since an interval-valued probability function will induce a linear ordering only on the subclasses of its domain composed of hypotheses whose 'values' are disjoint intervals.

It might even be argued that the cost incurred by departing from the point-probability model does not, despite the appearance of greater fidelity to the phenomenon of inexact belief-states, bring commensurate rewards of greater realism. If there are no exact point-valued degrees of belief, there are probably also no exact point-valued bounds to the range of imprecision of belief. At any rate, there is not much to be lost, and much to be gained, in working within the simpler theory of point-valued probabilities, which can be regarded as the ideal approximated sufficiently exactly in most cases. The situation is analogous to the case of physical magnitudes, where people are quite happy to invoke real-number values, even though, strictly speaking, lengths, volumes, masses, densities, and the like do not take exact values. Real-number theory, applied to the measurement of physical quantities, is so widely used because, as an idealisation which gives sufficiently accurate results within the ranges of imprecision in which we work, it is indispensable. The same, mutatis mutandis, is true here.

We shall now close this chapter with a brief account of some other arguments for supposing that subjective uncertainty, where it is considered to be measurable at all by real numbers, is measurable by probability functions.

## c OTHER ARGUMENTS FOR THE PROBABILITY CALCULUS

### c.1 The Standard Dutch Book Argument

Any reader of books or articles on philosophical probability will be struck by the variety of arguments produced for the thesis that strength of belief can be measured numerically and that those measures satisfying the axioms of the probability calculus are among the class of admissible measures. We have ourselves produced an argument for this conclusion, where we identified degrees of belief with subjectively fair betting-quotients. This argument exploits the Dutch Book theorem, which shows that a set of betting-quotients which fails to satisfy the probability calculus cannot all be fair. We then justified invoking the constraints imposed by the probability calculus on degrees of belief by identifying the latter as subjectively fair betting-quotients.

A rather different way in which the Dutch Book theorem has been used to justify obedience to the probability calculus is altogether more dubious, however, though it is the way adopted by almost everyone who invokes the theorem—hence our section heading, the "standard" Dutch Book argument. It proceeds from the postulate that to possess a degree of belief $p$ in $h$ *is actually to be prepared to bet indifferently on or against* h *at odds* p:1−p, so long as the stakes are kept small. The Dutch Book theorem clearly entails that if your degrees of belief, so characterised, do not satisfy the probability calculus, then there are positive and negative stakes (positive stakes mean you bet on, negative stakes mean you bet against) which you would accept in bets at the odds determined by your degrees of belief and which, once accepted, would cause you to lose money come what may. It is (plausibly) assumed that you would not voluntarily retain such a system of degrees of belief once their vulnerability to a Dutch Book had been brought to your attention, and it is then concluded that obedience to the probability calculus is a principle of *economic* rationality.

The trouble with this use of the Dutch Book theorem, which critics have not been slow to point out, is that the postulate, that degrees of belief entail willingness to bet at the odds based on them, is vulnerable to some telling objections. One is that there are hypotheses for which the wise choice of odds bears no relation to your real degree of belief: if $h$ is an unrestricted universal hypothesis over an infinite domain, for example, then while it may in certain circumstances be possible to falsify $h$, it is not possible to verify it. Thus the only sensible *practical* betting quotient to nominate on $h$ is 0; for you could never gain anything if your betting quotient was positive and $h$ was true, while you would lose if $h$ turned out to be false. Yet you might well believe that $h$ stands a non-zero chance of being true.

Our use of the Dutch Book theorem avoids this objection. We employed it in the context of an imaginary experiment, designed to bring out inconsistencies in your attributions of fair betting-quotients. In this thought-experiment two fictitious individuals $A$ and $B$ bet against each other at odds based on those betting quotients. *It was assumed throughout that the truth-value will be revealed to the bettors and the debts collected,* even though the hypotheses bet on may not be decidable in the real world we inhabit. The reason we consider this fictional state of affairs is that we measure your strength of belief in a hypothesis $h$ by the betting quotient you *would* consider fair on $h$ *were* the truth-value of $h$ to be revealed. And as the celebrated Stalnaker/Lewis theory of counterfactuals insists, to evaluate such counterfactuals we must look at imaginary worlds in which their antecedents are satisfied and in which as little else as possible differs from what obtains in the actual world (Stalnaker, 1975; Lewis, 1973).

Another decisive objection to the standard Dutch Book argument is one we have already alluded to, namely, that there may be all sorts of reasons why you might be unwilling to bet even at odds which faithfully represent your degree of belief. Most people, indeed, are induced to bet only at odds which they think are advantageous to them. Even if the stakes are kept very small, then as Ramsey observed (1931, p. 176), it is difficult to see why you should bother to be too realistic when giving or taking odds at all. It is the problematic nature of the postulate, that there are odds which you would equally give and take, which prompted C. A. B. Smith (1961) to abandon it and assume merely that there is a minimum of the odds which you would accept on a given hypothesis, and a maximum at which you would take the other side of the bet, and that these usually will not coincide (he then used Dutch Book considerations to show that these maxima and minima determine upper and lower probabilities).

De Finetti (1937, p. 102) and later Mellor (1971, p. 37) attempt to evade the difficulty by introducing an element of compulsion: you are now compelled to name betting-quotients, with your coercer free to appoint (i) the hypotheses to be bet on, (ii) which side of each of the bets you will take, and (iii) the stakes. The quotients you pick are then identified with your degrees of belief, and the Dutch Book Argument is invoked to show that they should satisfy the probability calculus. But the introduction of compulsion still leaves the identification of genuine degrees of belief with the betting quotients thus elicited open to serious, and in our opinion insuperable, objections. For example, there remains the problem of hypotheses which are only one-way decidable: there is still no justification for naming odds other than zero on these. In addition, there is no reason why the presumptive degrees of belief so elicited should obey the probability calculus. An example will make this clear. Suppose that somebody threatened to inflict on you the most hideous torture were you to deny that 2 + 2 = 5. It would be perfectly rational of you *in those circumstances* to agree that 2 + 2 = 5, but it certainly would not necessarily be rational to extend your agreement outside them; indeed, there is no reason at all why you should. The same holds for the forced-betting situation we are being asked to contemplate; it is simply not a good reason to give that just because there are situations like the one in which penalties are attached to a set of inconsistent betting-quotients, those betting-quotients should *in general* obey probabilistic constraints.

### c.2 Scoring Rules

Let $I_h$ be a function which takes the value 1 for those 'possible worlds' in which $h$ is true and 0 for those in which it is false (if

$h$ describes a generic outcome of some stochastic experiment $E$, then the 'possible worlds' on which $I_h$ is defined can be regarded simply as the distinct basic outcomes of $E$). $I_h$ is called the indicator function of $h$. Let $p$ be a real number between 0 and 1 inclusive. The payoffs of a bet on $h$ at odds $p:1-p$, with stake $S$, can be represented by the values of the function $S(I_h - p)$, for these are $S(I - p)$ when $h$ is true, and $-Sp$ when $h$ is false. By allowing $S$ to take negative values, the payoff function has the same form whichever side of the bet you take: if you are betting on $h$, $S$ is positive; if against, negative. Where $p$ represents your true degree of belief in $h$, this function is a so-called *scoring rule*. In general, a scoring rule is defined to be any function $f(I_h,p)$ of $I_h$ and $p$ which assigns a penalty depending on the value taken by $I_h$.

Different scoring rules have been considered. De Finetti, for example, (1974, p. 87) proposed a quadratic scoring rule $L(I_h,p)$, which exacts a penalty proportional to $(I_h - p)^2$. This scoring rule is actually used to improve the calibration of individual weather forecasters in the US, i.e., to decrease the difference between the proportion of days on which they predicted some weather-event like rain to occur with probability $p$, and $p$ itself. That the penalty is smaller the closer that proportion is to $p$ is left as an exercise at the end of the chapter.

Lindley (1982) contains a very striking result about (penalising) scoring rules. He defines a set of degrees of belief ('uncertainties') $p'_1, \ldots, p'_n$ to be admissible if for every other set $p'_1, \ldots, p'_n$ there are values of the indicator variables $I_{h_i}$ such that the sum of the scores $f(I_{h_i},p_i)$ is less than the sum of the scores $f(I_{h_i},p'_i)$, and for all values of the indicators the sum of the $f(I_{h_i},p_i)$ never exceeds that of the $f(I_{h_i},p'_i)$. Lindley shows that if $p_1, \ldots, p_n$ are admissible, then there is a continuous mapping of $p_1, \ldots, p_n$ into the closed unit interval which satisfies axioms 1.–4. of the probability calculus. In other words, if your degrees of belief are such that they could not be replaced by a different set which would give a lower net loss for some events, and is never higher on any, then they are either directly probabilities or they can be rescaled to become probabilities.

C. A. B. Smith, in the discussion to Lindley's paper, shows that by replacing the uniform penalties in Lindley's approach by positive and negative rewards, the scoring rule in that theory can be transformed into the standard form of bets in which the agent indifferently takes either side of the bets at the given odds determined by his degree of belief. But as Smith points out, this is just the unrealistically strong assumption which, we pointed out earlier, vitiates the force of the standard use of the Dutch Book argument.

### c.3 Using a Standard

Lindley (1985, pp. 17–20) advances a particularly simple argument for why degrees of belief can be represented by probabilities. The argument proceeds from observing that since the problem is one of the measurement of a quantity, namely belief, we should employ the same sort of procedure that is used in the measurement of physical magnitudes. This is to compare the quantity to be measured against some standard unit. The unit Lindley selects is provided by an urn containing a proportion $p$ of black balls. Suppose you are considering some arbitrary hypothesis $h$, say, that it will rain tomorrow. You are offered a choice of two options: to receive a prize if $h$ is true and to receive the same prize if a ball picked randomly from the urn is black. You are also told the value of $p$. Your preference for one or the other of the two options will clearly depend on the size of $p$: if $p$ is 1 you will almost certainly prefer the second, while if it is 0 you will prefer the first. At some value of $p$ you will be indifferent between the two, and Lindley makes this the measure of your belief in $h$. By increasing the number of balls in the urn, and hence increasing the number of possible values of $p$, you can refine your degree of belief in $h$ accordingly. A straightforward criterion of coherence then determines that degrees of belief so measured obey the probability calculus.

### c.4 The Cox-Good-Lucas Argument

Cox (1961), Good (1950, Appendix III), and Lucas (1970) propose variants of a quite different type of argument for supposing that subjective probabilities are technically probabilities in the sense of the probability calculus. They each show that if a real-valued function $f$ satisfies a few very general and plausible desiderata for a measure of reasonable degree of belief, then $f$ can be transformed ('regraduated') by a suitably increasing function $Q(f)$ into a conditional probability function. For example, all that Cox requires is that

$$f(c \,\&\, b \mid a) = G[f(c \mid b \,\&\, a), f(b \mid a)]$$

and that

$$f(\sim b \mid a) = H[f(b \mid a)]$$

for some functions $G$, $H$ satisfying certain differentiability conditions. He is then able to show that

$$0 \leq Q(f) \leq 1$$

and that

$$Q[f(c \,\&\, b \mid a)] = Q[f(c \mid b \,\&\, a)] \cdot Q[f(b \mid a)]$$

and

$$Q[f(\sim b \mid a)] = 1 - Q[f(b \mid a)].$$

From these properties of the transform $Q$ it is not difficult to show that it satisfies the axioms of the probability calculus.

### c.5 Introducing Utilities

Ramsey's pioneering work (1931) generated a great deal of subsequent interest in developing probability within a general theory of utility. Recall the standard Dutch Book argument. Here probabilities are elicited from you in terms of your preferences among bets at specified odds. As we pointed out, however, because of budgetary or other constraints, your propensities to bet or not, and the odds at which you will agree to bet, may bear little or no relation to your beliefs about the events in question. This sort of objection is standardly met in this behaviouralistic approach by measuring the prizes in units which are supposed to reflect the total value to you of winning and losing, a value which will after all reflect among other things your like or dislike of betting itself, the degree to which you are risk-averse, and how you rank the bet against various alternative courses of action open to you. Ramsey's fruitful insight, which initiated a programme of research that continues to flourish to this day, was to use an individual's preference-ordering over gambles to show that if the preferences obey certain consistency constraints, like transitivity, for example, then they can be represented, to a surprising degree of uniqueness, by real numbers, which are formally *expected utilities* relative to some probability and utility functions.

Savage (1954) and Jeffrey (1965) follow Ramsey's lead in developing simultaneously a theory of subjective probability and a theory of utility. Like Ramsey, Savage relates probabilities to preferences between gambles, i.e., acts with two possible outcomes, the receipt of a prize if some class of states of the world obtain, and nothing if not. Savage shows how from axioms determining, among other things, a linear (reflexive) preference-ordering between such acts, we may derive an ordering, called a *qualitative personal probability,* of the corresponding classes of states of the world. He then shows that, subject to certain additional conditions being satisfied, there is a unique probability function $P$ such that event $a$ is at least as probable as $b$ if and only if $P(a) \geq P(b)$. As had Ramsey, Savage also shows that his axioms determine a real-valued utility function defined on outcomes and unique up to a positive linear transformation, such that act $A$ is not less preferred to $B$ if the expected utility of $A$ is not less than that of $B$.

## d CONCLUSION

These arguments differ in the degree to which they are compelling. Nevertheless, it is a striking fact that, starting from often apparently very different assumptions, all plausible in their own way, so many arguments lead directly to the probability calculus. The latter seems, in other words, to be a sort of invariant of different ways of defining uncertainty, or as Lindley puts it, "inevitable", meaning that the choice of any plausible way of mathematically measuring uncertainty will lead to it. This convergence of arguments has a powerful cumulative effect and increases our conviction that the probability calculus corresponds to some quite objective feature of subjective uncertainty. The situation is rather analogous to that in the mathematical theory of effective computability, where a number of apparently distinct ways of characterising that notion turn out to define exactly the same class of functions, the so-called partial recursive functions. But enough, we feel, has been said about why degrees of belief should be formally probabilities; let us now see what methodological consequences flow from assuming that they are.

## EXERCISES

**1.** What are the odds corresponding to betting-quotients of

a. $\frac{1}{4}$
b. $\frac{3}{4}$
c. $\frac{7}{8}$
d. $\frac{14}{15}$

**2.** What are the betting-quotients corresponding to odds of

a. 1:2
b. 2:1
c. 3:4
d. 7:1

**3.** Show that if $t$ is a tautology, then a necessary condition of the fairness of a betting-quotient $p$ on $t$ is that $p = 1$.

**4.** I know that it will rain tomorrow, and hence I know that you must lose any bet with me in which you offer odds of more than 0 that it will not. Show that my expected gain from a bet at such odds is positive. Are non-zero odds unfair?

**5.** I have a large fortune, I am an atheist, my tastes are epicurean, and I am not noticeably risk-averse. Odds are offered me on a bet, which I decline. Does it follow that I believe those odds to be unfair?

**6.** $h$ is a universal hypothesis. No one can ever win a bet on the truth of $h$, but they might well lose one; so anyone's assessment of the fair odds on $h$, and hence their degree of belief in $h$, should be zero. Is this true?

**7.** Prove that if $h_1, h_2, \ldots$ are a denumerably infinite sequence of hypotheses and $h$ is true if and only if exactly one of the $h_i$ is true, then $P(h) \geq \Sigma\, P(h_i)$.

**8.** Where the bets (i), (ii), (iii) are as in the final part of the discussion in **a.3,** show that $A$ will have a non-zero net gain of $r^2 (p - \frac{q}{r})$ in all circumstances.

**9.** Suppose that $n$ days are examined on which forecaster $A$ predicts rain with probability $p$, and it rains on $m$ of them. Show that

a. the total penalty obtained on the quadratic scoring rule is $m(1 - p)^2 + (n - m)p^2$

b. this penalty takes a minimum value for $p = \frac{m}{n}$

# CHAPTER 6

# Updating Belief

## a BAYESIAN CONDITIONALISATION

We have shown that if your degrees of belief are consistent, then they obey axioms 1–4 of the probability calculus as presented in Chapter 2. These degrees of belief are degrees of belief at a particular time $t$, and for this reason axioms 1–4 and their consequences are often called the *synchronic* probability calculus. One of the consequences of that calculus is, of course, Bayes's Theorem, and the reader should by now be aware that Bayes's Theorem regulates the way in which beliefs are updated on the receipt of evidence. But updating beliefs means moving from one probability function $P$ relative to which that evidence has a probability usually less than 1 (if $e$ had probability 1, then $P(h \mid e) = P(h)$ and so $e$ would have no confirming power), to another $P'$ relative to which $e$ has probability 1. In the process, other propositions in the domain of $P$ will of course acquire new $P'$-values. But Bayes's Theorem tells us merely that $P(h \mid e)$ is equal to $\frac{P(e \mid h)P(h)}{P(e)}$. How then can it regulate the passage from $P$ to $P'$? The answer, which had until recently been regarded as too obvious to warrant special mention, is that $P'(h)$ is set equal to $P(h \mid e)$. This rule is known as the Principle of Bayesian Conditionalisation, and for future reference we shall highlight it.

> **Principle of Bayesian Conditionalisation:** When your degree of belief in e goes to 1, but no stronger proposition also acquires probability 1, set $P'(a) = P(a \mid e)$ for all $a$ in the domain of $P$, where $P$ is your probability function immediately prior to the change.

Although Bayesian conditionalisation was regarded for so long as intuitively obvious, no actual argument was ever advanced to justify it, presumably because it was thought to need none. But in an influential paper (1967) Ian Hacking observed that, unlike the axioms of probability themselves, there seemed to be no Dutch Book argument for it, for your degrees of belief both before and after learning *e* may be consistent, in the sense of being immune to a Dutch Book, without the latter being obtained by conditionalisation from the former. But other people have been unable to rest content with that answer—unsurprisingly, given the central importance of conditionalisation to Bayesian methodology—and they claim that there is indeed a Dutch Book argument for conditionalisation as the only consistent updating *rule*. They reason as follows (Skyrms, 1987, presents this line of thought in a lucid way). Suppose you follow a rule for changing your beliefs in the light of future evidence, according to which you decide what your degree of belief will be in the eventuality of $e_1$ being true, or $e_2$ being true,. . . . , or $e_n$ being true, where the $e_i$ form a partition of future possibilities into a set of mutually exclusive and exhaustive cases. Obviously, the $e_i$ cannot represent all the possible eventualities that may befall you through your life, or even those up to only fairly distant points in the future. But we can suppose that they might represent the results of some observation you are about to make. Now we can ask the question whether there is a betting strategy based on your current probabilities *and your updating rule* which would involve you in a sure loss or gain if for some $h$, $P'(h) \neq P(h \mid e_i)$, where $P'(h)$ is the probability your updating strategy dictates for $h$ if $e_i$ turns out to be true.

This is the question Paul Teller (1973) reports David Lewis having asked, and answered. Lewis's answer was yes. His argument, recounted by Teller, is presented here in a slightly simplified form. Let $e_i$ just be $e$. Suppose that your current conditional probability of $h$ given $e$ is $P'(h) = r \neq x = P(h \mid e)$. Consider the following betting strategy. You (i) buy, for $x$ dollars, a conditional bet on $h$ given $e$ paying one dollar, and (ii) buy a bet on $e$ paying $y = n - r$ dollars for the price $yP(e)$ dollars. The prices at which you buy and sell these bets are, by supposition, fair as far as you are concerned. If $e$ turns out to be false, bet (ii) ensures you lose overall the sum $yP(e)$ dollars. In the event of $e$'s truth, you (iii) sell a bet on $h$ paying one dollar for a dollar price equal to your new fair betting-quotient $r$ on $h$. Totting up, you will find that, whatever the truth-values of $h$

and $e$, bets (i)–(iii) will net you a loss of $yP(e)$ dollars. In other words, by following this particular betting strategy you could be made to suffer an inevitable loss or gain, depending on whether $y$ is positive or negative.

It is widely claimed that from the existence of such a "Dutch Strategy" (the terminology is van Fraassen's), it follows that you are indeed inconsistent in not having Bayesian conditionalisation as your updating rule, if you have a rule at all. But in fact that conclusion does not follow at all, as the following example shows. Suppose that $h$ is a proposition of which you are currently certain (it might, for example, describe some sensation you are having or recently had). Suppose also that you suspect that you have an incipient brain lesion, one of whose effects you believe will be tomorrow to diminish your confidence in things you were convinced of today. Let $e_i$ be the proposition that your fair betting-quotient on $h$ tomorrow will be $r_i$, where $r_1, r_2, \ldots, r_n$ represents the finest resolution of your belief spectrum into units between which you can personally distinguish (these will be intervals, but no harm is done by representing them by their midpoints). You express all your beliefs about your current state in the equations $P(h) = 1$ and $P[P_t(h) = r] > 0$, where $P_t$ is your probability function tomorrow, and $r$ is one of the numbers $r_i < 1$. It follows, if you are consistent, that $P[h \mid P_t(h) = r] = 1$. What should your updating rule be relative to the proposition $P_t(h) = r$? Well, you clearly have no choice, if you remain consistent, but to have $r$ as your updated probability of $h$ in the event of the truth of that proposition. But this conclusion is in conflict with the Lewis-Teller Dutch Book argument, for according to that the only consistent updating rule is conditionalisation, which demands that your updated probability $P_t(h)$ of $h$ at $t$ is $P(h \mid e) = 1$. Are you inconsistent for infringing conditionalisation in this way? and if so, how does the inconsistency arise?

The non-conditionalising updating rule, that if you discover that $P_t(h) = r$ is true, then your new degree of belief $P_t(h)$ in $h$ will be $r$, clearly cannot introduce inconsistencies, for it is in effect a tautology. Could the initial assignments $P(h) = 1$, $P[P_t(h) = r] > 0$ be inconsistent? Surely not; the situation they describe, is—unfortunately perhaps—quite possible. And yet those assignments have implicitly been charged with inconsistency by van Fraassen. On what ground? *That any infraction of what he calls the Reflection principle, namely that for all* r $P[h \mid P_t(h) = r] = r$ *for* $t > 0$ ($P_0$ *being* P)*, is vulnerable to that same Dutch Book argument that allegedly proves that departures from conditionalisation are inconsistent*. To see this, simply note that *P'(h)* in the Lewis-Teller argument must now be equal to *r,* since *r* is the betting-quotient you will have on *h* as a result of 'learning' that $P_t(h) = r$.

But this is to invoke that same Dutch Book argument to rebut an objection to its use in allegedly justifying conditionalisation. In other words, it is to argue in a circle, for it is precisely the conclusion drawn from that argument which is in doubt, namely, that probability assignments which are vulnerable to the diachronic Dutch strategy are inconsistent. And indeed, for a quite independent reason which Christensen (1991) was the first to point out, that conclusion does not follow. This is that a set of betting-quotients can only be shown inconsistent by a Dutch Book if they are ones which you *simultaneously* regard as fair. As Christensen remarks, I may have a fair betting-quotient *p* on some proposition today and an entirely different one *q* tomorrow. Were I to buy a bet paying one dollar on that proposition today for my currently fair price *p* dollars, and sell it tomorrow for my new fair price *q* dollars, I should make a certain loss if $p > q$. But to advance that as a Dutch Book argument to show that you are inconsistent in the way you assign probabilities is clearly absurd. You have simply at one time accepted *p* and at another *q* as fair betting-quotients, not simultaneously accepted both. Consider the analogy with deductive logic, an analogy we shall press, in its different aspects, on several future occasions: my set of sentences held to be true at one and the same time is inconsistent if it contains *c* and *~c,* for some *c*. But of course I can consistently believe *c* true today and *~c* true tomorrow. It is really a very obvious point.

If we now inspect the Lewis-Teller Dutch strategy against any non-conditionalising updating rule, we find that it is based on current fair betting-rates $P(h \mid e)$ and *P(e),* and also on a betting-rate *P'(h)* the agent—you, say—will believe fair in the event of some *e*'s being true. But *P'(h)* is not necessarily a betting-quotient you believe fair now, and in the brain lesion example it was one you certainly did *not* accept as your current fair betting-rate. Hence, though there is a Dutch strategy based on all those betting-rates, it is not one that can possibly convict you as inconsistent in your assignments, since they are not all accepted by you as fair at one and the same time.

This suggests that the scope of the Lewis-Teller argument should be restricted to those updating rules only in which the updated probabilities $P'(h)$ are those betting rates you would *currently* think fair on $h$ were $e$ to be true. Given that restriction, it now does seem reasonable to conclude from the Lewis-Teller argument that if the updated $P'(h)$ differs from $P(h \mid e)$, that is a ground for deeming $P'(h)$ and $P(h \mid e)$ inconsistent. Indeed so, but if the restriction to updated rates you currently think fair is imposed, then $P'(h)$ is *by definition* equal to the conditional probability $P(h \mid e)$, and the Lewis-Teller argument is otiose.

We can conclude, then, that the Lewis-Teller Dutch Book argument is either unsound or otiose. Either way, it does not show, and no diachronic Dutch Book argument can show, that a failure to conditionalise demonstrates inconsistency. It is of interest to note, however, that that argument does work for 'synchronic' reflection: $P_t[h \mid P_t(h) = r] = r$. If we make all the same Lewis-Teller bets here, we get a Dutch Book if $r$ is different from $s$ in $P[h \mid P(h) = r] = s$, as we see from the following slightly simplified Lewis-Teller strategy. Suppose you were to buy (i) a conditional bet on $h$ given $P(h) = r$ paying one dollar, for your fair price $s$; (ii) buy a bet on '$P(h) = r$' paying $y$ dollars, for your fair price $P[P(h) = r]$, where $y = s - r$; and (iii) if '$P(h) = r$' is true, sell a bet on $h$ paying one dollar for your fair price $r$. If you tot up the gains from these bets, you will find, as in the Lewis-Teller Dutch Book, that you will lose $yP[P(h) = r]$ dollars for sure. This Dutch Book does reveal inconsistency, since all the bets involved are ones you do now simultaneously think fair. It might be argued that one hardly needs to go through the argument, since it is intuitively obvious that your fair betting-quotient on $h$ conditional on that quotient's being $r$ must, if you are consistent, be $r$. Still, obvious things may still be incorrect, and it is useful to have an independent check.

Where does all this leave Bayesian conditionalisation? There is certainly no unconditional justification for the rule; in intimating this Hacking was quite correct. It does, however, become a valid rule if the following condition is met: on learning $e$, your conditional probabilities relative to $e$ remain unchanged when $P(e)$ changes to $P'(e) = 1$. Then you must conditionalise. For if $P'(e) = 1$ then $P'(h \mid e) = P'(h)$; so if $P'(h \mid e) = P(h \mid e)$, then $P'(h) = P(h \mid e)$. The intuitive reasonableness of this may not be apparent from the formulas, but it becomes so when translated into words: if on learning $e$ you see no reason to change your opinion that the fair betting-quotient on $h$ were $e$ to be true is $p$, say, then your only consistent value of that fair betting-quotient in the light of $e$'s having occurred is clearly $p$.

We can formulate the same condition in another way. Your unconditional probability $P(h)$ is your estimate of a presumptive fair betting-quotient on $h$ relative to your present stock $K$ of knowledge. Your conditional probability $P(h \mid e)$ is your estimate of the fair betting-quotient on $h$ relative to the augmented information $K \cup \{e\}$. On the supposition that learning that $e$ is true, i.e., adding $e$ to $K$, does not disturb this estimate, then your new fair betting-quotient $P'(h)$ on $h$ will clearly be equal to $P(h \mid e)$. If, however, $e$ deductively implies, by itself or in conjunction with $K$, that your degree of belief at a future time $t$ in $h$ is $r \neq P(h \mid e)$, as in our brain lesion example, then learning $e$ will disturb your prior conditional estimate $P(h \mid e)$ and the condition for conditionalisation will fail. We can confirm this by noting that in that example we had $P[h \mid P_t(h) = r] = 1$, while we also know that consistency requires that $P_t\, [h \mid P_t\, (h) = r] = r$. Hence the conditional probability of $h$ relative to $P_t(h) = r$ will not be the same for the functions $P$ and $P_t$.

On the other hand, scientific evidence is hardly likely to have deductive consequences imposing conditions on your future states of belief. Indeed, it is hardly likely, unless some of the more bizarre interpretations of quantum mechanics turn out to be correct, to have any deductive consequences regarding you at all. Granted this, the conditionalisation principle merely describes the way in which an ideal reasoner, who has worked out in full already the evidential impact of $e$ on $h$, relative to $K$, and expressed this in $P(h \mid e)$, will respond to the news of $e$'s truth. Of course, none of us is an ideal reasoner, able to see all the consequences of any proposition $e$ we consider, but that does not stop us using the rest of the Bayesian apparatus, whose correct application similarly requires a knowledge of deductive relationships, as in the additivity axiom, for example, which no human—or for that matter even non-human—reasoner could ever grasp in their entirety. We shall not comment here on the objections that have been raised against the Bayesian theory on this count, and which we believe misguided, but postpone them to the final chapter of the book, where they will be resolved.

## b JEFFREY CONDITIONALISATION

In his very influential book (1965) Richard Jeffrey inaugurated a new chapter of Bayesian research, into what has come to be called *probability kinematics*. Probability kinematics is about consistent changes in belief on receipt of new data. Now it might be thought that this is just what the Principle of Bayesian Conditionalisation (in the appropriate circumstances) adjudicates, and that that is the end of the story. Not so, as Jeffrey pointed out with the help of the following example (1965, p. 154).

You are observing a piece of cloth by candlelight. Before the observation you have degrees of belief $P(b)$,$P(g)$, and $P(v)$ that the cloth is blue, green, and violet respectively. Afterwards these change to $P'(b)$,$P'(g)$, and $P'(v)$, in such a way that the differences $P(b) - P'(b)$, $P(g) - P'(g)$, and $P(v) - P'(v)$ are all different and non-zero. Can we find a proposition $e$ such that we represent the change from $P$ to $P'$ as one mediated by Bayesian conditionalisation on $e$; i.e., such that

$$P'(b) = P(b \mid e),\ P'(g) = P(g \mid e),\ \text{and}\ P'(v) = P(v \mid e)?$$

Diaconis and Zabell (1982, theorem 2.1) provide the formal condition under which a probability *function* can be obtained from another in this way. But what could $e$ possibly be in the example above? It is not enough for $e$ to say merely that you observed the cloth in dim light, for how could this by itself give rise to the appropriate conditional degrees of belief? It has been suggested (Skyrms, 1985) that $e$ is simply the information that your new degrees of belief will take the values $P'(b)$, $P'(g)$, and $P'(v)$. While such an $e$ will do the trick from the purely logical point of view, it hardly seems a solution to the problem of finding a statement describing the content of the experience which *caused* the changes in belief. If we restrict our range of evidence statements to those describing data inputs that cause beliefs to change, then we seem forced to conclude that there is none which describes the process by which the candlelight observation causes the corresponding changes of belief. In other words, not all changes of belief induced by even localised experiences are representable as changes satisfying the principle of Bayesian conditionalisation.

Let us call changes of belief which are not conditionalisation-changes *exogenous*. So the exogenous change from $P(e) = p$ to $P'(e) = 1$ is required for Bayesian conditionalisation to determine the distribution $P'(c) = P(c \mid e)$ over all those other propositions $c$ which the agent contemplates. In the Jeffrey example we also have an exogenous change, taking the probabilities $P(b)$, $P(g)$, and $P(v)$ to $P'(b)$, $P'(g)$, and $P'(v)$, but one in which the new probabilities can take arbitrary values. Presumably this change will have some effect on the probabilities of other propositions in the system. Is there an analogue of Bayesian conditionalisation for updating $P$ in this case?

Jeffrey, who was the first to ask this question, also provided an answer in the special case where the propositions whose probabilities are exogenously changed are mutually exclusive. His answer is the *Rule of Jeffrey Conditionalisation*. We shall give the rule first for the simplest case where there is only one proposition, $d$ say, whose probability exogenously changes from $P(d)$ to $P'(d)$, where $0 < P(d) < 1$. According to Jeffrey's rule for this case, the posterior probability $P'(c)$ of any other proposition $c$ is given by

$$\textbf{(1)}\quad P'(c) = P(c \mid d)P'(d) + P(c \mid {\sim}d)P'({\sim}d),$$

where (since we are assuming consistency) $P'({\sim}d) = 1 - P'(d)$.

It is not difficult to see that (1) is a generalisation of Bayesian conditionalisation. For suppose that the exogenous change in $d$'s probability takes it to $P'(d) = 1$. Then $P'({\sim}d) = 0$, and so from (1) we infer that $P'(c) = P(c \mid d)$, which is of course just the rule of Bayesian conditionalisation. But it is a generalisation of Bayesian conditionalisation in another way also, for (1) is equivalent to the conjunction of the pair of equalities $P'(c \mid d) = P(c \mid d)$ and $P'(c \mid \sim d) = P(c \mid \sim d)$. In other words, Jeffrey conditionalisation is equivalent to the statement that the conditional probabilities of all sentences relative to $d, {\sim}d$

are unchanged. And we know that Bayesian conditionalisation is valid just when $P'(c \mid d) = P(c \mid d)$ for all $c$, where $d$ is the sentence whose $P'$ probability changes exogenously to 1. Since we can take the conjugate condition $P'(c \mid \sim d) = P(c \mid \sim d)$ to be vacuously satisfied where $P'(d) = 1$, the rules of Bayesian and Jeffrey conditionalisation are therefore essentially one and the same.

But they nevertheless have significant differences, reflecting the fact that one, Bayesian conditionalisation, is obtained as the limit of the other. For example, an important feature of Jeffrey conditionalisation not shared by Bayesian conditionalisation (except in extreme cases) is that it allows you to recover your original probability distribution. Suppose that, as a result of some non-linguistically expressible experience like the observation by candlelight, you change your probability of some sentence $d$ from $P(d)$ to $P'(d)$, where $P'(d) \neq 1$, maintaining your conditional probabilities on $d$. You then use (1) to obtain your posterior probability distribution $P'$ over all the other propositions you are contemplating. For some reason you subsequently decide that your first opinion was a more accurate assessment, and so you revert from $P'(d)$ to $P''(d) = P(d)$. Using (1) again you obtain a posterior distribution $P''$ on all the other propositions in your system, by setting $P''(c) = P'(c \mid d) P''(d) + P'(c \mid \sim d) P''(\sim d)$. It is straightforward to show that $P''(c) = P(c)$ for all $c$. In other words, you can change the probability of $d$ exogenously from $P(d)$ to $P'(d)$ and back to $P(d)$, and recover the original distribution $P$ over all propositions, by successive applications of (1). For Bayesian conditionalisation, where the new probability of $d$ goes to 1, you can recapture your original distribution only in the exceptional circumstance that it already assigned $d$ the probability 1.

The condition for the applicability of the Jeffrey rule, (1), as we saw, is that the conditional probabilities $P(\cdot \mid \pm d)$ remain unchanged. In a careful discussion of Jeffrey's rule, Pearl (1988, pp. 64–70) points out that verifying that this condition is satisfied requires considerable topological knowledge of one's belief structure, and he suggests that one way that knowledge can be acquired is by seeing whether the other sentences in the domain of $P$ are *conditionally independent* of $d$ given a sentence $e$ which describes the observation which caused the shift from $P(d)$ to $P'(d)$. For by the probability calculus

$$P(c \mid e) = P(c \mid d \,\&\, e)\, P(d \mid e) + P(c \mid \sim d \,\&\, e)\, P(\sim d \mid e),$$

so that if we define $P'(\cdot) = P(\cdot \mid e)$, then the equation above becomes formally identical with (1) if $c$ is independent of $e$ given $d$, i.e., $P(c \mid \pm d \,\&\, e) = P(c \mid \pm d)$. In other words, determining whether there is conditional independence given $e$ determines the applicability of Jeffrey's rule.

But this is precisely to ignore the fact which inspired Jeffrey to find an alternative to Bayesian conditionalisation, namely, that there may well be no such $e$. If there isn't one, then Pearl's suggestion cannot be implemented. On the other hand, if there is, then Jeffrey conditionalisation is redundant anyway, as Pearl himself observes (1988, p. 70). Moreover, the information you need in order to know whether Jeffrey's rule is applicable is just the same information that you need in order to know whether Bayesian conditionalisation is applicable, for, as we know, both rules require just the same condition of invariance of the relevant conditional probabilities. So it is as hard, or as easy, to verify the conditions of applicability for Bayesian conditionalisation as for it is Jeffrey conditionalisation.

## ■ c GENERALISING JEFFREY'S RULE TO PARTITIONS

We can, in certain conditions, generalise (1) to apply to situations where more than one proposition has its probability changed exogenously. (1) determines $P'$ as a weighted average of the conditional probabilities $P(\cdot \mid d)$, $P(\cdot \mid \sim d)$, with weights $P'(d)$, $P'(\sim d)$ respectively. Now $\{d, \sim d\}$ is a partition, i.e., a set of mutually exclusive and exhaustive alternatives, and (1) generalises straightforwardly to your simultaneously changing your degrees of belief from $P(d_i)$ to $P'(d_i)$ on the members of any $n$-fold partition $d_1, \ldots, d_n$:

**(2)** $P'(c) = \Sigma P(c \mid d_i) P'(d_i)$.

Observe that, analogously to (1), (2) is equivalent to the condition that all the $n$ identities $P'(c \mid d_i) = P(c \mid d_i)$ hold. The condition that the conditional probabilities remain invariant holds, as we saw above, for Bayesian conditionalisation also (though here the condition reduces to $P'(c \mid d) = P(c \mid d)$, as $P'(c \mid \sim d)$ is undefined when $P'(d) = 1$). This is important, for it is that very condition which determines the applicability of Bayesian conditionalisation. Likewise, we can infer that Jeffrey conditionalisation also is a valid rule in all cases where the relevant conditional probabilities remain unchanged.

## d DUTCH BOOKS AGAIN

The Jeffrey rule, like that of Bayesian conditionalisation, has only conditional validity: if the relevant conditional probabilities are maintained, the rule is valid; if not, not. Nevertheless, just as we saw that some people have attempted to show by a Dutch Book argument that Bayesian conditionalisation is a rule of quite general validity, so some other people have tried to do the same thing for the Jeffrey rule. In particular Armendt (1980) and Skyrms (1987) have attempted to construct Dutch Book arguments for Jeffrey conditionalisation. Both their discussions adopt the same assumption as did the Lewis-Teller one, namely, that the agent updates his/her probability function according to some specific rule. They then show that if their rule infringes the condition necessary and sufficient for Jeffrey's rule—that all the posterior conditional probabilities $P'(h \mid e)$ and $P'(h \mid \sim e)$ are equal to the corresponding prior conditional probabilities $P(h \mid e)$, $P(h \mid \sim e)$, where the change from $P$ to $P'$ originates in $e$—then there is a betting strategy that would generate a Dutch Book from those probabilities. Suppose, for example, that $P(h \mid e) > P'(h \mid e)$. The Dutch Book strategy adapts the Lewis-Teller one to the present case and consists of buying a conditional bet on $h$ given $e$ paying one dollar for the agent's current fair price $P(h \mid e)$ dollars, selling it back after $P$ has shifted to $P'$ for the agent's new fair price $P'(h \mid e)$ dollars, and selling a side-bet now on $e$ paying $P'(h \mid e) - P(h \mid e)$ dollars for the agent's current fair price $P(e)$ dollars. Totting up, we can see that the agent would lose $P(e)[P'(h \mid e) - P(h \mid e)]$ dollars come what may.

But the existence of such a Dutch Book strategy no more points up an inconsistency in the agent's beliefs than did the Lewis-Teller one, and for the same reason: it is constructed from betting quotients, some of which you believe fair now and some of which you will believe fair in the event of your probability of $e$ exogenously changing from $P(e)$ to $P'(e)$. In other words, it is based on a set of betting quotients not restricted to those which you believe currently fair. So, like the Lewis-Teller diachronic Dutch Book argument, this one also is unsound and does not establish the Jeffrey rule as a general consistency constraint. The thrust of these unsuccessful Dutch Book arguments for Jeffrey's rule as an absolute and unconditional rule of consistent behaviour is, we should now recognise, wrongly directed. Jeffrey's rule, like that of Bayesian conditionalisation, is *not* unconditionally valid. It is only valid provided that your conditional probabilities relative to the members of the partition whose probabilities have exogenously changed remain themselves unchanged.

But still they try: in a recent book (1989, pp. 331–37), van Fraassen attempts a different route to Jeffrey and Bayesian conditionalisation, this time by showing them to be consequences of certain symmetry conditions. In our opinion all that we should infer from these demonstrations is that symmetry principles, whatever their fertility in generating successful scientific theories, as in modern particle physics, should not be regarded as having axiomatic status. Symmetry arguments are popular among neo-Bayesians as means of determining probability distributions, and we shall have more to say about them in our final chapter, in particular about E. T. Jaynes's influential programme for determining prior probability distributions explicitly as the invariant distributions relative to the constraints imposed by background data.

## e THE PRINCIPLE OF MINIMUM INFORMATION

Yet another attempt to underwrite Jeffrey and Bayesian conditionalisation as rules of general validity also remedies an interesting formal defect in Jeffrey's rule, which is that it cannot be used to determine a posterior probability function when the propositions which have their probabilities exogenously changed do not form a partition. The problem cannot be got round by successively Jeffrey-conditioning on each of these in turn (i.e., applying (1) to each in some specified order), because the final distribution so obtained will in general depend on the order in which the propositions are taken. This is not true of successive Bayesian conditionalisations: it follows from a result in Chapter 2 (namely, (16) in section **c.3**) that successive Bayesian conditioning on $n$ sentences $a_1, \ldots, a_n$ gives the same result as conditioning on all simultaneously; hence the order of conditioning is unimportant.

A solution to the problem of choosing an updated probability function fitting $n$ exogenously determined probabilities $P'(a_1), \ldots, P'(a_n)$, where the $a_i$ can be arbitrarily chosen, and are not necessarily a partition, arises from exploring the analogy with the problem of fitting that curve to $n$ points which is closest to some initially specified curve. The initial 'curve' is the initial or prior probability function $P$, and closeness is to be measured now in the space of probability functions. But how? The favoured measure of closeness is that expressed in the so-called *Principle of Minimum Information,* also known as the *Principle of Minimum Cross-Entropy.* According to this, the closest function $P'$ to $P$ which satisfies the $n$ constraints $P'(a_1) = p_1', \ldots, P'(a_n) = p_n'$, is that which minimises the functional $I(P',P)$ subject to those constraints, where $I(P',P) = \Sigma p_i'\log(\frac{p_i'}{p_i})$ if the distributions $P,P'$ are discrete and none of the $p_i$ are zero when $p_i'$ is not. $I(P',P)$, read as *the information in* P' *relative to* P, is equal to the corresponding integral if $P,P'$ are continuous, where $\{p_i : i = 1, \ldots, n\}$ and $\{p_i' : i = 1, \ldots, n\}$ are replaced by probability densities $f(x), f'(x)$.

$I(P',P)$ has some interesting properties. For example, when the prior distribution $P$ is uniform, then minimising $I$ is equivalent to maximising the Shannon entropy $-\Sigma p_i'\log p_i'$. But whereas the Shannon entropy is, in the continuous case, not invariant under changes of random variable, $I$ is—because the Jacobian of the transformation from one variable $X$ to another $Y$ cancels top and bottom in the ratio $\frac{f(y)}{f'(y)}$. It is also not difficult to establish that Jeffrey's rule and hence Bayesian conditionalisation emerge as special cases of the rule to choose the $P'$ which minimizes $I$ subject to the relevant constraints. Does this amount to the sought-after demonstration of Jeffrey and Bayesian conditionalisation as general rules of correct reasoning? Our answer is, perhaps predictably, no.

First, a technical objection. A unique $I$-minimizing $P'$ exists in a very wide class of cases, including that where the constraint is of the form $P(a_1) = p_1', \ldots, P(a_n) = p_n'$, where the

$a_i$ can be any propositions, thus apparently solving the problem posed by the restriction of Jeffrey's rule to the case where the $a_i$ have to be disjoint. However, that is not the end of the story. $I(P',P)$ is supposed, recall, to represent the distance in function space between $P$ and $P'$. Now while $I(P',P)$ is indeed 0 when and only when $P=P'$, $I$ is not symmetric in $P$ and $P'$ and therefore *not* a true distance. Another functional, $\sup | P(A)-P'(A) |$, the so-called *variation distance* between $P$ and $P'$, is a distance in the sense of being a metric in function space, but minimising it does not always yield the same function $P'$ as minimising $I$ for the very same class of constraints (good accounts of these matters, with examples, are given by Diaconis and Zabell, 1982, and Williams, 1980). Hence the claim that the $P'$ which minimises $I$ is the closest to $P$, subject to the appropriate constraints, should not be taken too seriously.

But even if some better backing could be given to that claim, it would still leave open the question why closeness should be a relevant consideration in the choice of a posterior probability. We seem to be back in the shadowy, if not downright shady, world of the synthetic a priori. We want all our rules to be justified against the criterion of whether they are genuinely *logical* principles, and closeness in function space does not sound like a logical principle. However, in a well-known paper Shore and Johnson (1980) purport to demonstrate that it is. They list four quite plausible consistency criteria and then show that they determine the Principle of Minimum Information uniquely. However, they invoke an additional assumption, that the posterior probability $P'$ is also one which extremises (maximises or minimises) some functional $H(P',P)$. Now this is simply to fall back on a 'closeness' criterion again, so we cannot conclude with Shore and Johnson that they have shown that consistency criteria alone determine that the Principle of Minimum Information is the correct updating rule.

## f CONCLUSION

Bayesian and Jeffrey conditionalisation are at bottom the same principle, the latter subsuming the former as a special case. It is valid subject to the condition that the exogenous change from $P$ to $P'$ should not disturb the values of the existing $P$-conditional probabilities relative to the partition of propositions in which the change originates. This condition is less restrictive than it might sound, being no more than the assumption that the full bearing of the truth of $e$—where, let us suppose, $P(e)$ changes to $P'(e)$ exogenously—on each $h$ is already fully worked out in assigning the conditional probabilities $P(h \mid e)$, so that no further reflection causes you to change your mind once the change from $P$ to $P'$ on $e$ has taken place. This is a condition we can imagine satisfied by an ideal scientific reasoner, and indeed it was almost certainly because they had such a reasoner in mind that the pioneers of the Bayesian theory thought it unnecessary to provide an explicit justification for assuming conditionalisation to take place on the receipt of new data. We hope that we have vindicated their practice, at the very least.

## EXERCISES

1. Show that $P'$ as defined by (1) satisfies axioms 1–3 of the probability calculus.
2. Show that (1) is equivalent to $P'(c \mid d) = P(c \mid d)$ and $P'(c \mid \sim d) = P(c \mid \sim d)$. (Hint. The implication, from the conditional probabilities staying unchanged to (1), is straightforward. The converse follows from expanding $P'(c \,\&\, d)$ according to (1).)
3. Show that if $P'(c) = P(c \mid d)$, then $P'(c \mid d) = P(c \mid d)$.
4. Suppose that your probability of $d$ goes exogenously from $P(d)$ to $P'(d)$ and then from $P'(d)$ to $P''(d) = P(d)$. Show that if (1) is used to determine a posterior probability distribution $P'$, and then again to obtain a distribution $P''$, then the distribution $P''$ so obtained is identical to $P$.
5. Show that if $P'$ is obtained by Bayesian conditionalisation on $d$ from an original distribution $P$, i.e., that $P'(c) = P(c \mid d)$, then there is no proposition $e$ such that $P(c) = P'(c \mid e)$ for all $c$, unless $P(d) = 1$.
6. Construct an example where a posterior probability function obtained by successive Jeffrey-conditioning changes if the order of conditioning changes.

**7.** Suppose that $P', P$ are discrete and show that if $P(e)$ changes to $P'(e)$, then the function $P'$ which minimises $I(P', P)$ subject to the constraint $P'(e) = p'$ is given by (1); i.e., $P'$ is such that for all $a$, $P'(a) = P(a \mid e)p' + P(a \mid \sim e)(1 - p')$.

# PART II

# *Bayesian Induction: Deterministic Theories*

Philosophers of science have traditionally concentrated attention primarily on deterministic hypotheses, leaving statisticians to discuss the methods by which statistical or non-deterministic theories should be assessed. Accordingly, a large part of what would more naturally be regarded as philosophy of science is normally treated as a branch of statistics, going under the heading 'statistical inference'. So it is not surprising that philosophers and statisticians have developed distinct methods for their different purposes. We shall follow the tradition of treating deterministic and statistical theories separately. As will become apparent, however, we regard this separation as artificial and shall, in the course of the book, expound the unified treatment of scientific method afforded by Bayesian principles.

# CHAPTER 7

# Bayesian Versus Non-Bayesian Approaches

In this chapter we shall consider how, by attributing positive probabilities to hypotheses in the manner described in Chapter 2, one can account for many of the characteristic features of scientific practice, particularly as they relate to deterministic theories.

## a THE BAYESIAN NOTION OF CONFIRMATION

Information gathered in the course of observation is often considered to have a bearing on the acceptability of a theory or hypothesis (we use the terms interchangeably), either by confirming it or by disconfirming it. Such information may either derive from casual observation or, more commonly, from experiments deliberately contrived in the hope of obtaining relevant evidence. The idea that evidence may count for or against a theory, or be neutral towards it, is a central feature of scientific inference, and the Bayesian account will clearly need to start with a suitable interpretation of these concepts.

Fortunately, there is a suitable and very natural interpretation, for if $P(h)$ measures your belief in a hypothesis when you do not know the evidence $e$, and $P(h \mid e)$ is the corresponding measure when you do, $e$ surely confirms $h$ when the latter exceeds the former. So we shall adopt the following as our definitions:

$e$ **confirms or supports** $h$ when $P(h \mid e) > P(h)$

$e$ **disconfirms or undermines** $h$ when $P(h \mid e) < P(h)$

$e$ **is neutral with respect to** $h$ when $P(h \mid e) = P(h)$

One might reasonably take $P(h \mid e) - P(h)$ as measuring the degree of $e$'s support for $h$, though other measures have been suggested (e.g., Good, 1950); disagreements on this score will not need to be settled in this book. We shall refer, in the usual way, to $P(h)$ as 'the prior probability of $h$' and to $P(h \mid e)$ as $h$'s 'posterior probability' relative to, or in the light of, $e$. The reasons for this terminology are obvious, but it ought to be noted that the terms have a meaning only in relation to evidence: as Lindley (1970, p. 38) put it, "[t]oday's posterior distribution is tomorrow's prior". It should be remembered too that all the probabilities are evaluated in relation to accepted background knowledge.

## b THE APPLICATION OF BAYES'S THEOREM

Bayes's Theorem relates the posterior probability of a hypothesis, $P(h \mid e)$, to the terms $P(h)$, $P(e \mid h)$, and $P(e)$. Hence, knowing the values of these last three terms, it is possible to determine whether $e$ confirms $h$, and, more importantly, to calculate $P(h \mid e)$. In practice, of course, the various probabilities may only be known rather imprecisely; we shall have more to say about this practical aspect of the question later.

The dependence of the posterior probability on the three terms referred to above is reflected in three striking phenomena of scientific inference. First, other things being equal, the extent to which evidence $e$ confirms a hypothesis $h$ increases with the likelihood of $h$ on $e$, that is to say, with $P(e \mid h)$. At one extreme, where $e$ refutes $h$, $P(e \mid h) = 0$; hence, disconfirmation is at a maximum. The greatest confirmation is produced, for a given $P(e)$, when $P(e \mid h) = 1$, which will be met in practice when $h$ logically entails $e$. Statistical hypotheses, which will be dealt with in Parts III, IV, and V of this book, are more substantially confirmed the higher the value of $P(e \mid h)$.

Secondly, the posterior probability of a hypothesis depends on its prior probability, a dependence sometimes discernible in scientific attitudes to ad hoc hypotheses and in frequently expressed preferences for the simpler of two hypotheses. As we shall see, scientists always discriminate, in advance of any experimentation, between theories they regard as more-or-less credible (and, so, worthy of attention) and others.

Thirdly, the power of $e$ to confirm $h$ depends on $P(e)$, that is to say, on the probability of $e$ when it is not assumed that $h$ is true (which, of course, is not the same as assuming $h$ to be false). This dependence is reflected in the scientific intuition that the more surprising the evidence, the greater its confirming power. However, $P(e) = P(e \mid h)P(h) + P(e \mid \sim h)P(\sim h)$ (as we showed in Chapter 2, section **c.3**), so that really, the posterior probability of $h$ depends on the three basic quantities $P(h)$, $P(e \mid h)$, and $P(e \mid \sim h)$.

We shall deal in greater detail with each of these facets of inductive reasoning in the course of this chapter.

## c FALSIFYING HYPOTHESES

A characteristic pattern of scientific inference is the refutation of a theory, when one of a theory's empirical consequences has been shown to be false in an experiment. As we saw, this kind of reasoning, with its straightforward and unimpeachable logical structure, exercised such an influence on Popper that he made it the centrepiece of his scientific philosophy.

Although the Bayesian approach was not conceived specifically with this aspect of scientific reasoning in view, it has a ready explanation for it. The explanation relies on the fact that if, relative to background knowledge, a hypothesis $h$ entails a consequence $e$, then (relative to the same background knowledge) $P(h \mid \sim e) = 0$. Interpreted in the Bayesian fashion, this means that $h$ is maximally disconfirmed when it is refuted. Moreover, as we should expect, once a theory is refuted, no further evidence can ever confirm it, unless the refuting evidence or some portion of the background assumptions is revoked. (The straightforward proofs of these claims are suggested as an exercise.)

## d CHECKING A CONSEQUENCE

A standard method of investigating a deterministic hypothesis is to draw out some of its logical consequences, relative to a stock of background knowledge, and check whether they are true or not. For instance, the General Theory of Relativity was confirmed by establishing that light is deflected when it passes near the sun, as the theory predicts. It is easy to show, by means of Bayes's Theorem, why and under what circumstances a theory is confirmed by its consequences.

If $h$ entails $e$, then, as may be simply shown, $P(e \mid h) = 1$. Hence, from Bayes's Theorem: $P(e \mid h) = \frac{P(h)}{P(e)}$. Thus, if $0 < P(e) < 1$, and if $P(h) > 0$, then $P(h \mid e) > P(h)$. It follows that any evidence whose probability is neither of the extreme values must confirm every hypothesis with a non-zero probability of which it is a logical consequence.

Succeeding confirmations must eventually diminish in force, for the theory has an upper limit of probability beyond which no amount of evidence can push it. This too follows from Bayes's Theorem. Suppose $e_1, e_2, \ldots, e_n, \ldots$ are consequences of $h$. Then Bayes's Theorem asserts that

$$P(h \mid e_1 \,\&\, e_2 \,\&\, \ldots \,\&\, e_n) = \frac{P(h)}{P(h \mid e_1 \,\&\, e_2 \,\&\, \ldots \,\&\, e_n)}.$$

Now

$$P(e_1 \,\&\, e_2 \,\&\, \ldots \,\&\, e_n) = P(e_1)P(e_2 \,\&\, \ldots \,\&\, e_n \mid e_1)$$

and

$$P(e_2 \,\&\, \ldots \,\&\, e_n \mid e_1) = \mathrm{P}(\mathrm{e}_2 \mid e_1)P(e_3 \,\&\, \ldots \,\&\, e_n \mid e_1 \,\&\, e_2).$$

Thus, in general,

$$P(e_1 \,\&\, e_2 \,\&\, \ldots \,\&\, e_n) = P(e_1)P(e_2 \mid e_1) \ldots P(e_n \mid e_1 \,\&\, \ldots \,\&\, e_{n-1}).$$

Hence,

$$P(h \mid e_1 \,\&\, e_2 \,\&\, \ldots \,\&\, e_n) = \frac{P(h)}{P(e_1)P(e_2 \mid e_1) \ldots P(e_n \mid e_1 \,\&\, \ldots \,\&\, e_{n-1})}.$$

Provided $P(h) > 0$, the term $P(e_n \mid e_1 \,\&\, \ldots \,\&\, e_{n-1})$ must tend to 1 as $n$ increases. If it did not, the posterior probability of $h$ would at some point exceed 1, which is impossible (Jeffreys, 1961, pp. 43–44). This explains why it is not sensible to test a hypothesis indefinitely, though without more detailed information on the individual's belief-structure, in particular regarding the values of $P(e_n \mid e_1 \,\&\, \ldots \,\&\, e_{n-1})$, one could not know the precise point beyond which further predictions of the hypothesis were sufficiently probable not to be worth examining.

Specific categories of a theory's consequences also have a restricted capacity to confirm (Urbach, 1981). Suppose $h$ is the theory under discussion and that $h_r$ is a substantial restriction of that theory. A substantial restriction of Newton's theory might, for example, express the idea that freely falling bodies near the earth descend with a constant acceleration or that the period and length of a pendulum are related by the familiar formula. Since $h$ entails $h_r$, $P(h) \leq P(h_r)$—(*see* Chapter 2, section **c.3**)—and if $h_r$ is much less speculative than its progenitor, it will often be significantly more probable.

Now consider a series of predictions derived from $h$, but which also follow from $h_r$. If the predictions are verified, they may confirm both theories, whose posterior probabilities are given by Bayes's Theorem, thus:

$$P(h \mid e_1 \ \& \ e_2 \ \& \ldots \& \ e_n) = \frac{P(h)}{P(e_1 \ \& \ e_2 \ \& \ldots \& \ e_n)}$$

and

$$P(h_r \mid e_1 \ \& \ e_2 \ \& \ldots \& \ e_n) = \frac{P(h_r)}{P(e_1 \ \& \ e_2 \ \& \ldots \& \ e_n)}.$$

Combining these two equations to eliminate the common denominator, one obtains

$$P(h \mid e_1 \ \& \ e_2 \ \& \ldots \& \ e_n) = \frac{P(h)}{P(h_r)} P(h_r \mid e_1 \ \& \ e_2 \ \& \ldots \& \ e_n).$$

Since the maximum value of the last probability term in this equation is 1, it follows that however many predictions of $h_r$ have been verified, the main theory, $h$, can never acquire a posterior probability in excess of $\frac{P(h)}{P(h_r)}$. Hence, the type of evidence characterised by entailment from $h_r$ may well be limited in its capacity to confirm $h$.

This result explains the familiar phenomenon that repetitions of a particular experiment often confirm a general theory only to a limited extent, for the predictions verified by means of a given kind of experiment (that is, an experiment designed to a specified pattern) do normally follow from and confirm a much-restricted version of the predicting theory. When an experiment's capacity to generate confirming evidence has been exhausted through repetition, further support for $h$ would have to be sought from other experiments, experiments whose outcomes were predicted by different parts of $h$.

The arguments and explanations in this section rely on the possibility that evidence already accumulated from an experiment may increase the probability of further performances of the experiment producing similar results. Such a possibility is denied by Popperians on the grounds that the probabilities involved are subjective. How then do they explain the fact, attested by every scientist, that by repeating some experiment, one eventually (usually quickly) exhausts its capacity to confirm a given hypothesis? Alan Musgrave (1975) attempted an explanation designed on Popperian lines. He claimed that after a certain, unspecified number of repetitions of an experiment, the scientist would form a generalisation to the effect that whenever the experiment was performed, it would yield a similar result. Musgrave then proposed that the generalisation should be entered into 'background knowledge'. Relative to this newly augmented background knowledge, the experiment is certain to produce a similar result at its next performance. Musgrave then appealed to the principle that evidence confirms a hypothesis in proportion to the difference between its probability relative to the hypothesis together with background knowledge and its probability relative to background knowledge alone. That is, the degree to which $e$ confirms $h$ is proportional to $P(e \mid h \& b) - P(e \mid b)$, $b$ being background knowledge. Musgrave then inferred that even if the experiment did produce the expected result when next performed, the hypothesis would receive no new confirmation. Watkins (1984, p. 297) has endorsed this account.

A number of decisive objections may be raised against it, though. First, as we shall show in the next section, although it seems to be a fact and is an essential constituent of Bayesian reasoning, there is no basis in Popperian methodology for confirmation to depend on the probability of the evidence; Popper simply invoked the principle ad hoc. Secondly, Musgrave's suggestion takes no account of the fact that particular experimental results may be generalised in infinitely many ways. This is a substantial objection, since different generalisations give rise to different expectations about the outcomes of future experiments. Musgrave's account is incomplete without a rule to specify in each case the appropriate generalisation that should be formulated and adopted, and it is hard to imagine how such a rule could be justified within the confines of Popperian philosophy. Finally, the decision to designate the generalisation background knowledge, with the consequent effect on our evaluation of other theories and on our future conduct regarding, for example, whether to repeat certain experiments, is comprehensible only if we have invested some confidence in the theory. But then Musgrave's account tacitly calls on the same kind of inductive considerations as it was designed to circumvent, so its aim is defeated.

## e THE PROBABILITY OF THE EVIDENCE

The degree to which $h$ is confirmed by $e$ depends, according to Bayesian theory, on the extent to which $P(e \mid h)$ exceeds $P(e)$. An equivalent way of putting this is to say that confirmation is correlated with the difference between $P(e \mid h)$ and $P(e \mid \sim h)$, that is, with how much more probable the evidence is if the hypothesis is true than if it is false. This is obvious from the third form of Bayes's Theorem (*see* Chapter 2):

$$\frac{P(h \mid e)}{P(h)} = \frac{1}{P(h) + \dfrac{P(e \mid \sim h)}{P(e \mid h)} P(\sim h)}.$$

These facts are reflected in the everyday experience that information that is particularly unexpected or surprising, unless some hypothesis is assumed to be true, supports that hypothesis with particular force. Thus, if a soothsayer predicts that you will meet a dark stranger sometime and you do, your faith in his powers of precognition would not be much enhanced: you would probably continue to think his predictions were just the result of guesswork. However, if the prediction also gave the correct number of hairs on the head of that stranger, your previous scepticism would no doubt be severely shaken.

Cox (1961, p. 92) illustrated this point with an incident in *Macbeth*. The three witches, using their special brand of divination, predicted to Macbeth that he would soon become both Thane of Cawdor and King of Scotland. Macbeth finds both these prognostications almost impossible to believe:

> By Sinel's death, I know I am Thane of Glamis,
> But how of Cawdor?
> The Thane of Cawdor lives, a prosperous gentleman,
> And to be King stands not within the prospect of belief,
> No more than to be Cawdor.

But a short time later he learns that the Thane of Cawdor prospered no longer, was in fact dead, and that he, Macbeth, has succeeded to the title. As a result, Macbeth's attitude to the witches' powers is entirely altered, and he comes to believe in their other predictions and in their ability to foresee the future.

The following, more scientific, example was used by Jevons (1874, vol. 1, pp. 278–79) to illustrate the dependence of confirmation on the improbability of the evidence. The distinguished scientist Charles Babbage examined numerous logarithmic tables published over two centuries in various parts of the world. He was interested in whether they derived from the same source or had been worked out independently. Babbage (1827) found the same six errors in all but two and drew the "irresistible" conclusion that, apart from these two, all the tables originated in a common source.

Babbage's reasoning was interpreted by Jevons roughly as follows. The theory $t_1$, which says of some pair of logarithmic tables that they shared a common origin, is moderately likely in view of the immense amount of labour needed to compile such tables ab initio, and for a number of other reasons. The alternative, independence theory might take a variety of forms, each attributing different probabilities to the occurrence of errors in various positions in the table. The only one of these which seems at all likely would assign each place an equal probability of exhibiting an error and would, moreover, regard those errors as more-or-less independent. Call this theory $t_2$ and let $e^i$ be the evidence of $i$ common errors in the tables. The posterior probability of $t_1$ is inversely proportional to $P(e^i)$, which, under the assumption of only two rival hypotheses, can be expressed as $P(e^i) = P(e^i \mid t_1)\, P(t_1) + P(e^i \mid t_2)P(t_2)$. (This is the theorem of total probability—*see* Chapter 2, section **c.3.**) Since $t_1$ entails $e^i$, $P(e^i) = P(t_1) + P(e^i \mid t_2)P(t_2)$. The quantity $P(e^i \mid t_2)$ clearly decreases with increasing $i$. Hence $P(e^i)$ diminishes and approaches $P(t_1)$, as $i$ increases; and so $e^i$ becomes increasingly powerful evidence for $t_1$, a result which agrees with scientific intuition.

In fact, scientists seem to regard a few shared mistakes in different mathematical tables as so strongly indicative of a common source that at least one compiler of such tables attempted to protect his copyright by deliberately incorporating three minor errors "as a trap for would-be plagiarists" (L. J. Comrie, quoted by Bowden, 1953, p. 4).

The relationship between how surprising a piece of evidence is on background assumptions and its power to confirm a hypothesis is a natural consequence of Bayesian theory and was not deliberately built in. On the other hand, methodologies that eschew probabilistic assessments of hypotheses seem constitutionally incapable of accounting for the phenomenon. Such approaches would need to be able, first, to discriminate between items of evidence on grounds other than their deductive or probabilistic relation to a hypothesis. And having established such a basis for discriminating, they must show a connection with confirmation. The objectivist school has more-or-less dodged this challenge. An exception is Popper. In tackling the problem, he moved partway towards Bayesianism; however, the concessions he made were insufficient. Thus Popper conceded that, in regard to confirmation, the significant quantities are $P(e \mid h)$ and $P(e)$, and as we have already reported, he even measured the amount of confirmation (or "corroboration", to use Popper's preferred term) which $e$ confers on $h$ by the difference between these quantities (Popper, 1959a, appendix *ix).

But Popper never stated explicitly what he meant by the probability of evidence. On the one hand, he would never have allowed it to have a subjective connotation, for that would have compromised the supposed objectivity of science; on the other hand, he never worked out what objective significance the term could have. His writings suggest that he had in mind some purely logical notion of probability, but as we saw in Chapter 4, there is no adequate account of logical probability. Popper also never explained satisfactorily why a hypothesis benefits from improbable evidence or, to put the objection another way, he failed to provide a foundation in non-Bayesian terms for the

Bayesian confirmation function which he appropriated. (For a discussion and decisive criticism of Popper's account, see Grünbaum, 1976.)

The Bayesian position has recently been misunderstood to imply that if some evidence is known, then it cannot support any hypothesis, on the grounds that known evidence must have unit probability. That the objection is based on a misunderstanding is shown in Chapter 15, where a number of other criticisms of the Bayesian approach will be rebutted.

## f THE RAVENS PARADOX

That evidence supports a hypothesis more the greater the ratio $\frac{P(e \mid h)}{P(e)}$ scotches a famous puzzle first posed by Hempel (1945) and known as the *Paradox of Confirmation* or sometimes as the *Ravens Paradox*. It was called a paradox because its premisses were regarded as extremely plausible, despite their counter-intuitive, or in some versions contradictory, implications, and the reference to ravens stems from the paradigm hypothesis ('All ravens are black') which is frequently used to expound the problem. The difficulty arises from three assumptions about confirmation. They are as follows:

1. Hypotheses of the form 'All *R*s are *B*' are confirmed by the evidence of something that is both *R* and *B*. For example, 'All ravens are black' is confirmed by the observation of a black raven. (Hempel called this Nicod's condition, after the philosopher Jean Nicod.)
2. Logically equivalent hypotheses are confirmed by the same evidence. (This is the Equivalence condition.)
3. Evidence of some object not being *R* does not confirm 'All *R*s are *B*'.

We shall describe an object that is both black and a raven with the term *RB*. Similarly, a non-black, non-raven will be denoted $\bar{R}\,\bar{B}$. A contradiction arises for the following reasons: an *RB* confirms 'All *R*s are *B*', on account of the Nicod condition. According to the Equivalence condition, it also confirms 'All non-*B*s are non-*R*s', since the two hypotheses are logically equivalent. But contradicting this, the third condition implies that *RB* does not confirm 'All non-*B*s are non-*R*s'.

The contradiction may be avoided by revoking the third condition, as is sometimes done. (We shall note later another reason for not holding on to it.) However, although the remaining conditions are compatible, they have a consequence which many philosophers have regarded as blatantly false, namely, that by observing a non-black, non-raven (say, a red herring or a white shoe) one confirms the hypothesis that all ravens are black. (The argument is this: 'All non-*B*s are non-*R*' is equivalent to 'All *R*s are *B*'; according to the Nicod condition, the first is confirmed by $\bar{R}\,\bar{B}$; hence, by the Equivalence condition, so is the second.)

If non-black, non-ravens support the raven hypothesis, this seems to imply the paradoxical result that one could investigate that and other generalisations of a similar form just as well by observing white paper and red ink from the comfort of one's writing desk as by studying ravens on the wing. However, this would be a non sequitur. For the fact that *RB* and $\bar{R}\,\bar{B}$ both confirm a hypothesis does not imply that they do so with equal force. Once it is recognised that confirmation is a matter of degree, the conclusion is no longer so counter-intuitive, because it is compatible with $\bar{R}\,\bar{B}$ confirming 'All *R*s are *B*', but to a minuscule and negligible degree.

Indeed, most people do have a strong intuition that an *RB* confirms the ravens hypothesis ($h$) more than an $\bar{R}\,\bar{B}$. We can appreciate why that might be by consulting Bayes's Theorem as it applies to the two types of datum:

$$\frac{P(h \mid RB)}{P(h)} = \frac{P(RB \mid h)}{P(RB)} \quad \& \quad \frac{P(h \mid \bar{R}\,\bar{B})}{P(h)} = \frac{P(\bar{R}\,\bar{B} \mid h)}{P(\bar{R}\,\bar{B})}$$

These expressions can be simplified. First, $P(RB \mid h) = P(B \mid h \,\&\, R)P(R \mid h) = P(R \mid h) = P(R)$. We arrived at the last equality by assuming that whether some arbitrary object is a raven is independent of the truth of $h$, which seems plausible to us, at any rate as a good approximation, though Horwich (1982, p. 59) thinks it has no plausibility. By similar reasoning, $P(\bar{R}\,\bar{B} \mid h) = P(\bar{B} \mid h) = P(\bar{B})$. Also $P(RB) = P(B \mid R)P(R)$, and $P(B \mid R) = \sum P(B \mid R \,\&\, \theta)P(\theta \mid R) =$ (assuming independence between $\theta$ and $R$) $\sum P(B \mid R \,\&\, \theta)P(\theta)$, where $\theta$ represents possible values of the percentage of ravens in the universe that are black (according to $h$, of course, $\theta = 1$). Finally, $P(B \mid R \,\&\, \theta) = \theta$, for if the percentage of black ravens in the universe is $\theta$, the probability of an arbitrary raven being black is also $\theta$. (This is intuitively correct and is formalised in the so-called Principal Principle, which we shall discuss later.)

Combining all these considerations with the above forms of Bayes's Theorem yields

$$\frac{P(h \mid RB)}{P(h)} = \frac{1}{\Sigma\theta P(\theta)} \quad \& \quad \frac{P(h \mid \overline{R}\,\overline{B})}{P(h)} = \frac{1}{P(\overline{R} \mid \overline{B})}.$$

Consider first the term $P(\overline{R} \mid \overline{B})$. Presumably there are vastly more non-black things in the universe than ravens. So even if no ravens are black, the probability of some object about which we know nothing, except that it is not black, being a non-raven must be very high, indeed, practically 1. Hence, $P(h \mid \overline{R}\,\overline{B}) \simeq P(h)$, and, so, the observation that some object is neither a raven nor black provides very little confirmation for $h$.

According to the equation above, the degree to which $RB$ confirms $h$ is inversely proportional to $\Sigma\theta P(\theta)$. This means, for example, that if it is initially very probable that all or virtually all ravens are black, then $\Sigma\theta P(\theta)$ would be large and $RB$ would confirm $h$ rather little. While if it is initially relatively probable that most ravens are not black, confirmation could be substantial. Intermediate levels of uncertainty about the proportion of ravens that are black would bring their own levels of confirmation. By contrast, because the class of non-black objects is so much larger than the class of ravens, $\overline{R}\,\overline{B}$ confirms 'All ravens are black' to only a tiny extent, irrespective of $P(\theta)$. Mackie's well-known Bayesian solution to the ravens paradox, which is given in the Exercises section at the end of this chapter, is similar and also depends on an assumed large disparity in the number of non-black objects and ravens.

Our Bayesian working of the raven example appears to support the Nicod condition, with the minor limitation that no confirmation is possible, even with positive instances, when the hypothesis has a prior probability of 1. But a Bayesian approach anticipates the violation of Nicod's condition in other circumstances too. And numerous examples have been suggested as plausible instances of such violations. The first of these seems to be due to Good (1961). We shall use an example that is taken, with some modification, from Swinburne (1971). The hypothesis under examination is 'All grasshoppers are located outside the County of Yorkshire'. The observation of a grasshopper just beyond the county border is an instance of this generalisation and, according to Nicod, confirms it. But it might be more reasonably argued that since there are no border controls or other obstacles restricting the movement of grasshoppers in that area, the observation of one on the edge of the county increases the probability that others have actually entered and hence undermines the hypothesis. In Bayesian terms, this is a case where, relative to background information, the probability of some datum is reduced by a hypothesis —that is, $P(e \mid h) < P(e)$—which is therefore disconfirmed—in other words, $P(h \mid e) < P(h)$.

A much more striking example where Nicod's conditions break down was invented by Rosenkrantz (1977, p. 35). Three people leave a party, each with a hat. The hypothesis that none of the three has his own hat is confirmed, according to Nicod, by the observation that person 1 has person 2's hat and by the observation that person 2 has person 1's hat. But since there are only three people, the second observation must *refute* the hypothesis, not confirm it.

Our grasshopper example provides an instance where a datum of the type $\bar{R}B$ confirms a generalisation of the form 'All *R*s are *B*'. Imagine that an object which looked for all the world like a grasshopper had been found hopping about just outside Yorkshire and that it turned out to be some other sort of insect. The discovery that the object was not a grasshopper would be relatively unlikely unless the grasshopper hypothesis was true (hence, $P(e) < P(e \mid h)$); thus it would confirm that hypothesis. If the deceptively grasshopper-like object were within the county boundary, the same conclusion would follow, though the degree of confirmation would be greater. This shows that 'All *R*s are *B*' may also be confirmed by a datum of the $\bar{R}\bar{B}$ type. Hence, the impression that non-*R*s never confirm such hypotheses may be dispelled.

Horwich (1982) has argued that the raven hypothesis may be differently confirmed, depending on how the black raven was chosen, either by randomly selecting an object from the population of ravens or by making the selection from the population of black objects. (Horwich denotes the evidence that some object is a black raven as either $R^*B$ or $RB^*$, depending on whether it was discovered by the first selection process or the second.) Prompted by an unpublished paper by Kevin Korb ("Infinitely Many Resolutions of Hempel's Paradox", 1993), we agree with Horwich that this is so; the Bayesian explanation which Horwich gives is recapitulated, in a slightly different context, in Chapter 14, section **f.**

But Horwich offers another explanation, which fits poorly with his Bayesian one. For he claims that the datum $R^*B$ is always more powerfully confirming than $RB^*$, because, he says, only it subjects the raven hypothesis to the risk of falsification. But this surely conflates the process of collecting evidence, which may indeed subject the hypothesis to different risks of refutation, with the evidence itself, which either refutes the hypothesis or does not refute it, and in the case of $R^*B$ and $RB^*$, it does not. (For a fuller discussion of this point, the reader is referred to Chapter 15, section **g.**)

Our conclusions are, first, that the supposedly paradoxical consequences of Nicod's condition and the Equivalence condition are not problematic, and, secondly, that there are separate reasons for rejecting Nicod's condition, which, moreover, conform to Bayesian principles.

## g THE DESIGN OF EXPERIMENTS

Why should anyone go to the trouble and expense of performing new experiments and seeking more evidence for hypotheses? The question has been debated recently and is sometimes felt to be something of a problem. Maher (1990) argues that since evidence can neither conclusively verify nor conclusively refute a theory, Popper's scientific aims cannot be served by gathering new data. Since a large part of scientific activity is devoted to the acquisition of new evidence, if Mayer were right, there would appear to be a serious gap in Popper's philosophy. Miller (1991, p. 2) claims that the same difficulty appears in Bayesian philosophy:

> If e is the agent's total evidence, then $P(h \mid e)$ is the value of his probability and that is that. What incentive does he have to change it, for example by obtaining more evidence than he has already? He might do so, enabling his

total evidence to advance from e to e+; but in no clear way would $P(h \mid e+)$ be a better evaluation of probability than $P(h \mid e)$ was.

There seems to us, on the contrary, a quite straightforward reason why a Bayesian might seek new evidence, namely, in order to diminish uncertainty about some aspect of the world, in a desire to find out the truth. Suppose, for instance, that the question of interest concerns a particular parameter. You might start out fairly uncertain about its value, in the sense that your probability distribution over its possible values is rather diffuse. A suitable experiment, if successful, would furnish evidence to lessen that uncertainty by changing the probability distribution, via Bayes's Theorem, so that it was now more concentrated in a particular region, the greater the concentration and the smaller the region the better. This criterion has been given a precise, quantitative expression by Lindley (1956), in terms of Shannon's characterisation of information. Lindley showed that in the case where knowledge of a parameter, $\theta$, is sought, provided the density of $x$ varies with $\theta$, any experiment in which $x$ is measured has an expected yield in information. But, of course, this result is compatible with a good experiment (with a high expected information yield) being relatively uninformative in a particular case; similarly, a poor experiment may to one's surprise be relatively informative.

In deciding which experiment to perform, one must also take at least three other factors into account: the cost of the experiment; the morality of carrying it out; and the value, both theoretical and practical, of the hypotheses one is interested in. Bayes's Theorem, of course, implies nothing about how these separate factors should be balanced.

## h THE DUHEM PROBLEM

### h.1 The Problem

The so-called Duhem (or Duhem-Quine) problem is a problem for theories of science of the type associated with Popper, which emphasise the power of certain evidence to refute a hypothesis. According to Popper's influential views, the characteristic of a theory which makes it 'scientific' is its falsifiability: "Statements or systems of statements, in order to be ranked as scientific, must be capable of conflicting with possible, or conceivable, observations" (Popper, 1963, p. 39). And, claiming to apply this criterion, Popper (1963, ch. 1) judged Einstein's gravitational theory to be scientific and Freud's psychology, unscientific. There is a strong flavour of commendation about the term *scientific* which has proved extremely misleading. For a theory that is scientific in Popper's sense is not necessarily true, or even probably true or so much as close to the truth, nor can it be said definitely that it is likely to lead to the truth. In fact, there seems to be no conceptual connection between a theory's capacity to pass Popper's test of scientificness and its having any epistemic or inductive value. There is little alternative, then, so far as we can see, to regarding Popper's demarcation between scientific and unscientific statements as part of a theory about the content and character of what is usually termed science, not as having any normative significance.

Yet as an attempt at understanding the methods of science, Popper's ideas bear little fruit. His central claim was that scientific theories are falsifiable by "possible, or conceivable, observations". This poses a difficulty, for an observation can only falsify a theory (that is, conclusively demonstrate its falsity) if it is itself conclusively certain. But observations cannot be conclusively certain. Popper himself recognised this but seems not to have appreciated its incongruity with his falsificationist thesis. He held every observation report to be fallible; but, reluctant to admit degrees of fallibility or anything of the kind, he concluded that observation reports that are admitted as evidence "are accepted as the result of a decision or agreement; and to that extent they are *conventions*" (Popper, 1959a, p. 106; our italics). It is unclear to us to what psychological attitude this sort of acceptance corresponds, but whatever it is, Popper's view of evidence statements seems to pull the rug from under falsificationism: it implies that no theory can really be falsified by evidence. The nearest thing to a refutation would occur when 'conventionally accepted' evidence was inconsistent with a theory, which could then, at best, be described as 'conventionally' rejected. Indeed, Popper conceded this much: "From a logical point of view, the testing of a theory depends upon basic statements whose acceptance or rejection, in its turn, depends upon our *decisions*. Thus it is *decisions* which settle the fate of theories" (Popper, 1959a, p. 108).

Watkins is one of those who saw that falsificationism presupposes the existence of some infallibly true observation statements, and he attempted to restore the Popperian position by advancing the claim that such statements do in fact exist. He would agree that statements like 'The hand on this dial is pointing to the numeral 6' are fallible—it is unlikely, but possible, that the person reporting it missaw the position of the hand. But he claimed that introspective perceptual reports, such as 'In my visual field there is now a silvery crescent against a dark blue background', "may rightly be regarded by their authors when they make them as infallibly true" (Watkins, 1984, pp. 79 and 248). But in our view Watkins is wrong, and the statements he regards as infallible are open to exactly the same sceptical doubts as any other observation report. We can illustrate this through Watkins's example: clearly, it is possible, though admittedly not very probable, that the introspector has misremembered and mistaken the shape he usually describes as a crescent or the sensation he usually receives on reporting blue and silvery images. These and other sources of error ensure that introspective reports are not exempt from the rule that non-analytic statements are fallible.

Of course, the kinds of observation statements we have mentioned, if asserted under appropriate circumstances, would never be seriously doubted. That is, although they could be false, they have a force and immediacy that carries conviction; they are 'morally certain', to use the traditional phrase. But if observation statements are merely indubitable, then whether a theory is regarded as refuted by observational data or not must rest ultimately on a subjective feeling of certainty. The fact that such convictions are so strong and uncontroversial may disguise their fallibility, but cannot undo it. Hence, no theory is strictly falsifiable, for none could be conclusively shown to be false by empirical observations. In practice the closest one could get to a refutation would be arriving at the conclusion that a theory that clashes with almost certainly true observations is almost certainly false.

A second objection to Popper's falsifiability criterion, and the one upon which we shall focus for its more general interest, is that it describes as unscientific most of those theories which are usually deemed science's greatest achievements. This is the chief aspect of the well-known criticisms advanced by Polanyi (1962), Kuhn (1970), and Lakatos (1970), amongst others. They have pointed out that, as had already been established by Duhem (1905), many notable theories of science are not falsifiable by what would generally be regarded as observation statements, even if those statements were infallibly true. Predictions drawn from Newton's laws or from the Kinetic Theory of Gases turn out to depend not only on those theories but also on certain auxiliary hypotheses. Hence, if such predictions fail, one is not compelled by logic to infer that the main theory is false, for the fault may lie with one or more of the auxiliary assumptions. The history of science has many occasions when an important theory led to a false prediction and where that theory, nevertheless, was not blamed for the failure. In such cases we find that one or more of the auxiliary assumptions used to derive the prediction was taken to be the culprit. The problem that arose from Duhem's investigations was which of the several distinct theories involved in deriving a false prediction should be regarded as the false element or elements in the assumptions.

### h.2 Lakatos and Kuhn on the Duhem Problem

Lakatos examined in detail the way that scientists react to anomalies; indeed, he made it a central feature of what he referred to as his "methodology of scientific research programmes". Lakatos claimed that scientific research of the most significant kind usually proceeds in what he called "research programmes". A research programme takes the form of a central or "hard core" theory, together with an associated "protective belt" of auxiliary assumptions. The function of the latter is to combine with the hard core to allow the drawing out of specific predictions, which can then be checked by experiment. The auxiliary assumptions are described as protective because during a research programme's lifetime they, not the central theory, are revised when a prediction is shown to be false.

Lakatos suggested Newtonian physics as an example of a research programme, the three laws of mechanics and the law of gravitation constituting the hard core, while various optical theories, assumptions about the number and positions of the planets, and so forth, he included in the protective belt. He also described a set of heuristic rules by which the research programme dealt with anomalies and advanced into new areas.

Kuhn's famous theory of scientific paradigms is similar to the methodology we have just described and was probably part of its inspiration. Both Lakatos and Kuhn were impressed that scientists tend to give the benefit of the doubt to some, especially fundamental, theories when these encounter anomalies—and both argued that such theories exerted a commanding influence over whole areas of scientific research. Lakatos's methodology, however, has two advantages. First, it describes scientific research programmes in some detail and analyses their modes of action; whereas Kuhn left his corresponding notion of a paradigm somewhat vague in comparison.

Secondly, Lakatos also outlined criteria of success for a research programme. He held that it was perfectly legitimate to treat the hard core systematically as the innocent party in a refutation, provided the research programme occasionally leads to successful novel predictions or to successful or "non–ad hoc" explanations of existing data. Lakatos called such programmes "progressive".

> The sophisticated falsificationist [which Lakatos counted himself as] . . . sees nothing wrong with a group of brilliant scientists conspiring to pack everything they can into their favourite research programme ('conceptual framework', if you wish) with a sacred hard core. As long as their genius—and luck—enables them to expand their programme *'progressively'*, while sticking to its hard core, they are allowed to do it. (Lakatos, 1970, p. 187)

If, on the other hand, the programme persistently produced false predictions, or if its explanations were habitually ad hoc, Lakatos called it "degenerating". (We shall devote the next section to the notion of adhocness.) Lakatos employed these tendentious terms even though he never succeeded in substantiating their intimations of approval and disapproval, and in the end he seems to have abandoned the attempt and settled on the more modest claim that, as a matter of historical fact, progressive programmes have usually been well regarded by scientists, while degenerating ones were distrusted and eventually dropped.

This last claim has, it seems to us, some truth to it, as evidenced, for example, by the case studies in the history of science included in Howson (1976). But although Lakatos and Kuhn identified and described an important aspect of scientific work, they provided no rationale or explanation for it. For instance, Lakatos was never able to explain why a research programme's occasional predictive or explanatory success could compensate for numerous failures, nor was he prepared to specify how many of such successes are needed to convert a degenerating programme into a progressive one. (They should occur "now and then", he said.) Hence, although the methodology of scientific research programmes points to some of the factors relevant to scientific change, it affords no explanation.

Lakatos was also unable to explain why some theories are raised to the status of the hard core of a research programme and are defended by a protective belt of hypotheses, while others are left to their own devices. From Lakatos's writings, one could think that the question was decided by the scientist at will (Lakatos called it a "methodological fiat"). Unfortunately, this suggests that it is a perfectly canonical scientific practice to set up any theory whatever as the hard core of a research programme, or as the central pattern of a paradigm, and to blame all empirical difficulties on auxiliary theories. This is far from being the case.

### h.3 The Duhem Problem Solved by Bayesian Means

The questions left unanswered by Lakatos may be resolved with the help of Bayes's Theorem, as Dorling (1979) has shown, by considering how the individual probabilities of several theories are altered when, as a group, they have been refuted.

Suppose a theory, $t$, and an auxiliary hypothesis, $a$, together imply an empirical consequence which is shown to be false by the observation of the outcome $e$. Let us assume that while the combination $t$ & $a$ is refuted by $e$, the two components taken individually are not refuted. We wish to consider the separate effects wrought on the probabilities of $t$ and $a$ by the adverse evidence $e$. The comparisons of interest here are between $P(t \mid e)$ and $P(t)$, and between $P(a \mid e)$ and $P(a)$. The conditional probabilities can be expressed using Bayes's Theorem, as follows:

$$P(t \mid e) = \frac{P(e \mid t)P(t)}{P(e)} \qquad P(a \mid e) = \frac{P(e \mid a)P(a)}{P(e)}.$$

In order to evaluate the posterior probabilities of $t$ and of $a$, one must first determine the values of the various terms on the right-hand sides of these equations. Before attempting this, it is worth noting that the equations convey no expectation that the refutation of $t$ & $a$ jointly considered will in general have a symmetrical effect on the separate probabilities of $t$ and of $a$, nor any reason why the degree of asymmetry may not be considerable in some cases. It is evident that the probability of $t$ changes very little if $P(e \mid t) \approx P(e)$, while that of $a$ is reduced substantially just in case $P(e \mid a)$ is substantially less than $P(e)$. The equations also allow us to discern the factors that determine which hypothesis suffers most in the refutation.

A historical example might best illustrate how a theory that produces a false prediction may still remain very probable; we shall, in fact, use an example that Lakatos (1970, pp. 138–40, and 1968, pp. 174–75) drew heavily on. In 1815, William Prout, a medical practitioner and chemist, advanced the hypothesis that the atomic weights of all the elements are whole number multiples of the atomic weight of hydrogen, the underlying assumption being that all matter is built out of different combinations of some basic element. Prout believed hydrogen to be that fundamental building-block, though the idea was entertained by others that a more primitive element might exist out of which hydrogen itself was composed. Now the atomic weights recorded at the time, though approximately integral when expressed as multiples of the atomic weight of hydrogen, did not match Prout's hypothesis exactly. Those deviations from a perfect fit failed to convince Prout that his hypothesis was wrong however; he instead took the view that there were faults in the methods that had been used to measure the relative weights of atoms. The noted chemist Thomas Thomson drew a similar conclusion. Indeed, both he and Prout went so far as to adjust several reported atomic weights in order to bring them into line with Prout's hypothesis. For instance, instead of accepting 0.829 as the atomic weight (expressed as a proportion of the weight of an atom of oxygen) of the element boron, which was the experimentally reported value, Thomson (1818, p. 340) preferred 0.875 "because it is a multiple of 0.125, which all the atoms seem to be". (Thomson erroneously took 0.125 as the atomic weight of hydrogen, relative to that of oxygen.) Similarly, Prout adjusted the measured atomic weight of chlorine (relative to hydrogen) from 35.83 to 36, the nearest whole number.

Thomson's and Prout's reasoning can be explained as follows: Prout's hypothesis *t*, together with an appropriate assumption *a* asserting the accuracy (within specified limits) of the measuring technique, the purity of the chemicals employed, and so forth, implies that the measured atomic weight of chlorine (relative to hydrogen) is a whole number. Suppose, as was the case in 1815, that chlorine's measured atomic weight was 35.83, and call this the evidence *e*. It seems that chemists of the early nineteenth century, such as Prout and Thomson, were fairly certain about the truth of *t*, but less so of *a*, though more sure that *a* is true than that it is false. Contemporary near-certainty about the truth of Prout's hypothesis is witnessed by the chemist J. S. Stas. He reported (1860, p. 42) that "In England the hypothesis of Dr Prout was almost universally accepted as absolute truth", and he confessed that when he started researching into the matter, he himself had "had an almost absolute confidence in the exactness of Prout's principle" (1860, p. 44). (Stas's confidence eventually faded after many years' experimental study, and by 1860 he had "reached the complete conviction, the entire certainty, as far as certainty can be attained on such a subject, that Prout's law . . . is nothing but an illusion", 1860, p. 45.) It is less easy to ascertain how confident Prout and his contemporaries were in the methods by which atomic weights were measured, but it is unlikely that this confidence was very great, in view of the many clear sources of error and the failure of independent measurements generally to produce identical results. On the other hand, chemists of the time must have felt that their methods for determining atomic weights were more likely to be accurate than not, otherwise they would not have used them. For these reasons, we conjecture that $P(a)$ was of the order of 0.6 and that $P(t)$ was around 0.9, and these are the figures we shall work with. It should be stressed that these numbers and those we shall assign to other probabilities are intended chiefly to illustrate how Bayes's Theorem resolves Duhem's problem; nevertheless, we believe them to be sufficiently accurate to throw light on the progress of Prout's hypothesis. As will become apparent, the results we obtain are not very sensitive to variations in the assumed prior probabilities.

In order to evaluate the posterior probabilies of $t$ and of $a$, one must fix the values of the terms $P(e \mid t)$, $P(e \mid a)$, and $P(e)$. These can be expressed, using the theorem on total probability (Chapter 2, (11) in section **c.3**), as follows:

$$P(e) = P(e \mid t)P(t) + P(e \mid {\sim}t)P({\sim}t)$$

$$\begin{aligned} P(e \mid t) &= P(e \,\&\, a \mid t) + P(e \,\&\, {\sim}a \mid t) \\ &= P(e \mid t \,\&\, a)P(a \mid t) + P(e \mid t \,\&\, {\sim}a)P({\sim}a \mid t) \\ &= P(e \mid t \,\&\, a)P(a) + P(e \mid t \,\&\, {\sim}a)P({\sim}a) \end{aligned}$$

Since $t \,\&\, a$, in combination, is refuted by $e$, the term $P(e \mid t \,\&\, a)$ is zero. Hence:

$$P(e \mid t) = P(e \mid t \,\&\, {\sim}a)P({\sim}a).$$

It should be noted that in deriving the last equation but one, we have followed Dorling in assuming that $t$ and $a$ are independent, that is, that $P(a \mid t) = P(a)$ and, hence, $P({\sim}a \mid t) = P({\sim}a)$. This seems to accord with many historical cases and is clearly right in the present case. By parallel reasoning to that employed above, we may derive the results:

$$P(e \mid a) = P(e \mid {\sim}t \,\&\, a)P({\sim}t)$$

$$P(e \mid {\sim}t) = P(e \mid {\sim}t \,\&\, a)P(a) + P(e \mid {\sim}t \,\&\, {\sim}a)P({\sim}a)$$

Provided the following terms are fixed, which we have done in a tentative way, to be justified presently, the posterior probabilities of $t$ and of $a$ can be determined:

$$P(e \mid {\sim}t \,\&\, a) = 0.01$$

$$P(e \mid {\sim}t \,\&\, {\sim}a) = 0.01$$

$$P(e \mid t \,\&\, {\sim}a) = 0.02$$

The first of these gives the probability of the evidence if Prout's hypothesis is not true but if the method of atomic weight measurement is accurate. Such probabilities were explicitly considered by some nineteenth-century chemists, and they typically took a theory of random assignment of atomic weights as the alternative to Prout's hypothesis (e.g., Mallet, 1880); we shall follow this. Suppose it had been established for certain that the atomic weight of chlorine lay between 35 and 36. (The final results we obtain respecting the posterior probabilities of $t$ and $a$ are, incidentally, not affected by the width of this interval.) The random-allocation theory would assign equal probabilities to the atomic weight of an element lying in any 0.01-wide band. Hence, on the assumption that $a$ is true, but $t$ false, the probability that the atomic weight of chlorine lies in the interval 35.825 to 35.835 is 0.01. We have assigned the same value to $P(e \mid \sim t \;\&\; \sim a)$ on the grounds that if $a$ were false because, say, some of the chemicals were impure or the measuring techniques faulty, then, still assuming $t$ to be false, one would not expect atomic weights to be biased towards any particular part of the interval between adjacent integers.

We have set the probability $P(e \mid t \;\&\; \sim a)$ rather higher, at 0.02. The reason for this is that although some impurities in the chemicals and some degree of inaccuracy in the method of measurement were moderately likely in the early nineteenth century, chemists would not have considered their techniques entirely haphazard. Thus if Prout's hypothesis were true, but the measuring technique imperfect, the measured atomic weights would have been likely to deviate somewhat from integral values; but the greater the deviation, the less likely, on these assumptions, so the probability of an atomic weight lying in any part of the 35–36 interval would not be distributed uniformly over the interval, but would be more concentrated around the whole numbers.

Let us proceed with the figures we have assumed for the crucial probabilities, noting however that the particular values of the three probability terms are unimportant, only their relative values need be taken into account in the calculation. Thus we would arrive at the same posterior probabilities for $a$ and $t$ with the weaker assumptions that $P(e \mid \sim t \;\&\; a) = P(e \mid \sim t \;\&\; \sim a) = \frac{1}{2} P(e \mid t \;\&\; \sim a)$. We thus obtain

$$P(e \mid \sim t) = 0.01 \times 0.6 + 0.01 \times 0.4 = 0.01$$

$$P(e \mid t) = 0.02 \times 0.4 = 0.008$$

$$P(e \mid a) = 0.01 \times 0.1 = 0.001$$

$$P(e) = 0.008 \times 0.9 + 0.01 \times 0.1 = 0.0082$$

Finally, Bayes's Theorem enables us to derive the posterior probabilities in which we were interested:

$P(t \mid e) = 0.878$ (Recall that $P(t) = 0.9$.)

$P(a \mid e) = 0.073$ (Recall that $P(a) = 0.6$.)

These striking results show that evidence of the kind we have described may exert a sharply asymmetric effect on the probabilities of $t$ and of $a$. The initial probabilities we assumed seem appropriate for chemists such as Prout and Thomson, and if they are correct, the results deduced from Bayes's Theorem explain why those chemists regarded Prout's hypothesis as being more-or-less undisturbed when certain atomic-weight measurements diverged from integral values, and why they felt entitled to adjust those measurements to the nearest whole number. Fortunately, these results are relatively insensitive to changes in our assumptions, so the accuracy of those assumptions is not a vital matter as far as our explanation is concerned. For example, if one took the initial probability of Prout's hypothesis ($t$) to be 0.7, instead of 0.9, keeping the other assignments, we find that $P(e \mid t) = 0.65$, while $P(a \mid e) = 0.21$. Hence, as before, after the refutation, Prout's hypothesis is still more likely to be true than false, and the auxiliary assumptions are still much more likely to be false than true. Other substantial variations in the initial probabilities produce similar results, though with so many factors at work, it is difficult to state concisely the conditions upon which these results depend without just pointing to the equations above.

Thus Bayes's Theorem provides a model to account for the kind of scientific reasoning that gave rise to the Duhem problem. And the example of Prout's hypothesis, as well as others that Dorling (1979 and 1982) has described, show, in our view, that the Bayesian model is essentially correct. By contrast, non-probabilistic theories seem to lack entirely the resources that could deal with Duhem's problem.

A fact that emerges when slightly different values are assumed for the various probabilities in the Prout's hypothesis example is that one or other of the theories may actually become more probable after the conjunction $t$ & $a$ has been refuted. For instance, when $P(e \mid t \& \sim a)$ equals 0.05, the other probabilities being assigned the same values as before, the posterior probability of $t$ is 0.91, which exceeds its prior probability. This may seem bizarre, but, as Dorling (1982) has argued, it is not so odd when one bears in mind that the refuting evidence normally contains a good deal more information than is required merely to disprove $t$ & $a$ and that this extra information may be confirmatory. In general, such confirmation occurs when $P(e) < P(e \mid t)$, which is easily shown to be equivalent to the condition $P(e \mid t) > P(e \mid \sim t)$. In other words, when evidence is easier to explain (in the sense that it receives a higher probability) if a given hypothesis is true than if it is not, then that theory is confirmed by the evidence.

## i GOOD DATA, BAD DATA, AND DATA TOO GOOD TO BE TRUE

**Good data.** The marginal influence which we have seen an anomalous observation may exert on the probability of a theory is to be contrasted with the dramatic effect that a confirmation can have. For instance, if the measured atomic weight of chlorine had been a whole number, in line with Prout's hypothesis, so that now $P(e \mid t \& a)$ is 1 instead of 0, and if the probabilities we assigned were kept, the probability of the hypothesis would have shot up from a prior of 0.9 to a posterior value of 0.998. And, even more dramatically, if the prior probability of $t$ had been 0.7, its posterior probability would have risen to 0.99.

The existence of this asymmetry between anomalous and confirming instances was highlighted with particular vigour by Lakatos, who regarded it as being of the greatest significance in science and as one of the characteristic features of a research programme. Lakatos maintained that a scientist involved in such a programme typically "forges ahead with almost complete disregard of 'refutations'", provided he is occasionally rewarded with successful predictions (1970, p. 137): he is "encouraged by Nature's YES, but not discouraged by its NO" (1970, p. 135). As we have indicated, we believe there to be much truth in Lakatos's observations; however, they are merely incorporated without explanation into his methodology, while the Bayesian has a simple and plausible explanatory model.

**Bad data.** An interesting fact that emerges from the Bayesian analysis is that a successful prediction derived from a combination of two theories, say $t$ and $a$, does not always redound to the credit of $t$, even if the prior probability of the evidence is small; indeed, it can even undermine it. We may illustrate this by referring again to the example of Prout's hypothesis.

Suppose the atomic weight of chlorine were 'measured', not in the old-fashioned chemical way, but by concentrating hard on the element in question and picking a number in some random fashion from a given range of numbers. And let us assume that this method assigns a whole-number value to the atomic weight of chlorine. This is just what one would predict on the basis of Prout's hypothesis, if the outlandish measuring technique were reliable. But reliability is obviously most unlikely, and it is equally obvious that, as a result, the measured atomic weight of chlorine adds practically nothing to the probability of Prout's hypothesis, notwithstanding its integral value. This intuition is upheld by Bayes's Theorem, as a simple calculation based on the above formulas shows. (As before, let $t$ be Prout's hypothesis and $a$ the assumption that the measuring technique is accurate. Then set $P(e \mid t \; \& \sim a) = P(e \mid \sim t \; \& \sim a) = P(e \mid \sim t \; \& \; a) = 0.01$, for reasons similar to those stated earlier, and let $P(a)$ be very small, say 0.0001, for obvious reasons. It then follows that $P(t)$ and $P(t \mid e)$ are equal to two decimal places.)

This example shows that Leibniz was wrong to declare as a general principle that "It is the greatest commendation of an hypothesis (next to truth) if by its help predictions can be made even about phenomena or experiments not tried". Leibniz and Lakatos, who quoted these words with approval (1970, p. 123), seem to have overlooked the fact that if a prediction can be deduced from a hypothesis only with the assistance of highly questionable auxiliary claims, then that hypothesis may accrue very little credit. This explains why the various sensational predictions which Velikovsky drew from his theory of planetary collisions failed to impress most serious astronomers, even when some of those predictions were to their amazement fulfilled. For instance, Velikovsky's prediction of the existence of large quantities of petroleum on the planet Venus relied not only on his pet theory that various natural disasters in the past had been caused by collisions between the earth and a comet, but also on a string of unsupported and not very plausible assumptions, such as that the comet in question originally carried hydrogen and carbon, that these had been converted to petroleum by electrical discharges supposedly created in the violent impact with the earth, that the comet had later evolved into the planet Venus, and some others (Velikovsky, 1950, p. 351). (More details of Velikovsky's theory are given in the next section.)

**Data too good to be true.** Data are sometimes said to be 'too good to be true' when they fit a favoured hypothesis more perfectly than seems reasonable to expect. For instance, suppose all the atomic weights listed in Prout's paper had been whole numbers, exactly. Such a result almost looks as if it was designed to impress, and it is just for this reason that it fails to.

We may analyse this response as follows. Let $e$ be the evidence of, say, 20 atomic-weight measurements, each a perfect whole number. No one could have regarded precise atomic weights measured at the time as absolutely reliable. The most natural view would have been that such measurements are subject to experimental error and, hence, that they would give a certain spread of results about the true value. On this assumption, which we shall label $a'$, it is extremely unlikely that numerous independent atomic-weight measurements would all produce whole numbers, even if Prout's hypothesis were true. So $P(e \mid t \;\&\; a')$ is extremely small and, clearly, $P(e \mid \sim t \;\&\; a')$ would be no larger. Now $a'$ has many possible alternatives, one of the more plausible (though initially it might not be very plausible) being that the experiments were consciously or unconsciously rigged in favour of Prout's hypothesis. If this were the only significant alternative (and so, in effect, equivalent to $\sim a'$), $P(e \mid t \;\&\; \sim a')$ would be very high, as would $P(e \mid {\sim}t \,\&\, {\sim}a')$. It follows from the equations on page 139 above that

$$P(e \mid t) \approx P(e \mid t \,\&\, {\sim}a')P({\sim}a') \text{ and}$$
$$P(e \mid {\sim}t) \approx P(e \mid {\sim}t \,\&\, {\sim}a')P({\sim}a'),$$

and, hence,

$$P(e) \approx P(e \mid t \,\&\, {\sim}a')P({\sim}a')\,P(t) + P(e \mid {\sim}t \,\&\, {\sim}a')\,P({\sim}a')\,P({\sim}t).$$

Now, presumably the rigging of the results to produce whole numbers, if it took place, would produce whole numbers equally effectively whether $t$ was true or not; in other words,

$$P(e \mid t \,\&\, {\sim}a') = P(e \mid {\sim}t \,\&\, {\sim}a');$$

hence

$$P(e) \approx P(e \mid t \,\&\, {\sim}a')P({\sim}a').$$

Therefore,

$$P(t \mid e) = \frac{P(e \mid t)P(t)}{P(e)} \approx \frac{P(e \mid t \,\&\, {\sim}a')P({\sim}a')\,P(t)}{P(e \mid t \,\&\, {\sim}a')P({\sim}a')} = P(t).$$

Thus $e$ does not confirm $t$ significantly, even though, in a misleading sense, it fits the theory perfectly. This is why it is said to be too good to be true. A similar calculation shows that the probability of $a'$ is diminished and, on the assumptions that we made, this implies that the probability of the experiments having been fabricated is enhanced. (The above analysis is essentially the same as given in Dorling, 1982).

A famous case of data that were allegedly too good to be true is that of Mendel's plant-breeding results. Mendel's genetic theory of inheritance allows one to calculate the probabilities with which certain plants would produce specific kinds of offspring. For instance, under certain circumstances, pea plants of a particular strain may be calculated to yield round and wrinkled seeds with probabilities 0.75 and 0.25, respectively. Mendel obtained seed-frequencies that matched the corresponding probabilities in this and in similar cases remarkably well, suggesting (misleadingly, Fisher contended) substantial support for the genetic theory. Fisher did not believe that Mendel had deliberately falsified his results to appear in better accord with his theory than they really were.

To do so, Fisher claimed, would "contravene the weight of the evidence supplied in detail by . . . [Mendel's] paper as a whole". But Fisher thought it a "possibility among others that Mendel was deceived by some assistant who knew too well what was expected" (1936, p. 132), an explanation he backed up with some (rather meagre) evidence. Dobzhansky (1967, p. 1589), on the other hand, thought it "at least as plausible" that Mendel had himself discarded results that deviated much from his ideal, in the sincere belief that they were contaminated or that some other accident had befallen them. (For a comprehensive review see Edwards, 1986.)

The argument put forward earlier to show that too-exactly whole-number atomic-weight measurements would not have supported Prout's hypothesis depends on the existence of some sufficiently plausible alternative hypothesis that would explain the data better. We believe that, in general, data are too good to be true relative to one hypothesis only if there are such alternatives. This principle accords with intuition; for if the technique for eliciting atomic weights had long been established as precise and accurate, and if careful precautions had been taken against experimenter bias and deception, all the natural alternatives to Prout's hypothesis could be discounted and the data would no longer seem suspiciously good; they would be straightforwardly good. Fisher, however, did not subscribe to the principle, at least, not explicitly; he believed that Mendel's results told against the genetic theory whatever alternative explanations might suggest themselves. Nevertheless, as just indicated, the consideration of such alternatives played a part in his argument. We shall refer again to Fisher's case against Mendel in the next chapter.

## ■ j AD HOC HYPOTHESES

As we have seen, an important scientific theory which, in combination with other hypotheses, has made a false prediction may nevertheless emerge relatively unscathed, while one or more of the auxiliary hypotheses are largely discredited. (We are using such expressions in the normal way to describe how hypotheses are received, regarding them as harmless metaphors for obvious and more-or-less precise probabilistic notions. Thus, a hypothesis that is unscathed by negative evidence is one whose posterior and prior probabilities are similar. On the other hand, it is difficult to see what opponents of the Bayesian approach could have in mind when they talk of theories being 'accepted' or 'retained', or 'put forward' or 'saved' or 'vindicated'.) When a set of auxiliary assumptions is discredited in a test, scientists frequently think up new assumptions which assist the main theory to explain the previously anomalous data. Sometimes these new assumptions give the impression that their role is simply to 'patch up' the theory, and in such cases Francis Bacon called them "frivolous distinctions" (1620, Book I, aphorism xxv). More recently they have been tagged 'ad hoc hypotheses', presumably because they would not have been introduced if the need to bring theory and evidence into line had not arisen.

But although particular ad hoc theories are fairly easy to evaluate intuitively, there is controversy over what general criteria apply. We shall see that the Bayesian approach clarifies the question. First let us consider a few examples of ad hoc theories.

### j.1 Some Examples of Ad Hoc Hypotheses

**Velikovsky's theory of collective amnesia.** Immanuel Velikovsky, in a daring book called *Worlds in Collision* that attracted a great deal of attention some years ago, put forward the theory that the earth has been subject, at various stages in its history, to cosmic disasters produced by near collisions with massive comets. One of these comets, which went on to make a distinguished career as the planet Venus, is supposed to have passed close by the earth during the Israelites' captivity in Egypt and to have caused many of the various remarkable events of the time, such as the ten plagues and the parting of the Red Sea, related in the Bible. One of the theory's predictions, apparently, is that since no group of people could have missed these tremendous goings-on, if they kept records at all, they would have recorded them. However, many communities failed to note in their writings anything out of the ordinary at that time. But Velikovsky, still convinced by his main theory, put this exceptional behaviour down to what he called a "collective amnesia". He argued that the cataclysms were so terrifying that whole peoples behaved "as if [they had] obliterated impressions that should be unforgettable". There was a need, Velikovsky said, to "uncover the vestiges" of these events, "a task not unlike that of overcoming amnesia in a single person" (1950, p. 288).

Individual amnesia is the issue in the next example.

**Dianetics.** Dianetics is a theory that purports to analyse the causes of insanity and mental stress, which it sees as the 'misfiling' of information in unsuitable locations in the brain. By refiling these 'engrams', it claims, sanity may be restored, composure enhanced, and, incidentally, the memory vastly improved. Not surprisingly, the therapy is long and expensive, and few people have been through it and borne out the theory's claims. However, one triumphant success, a young student, was announced by the inventor of Dianetics, L. Ron Hubbard, and in 1950 he exhibited this person to a large audience, claiming that she had a "full and perfect recall of every moment of her life". But questions from the floor ("What did you have for breakfast on October 3, 1942?"; "What colour is Mr. Hubbard's tie?", and the like) soon demonstrated that the hapless young woman had a most imperfect memory. Hubbard accounted for this to what remained of the assembly by saying that when the woman first appeared on the stage and was asked to come forward "now", the word "now" had frozen her in "present time" and paralysed her ability to recall the past. (An account of the incident and of the history of Dianetics is given by Miller, 1987.)

**An example from psychology.** Investigations into the IQs of different groups of people show that average levels of measured intelligence vary. Some environmentalists, so-called, attribute low scores primarily to poor social and educational conditions, an explanation that ran into trouble when it was discovered that a large group of Eskimos, leading a feckless, poor, and drunken existence, scored very highly on IQ tests. The distinguished biologist Peter Medawar (1974), in an effort to deflect the difficulty away from the environmentalist thesis, tried to explain this unexpected observation by saying that an "upbringing in an igloo gives just the right degree of cosiness, security and mutual contact to conduce to a good performance in intelligence tests."

In each of these examples, the theory which was proposed in place of the refuted one seems rather unsatisfactory. It is not likely that they would have been put forward except in response to particular empirical anomalies, hence the label "ad hoc", which suggests that the theory was advanced for the specific purpose of evading a difficulty. However, some theories of this kind cannot be condemned so readily. For instance, an ad hoc alteration which rescued Newtonian theory from a difficulty led directly to the discovery of a new planet and was generally deemed a shining success.

**The discovery of the planet Neptune.** The planet Uranus was discovered by Sir William Herschel in 1781. Astronomers quickly sought to describe the orbit of the new planet, using Newtonian theory and taking account of the perturbing influence of other known planets, so that predictions could be made concerning its future positions. But discrepancies between predicted and observed positions of Uranus substantially exceeded the admitted limits of experimental error and grew year by year. The possibility that the fault lay with Newton's laws was mooted by a few astronomers, but the prevailing opinion was that there must be some unknown planet providing an extra source of gravitational attraction on Uranus, which ought to be included in the Newtonian calculations. Two astronomers in particular, John Couch Adams and U. J. J. Le Verrier, working independently, were convinced of this, and using all the known sightings of Uranus, they estimated where the hypothetical planet should be. This was a remarkable mathematical achievement, but more importantly, careful telescopic observations and studies of old astronomical charts revealed in 1846 the presence of a planet with the anticipated characteristics. The planet was later called Neptune. Newton's theory was saved, for the time being. (The fascinating story of this episode is told by W. M. Smart, 1947.)

### j.2 A Standard Account of Adhocness

The common features of the examples we are considering are that a theory $t$, which we can call the main theory, was combined with an auxiliary hypothesis $a$, to predict $e$, when in fact $e'$ occurred, $e'$ being incompatible with $e$. And in order to retain the main theory in its desired explanatory role, a new auxiliary, $a'$, was proposed which, with $t$, implies $e'$. The new theories are ad hoc in the sense that they were advanced "for the sole purpose of saving a hypothesis seriously threatened by adverse evidence" (Hempel, 1966, p. 29). However, many philosophers have distinguished two kinds of ad hoc theory. Theory $t$ & $a'$ is of the first kind if it possesses no independent test implications—independent, that is, from the evidence that refuted its predecessor $t$ & $a$. It is ad hoc in the second sense if it does have such test implications but none has been verified. Lakatos (1970, p. 175) called the first kind of theory ad $hoc_1$, the second kind ad $hoc_2$. Often, the designation *ad hoc* is applied just to the new theory, $a'$, rather than to its conjunction with $t$.

The term *ad hoc* for hypotheses that do not meet one or other of these conditions seems not to be an old one; its earliest occurrence in English that we know of was in 1936, in a critical review of a book of psychology. The reviewer, W. J. H. Sprott, commented on some explanations offered in the book of certain aspects of childish behaviour:

> There is a suspicion of '*ad-hoc*-ness' about the 'explanations'. The whole point is that such an account cannot be satisfactory until we can predict the child's movements from a knowledge of the tensions, vectors and valences which are operative, *independent of our knowledge of how the child actually behaved*. So far we seem reduced to inventing valences, vectors and tensions from a knowledge of the child's behaviour. (Sprott, 1936, p. 249; our emphasis)

Sprott clearly regarded ad hoc theories as unsatisfactory, a view which many philosophers nowadays share. For example, Popper states it as one of his 'requirements' that a theory should not be ad $hoc_1$:

> We require that the new theory should be *independently testable*. That is to say, apart from explaining all the *explicanda* which the new theory was designed to explain, it must have new and testable consequences (preferably consequences of a *new kind)*. (Popper, 1963, p. 241)

A further requirement laid down by Popper is that the new theory "should pass the independent tests in question", that is, they should not be ad $hoc_2$. Lakatos (1970) agreed with Popper that a theory is unacceptable if it is ad hoc in either sense; others such as Hempel (1966, p. 29) emphasise only the first sense. Disapproval of ad hoc theories is not new; in the early seventeenth century, Bacon criticized as a "frivolous distinction" the type of hypothesis that is "framed to the measure of those particulars only from which it is derived" (i.e., ad $hoc_1$ hypotheses). Bacon argued that a hypothesis ought to be "larger and wider" than the observations that gave rise to it and, moreover, that "that largeness and wideness" should be confirmed "by leading us to new particulars" (i.e., the theory should not be ad $hoc_2$).

The theories advocated by Velikovsky, Medawar, and Hubbard in response to anomalous data are probably ad $hoc_1$, since they seem to make no independent predictions, though, of course, a closer study of those theories might reverse that judgment. According to the criteria we have discussed, the theories appear therefore to represent unsatisfactory scientific developments, which is intuitively right. The Adams–Le Verrier hypothesis, on the other hand, is not ad hoc in either sense, because it did make new predictions, some of which were verified by telescopic sightings of Neptune. Again, philosophical and intuitive judgment coincides.

Despite this seeming success, we believe the adhocness criterion to be misconceived and unfounded. In setting out our position, we shall first review some positive arguments that have been advanced in favour of the criterion, we shall then show why the criterion must be wrong, and finally, we shall present the Bayesian view on ad hoc theories.

### j.3 Popper's Defence of the Adhocness Criterion

Popper claimed that ad hoc theories are unscientific or, at any rate, less scientific than non–ad hoc theories. Popper's argument for this position, which is by no means simple or straightforward, appears to be based on his well-known preference for falsifiable theories (*see* Urbach, 1991):

> As regards auxiliary hypotheses we decide to lay down the rule that only those are acceptable whose introduction does not diminish *the degree of falsifiability or testability* of the system in question, but, on the contrary, increases it. . . . (1959a, pp. 82–83; original italics replaced)

By connecting this rule with the criterion of adhocness, Popper implied that the trouble with ad $hoc_1$ theories is that they represent reductions in the degree of falsifiability of the system in question. Before examining this claim, we need to enquire how, in Popper's scheme, degrees of falsifiability are measured and why a lower falsifiability should, in his view, make a new theory an "unacceptable" replacement for an established one.

Let us first describe what Popper said concerning degrees of falsifiability. In the special case where one theory implies another, it is uncontentious to say that the implied theory is not more falsifiable than the implying theory, since any statement that refutes the former also refutes the latter, but not necessarily vice versa. However, the type of theory under discussion here is not covered by this special case, since there is no implication between the falsified theory $t$ & $a$ and the theory $t$ & $a'$, which takes its place; in fact the two theories are logically incompatible. Popper, therefore, had to find a measure of falsifiability applicable to theories that are either logically independent or mutually contradictory. On the surface, this seems impossible, since, as Popper often pointed out, falsifiable theories have infinitely many "potential falsifiers". Comparing the falsifiability of theories by counting up their respective potential falsifiers would therefore not serve Popper's goal, but would put all universal theories on a par.

Popper believed he had a way out of this difficulty: "It may be possible", he said, "to compare theories as to their degree of testability [i.e., falsifiability] by ascertaining the minimum *degree of composition* which a basic statement must have if it is to be able to contradict the theory". The degree of composition of a statement, in Popper's account, is a measure of the number of "simpler statements" of which it is composed. To compute this number, Popper said, "it might be suggested that we should choose a certain class of statements [that are] . . . *elementary or atomic* ones, from which all other statements could then be obtained by conjunction and other logical operations". But such a choice, Popper judged, would "impose serious restrictions upon the free use of scientific language". Instead, he proposed "to compare the degrees of composition of basic statements . . . by selecting arbitrarily *(sic)* a class of *relatively* atomic statements". He defined a relatively atomic statement as a substitution instance of "a generating schema or matrix (for example, 'There is a measuring apparatus for . . . at the place . . . , the pointer of which lies between the gradation mark . . . and . . . ')". Popper called the "class of these statements, together with all the conjunctions that can be formed from them . . . a *'field'*".

This already complicated account needs yet further elaboration, we are told, because "in order to avoid inconsistencies which might arise through the use of different fields, it is necessary to use a somewhat narrower concept than that of a field, namely that of a *field of application*". Popper then defined the dimension of a theory as the number *d*, such that the theory cannot be falsified by any *d*-tuple (the conjunction of *d* relatively atomic statements) of the field of application, although it can be falsified by certain $d+1$-tuples. And $d+1$ is then taken as measuring the theory's degree of falsifiability. But before this account is complete, it is necessary to clarify the notion of a field of application, which Popper attempted in an appendix to the book in which the above is expounded.

> A theory *t* is called '*d*-dimensional with respect to the field of application *F*' if and only if the following relation holds between *t* and *F*: there is a number *d* such that (a) the theory does not clash with any *d*-tuple of the field and (b) any given *d*-tuple in conjunction with the theory divides all the remaining relatively atomic statements uniquely into two infinite sub-classes *A* and *B*, such that the following conditions are satisfied: ($\alpha$) every statement of the class *A* forms, when conjoined with the given *d*-tuple, a 'falsifying $d+1$-tuple' i.e. a potential falsifier of the theory; ($\beta$) the class *B* on the other hand is the sum of one or more, but always a finite number, of infinite sub-classes [$B_i$] such that the conjunction of any number of statements belonging to any one of these sub-classes [$B_i$] is compatible with the conjunction of the given *d*-tuple and the theory. (Popper, 1959a, pp. 285–86)

We do not pretend fully to understand Popper's Byzantine falsifiability measure, and, although he regarded it as "only provisional" (1959a, p. 285), so far as we know, he never revised or discussed it further, save in regard to the simple case of theories expressible as polynomials. To illustrate, consider two theories having the forms $y = ax + b$ and $y = ax^2$

$+ bx + c$, espectively, where $x$ and $y$ are variables and $a$, $b$, and $c$ are constants, unspecified except that the coefficient of $x^2$ in the second theory is non-zero (otherwise the first would entail the second). Referring to the above discussion, the field of application for these theories is taken by Popper to consist of statements of the form 'when $x = m, y = n$', where $m$ and $n$ are particular numbers derived from observation. Clearly, with $d$ unspecified parameters in the equation, $d+1$ is the least number of statements in the field needed to refute the corresponding theory, and this number is considered by Popper to be the theory's degree of falsifiability.

Let us return to the idea of adhocness. Any defence of the criterion in terms of falsifiability requires at least two steps. First, a theory that is ad hoc in sense one must be shown to necessarily reduce the falsifiability of the system in which it occurs. Secondly, it must be demonstrated that increased falsifiability, as defined by Popper, is an index of 'acceptability' or 'satisfactoriness'.

Consider the first point. It is easy to see that the above paradigm does not apply to our examples, since there are no unfilled parameters in any of the theories concerned. Velikovsky's modified theory tells us precisely which tribes suffered from amnesia (they are the ones that failed to record the catastrophes), just as the Neptune hypothesis specified exactly the orbit on which the planet was supposed to move. Moreover, no other theories that are said to be ad hoc that have come to our attention fit the Popperian pattern. On the second point, Popper has never demonstrated nor, as far as we know, has he ever tried to demonstrate that higher falsifiability in his specialised sense indicates greater epistemic merit or "cognitive value". Indeed, the two ideas seem not to overlap at all.

### j.4 Why the Standard Account Must Be Wrong

According to the standard account, all ad hoc hypotheses are unsatisfactory, though ad $\text{hoc}_1$, not surprisingly, is often regarded as worse than ad $\text{hoc}_2$. As we explained, we do not think any of the attempts to justify the adhocness criterion a priori have been successful. And we shall argue that this is to be expected, since there are positive reasons to reject the criterion, which we shall now set out. Our argument will appeal to counter-examples and to some more general considerations.

An attraction of the adhocness criterion, no doubt, is its apparent objectivity and its avoidance of subjective probability, but, as we shall show, the non-Bayesian account has its own subjective aspect, one which, in our view, is very inappropriate.

Consider first some counter-examples to the standard account. Suppose one were examining the hypothesis that a particular urn contains only white counters. Next, imagine that a counter is withdrawn from the urn at random, that after its colour has been noted, it is replaced, and that this operation is repeated 10,000 times. If 4950, say, of the selected counters were red and the rest white, the initial hypothesis and the various necessary auxiliary assumptions, taken together, would be refuted; and it is then natural to conclude that, contrary to the original assumption, the urn contains both red and white counters in approximately equal numbers. This seems a perfectly reasonable inference, the revised hypothesis appears well justified by the evidence, yet there is no independent evidence for it. And if we complicate the example by letting the urn vapourise just after the last counter has been inspected, there will be no possibility of such independent evidence. So the hypothesis about the (late) urn's contents is ad $hoc_{1\,\&\,2}$; but for all that, it seems plausible and satisfactory (Howson, 1984; Urbach, 1991).

Speculating on the contents of an urn is but a humble form of enquiry, which we cite for the simple way it illustrates that a theory can be acceptable even when we have no evidence independent of the observations which caused the theory to be proposed, nor any possibility of such evidence. Hence, the two adhocness criteria are misguided. Examples from the higher sciences confirm this. Take the following case from the science of genetics: suppose it was initially assumed or believed that two characteristics of a certain plant are inherited in accordance with Mendel's principles through the agency of a pair of independently acting genes located on different chromosomes. Imagine now that plant-breeding experiments throw up a surprising number of plants carrying both characteristics, so that the original assumption that the genes act independently is revised in favour of a theory that they are linked on the same chromosome. Again, the revised theory would be strongly confirmed and established as acceptable merely on the evidence that stimulated its formulation and without the necessity of further, independent evidence. (An example of this sort is worked out by Fisher, 1970, ch. IX.)

The discovery of the planet Neptune illustrates the same point. Adams arrived at what he regarded as the most likely mass and elements of the orbit of the hypothetical planet by the mathematical technique of least squares applied to all the observations that had hitherto been collected on the positions of Uranus. Adams's hypothesis fitted these observations so well that *even before Neptune had been seen through the telescope or detected on astronomical charts,* its existence was contemplated with the greatest confidence by the leading astronomers of the day. For instance, in his retirement address as president of the British Association, Sir John Herschel, after remarking that the previous year had seen the discovery of a new minor planet, went on: "It has done more. It has given us the probable prospect of the discovery of another. We see it as Columbus saw America from the shores of Spain. Its movements have been felt, trembling along the far-reaching line of our analysis, *with a certainty hardly inferior to that of ocular demonstration*" (quoted in Smart, 1947, p. 61; our italics). And the Astronomer Royal, Sir George Airy, who was initially inclined to believe that the problem with Uranus would be resolved by introducing a slight adjustment to the inverse-square law, spoke of *"the extreme probability* of now discovering a new planet in a very short time" (also quoted in Smart, 1947, p. 61; our italics). Neptune was discovered a very short time later.

We turn now to a more general objection to the idea that hypotheses are acceptable only if corroborated by independent evidence. Imagine a scientist who is interested in the conjunction of hypotheses $t$ & $a$, whose implication $e$ can be checked in an experiment. The experiment is performed with the result $e'$, incompatible with $e$, and the scientist advances a new theory, $t$ & $a'$, which is consistent with the observations but is ad hoc in one or other of the two senses, that is, there is either no fresh evidence for $a'$ or no possibility of such evidence. The new theory therefore is unacceptable according to the view we are considering.

Suppose, next, that another scientist, working without knowledge of his colleague, also wished to test $t$ & $a$ but that he chose a different experiment for this purpose, one with only two possible outcomes: either $e$ or $\sim e$. Of course, he would obtain the latter, and having done so, he would be obliged to revise the refuted theory, to $t$ & $a'$, say. This scientist now notices that $e'$ follows from the new theory, and he performs the orthodox experiment to verify $e'$. The new theory can then count a successful prediction to its credit, and so is not ad hoc. Hence, according to the standard view, it is perfectly acceptable.

This is strange, to say the least, because we have arrived at opposite evaluations of the very same theory, breaching at the same time what we previously called the Equivalence condition and showing that the standard adhocness criterion is inconsistent. Whatever measures might be taken to resolve the inconsistency, it seems to us that one element of the criterion ought to be removed, namely, the significance it attaches to whether the theory concerned was thought up before or after the evidence was known. This introduces into the principles of theory-evaluation considerations concerning the state of the experimenters' minds, which are intuitively irrelevant and incongruous in a methodology with pretentions to objectivity. No such considerations enter the corresponding Bayesian evaluations.

#### j.5 The Bayesian View of Ad Hoc Theories

We have argued, contrary to the standard view, that a theory could be scientific and plausible even if it is ad hoc. An acceptable ad hoc theory is a possibility allowed for by the Bayesian principle that theories should be evaluated according to their probabilities. To illustrate, consider the ad hoc theory $a'$, which we have supposed was put forward in response to some refuting evidence $e'$. The probability of this theory must be reckoned relative to $e'$ and any other available relevant information, $b$. The probability calculus places no restrictions on the value of $P(a' \mid e' \,\&\, b)$; it might, for example, be below 0.5, so that $a'$ would be more likely false than true, or greater than 0.5, when the reverse would be the case. Hence $a'$ does not need the support of new independent predictions in order to be quite plausible and acceptable (Horwich, 1982, pp. 105–108).

Scientists are also interested in whether $t$ in the presence of the newly thought-up $a'$ provides a competent explanation of the previously anomalous $e'$. It would only do so if $t$ & $a'$ was sufficiently credible; since $P(t \,\&\, a' \mid e' \,\&\, b) \leq P(a' \mid e' \,\&\, b)$, this would be the case only if $a'$ was itself acceptable, in the sense indicated.

The Bayesian approach, incidentally, explains why people often respond immediately with incredulity, even derision, on first hearing certain ad hoc hypotheses. It is hardly likely that their amusement stems from perceiving, or even thinking that they perceive, that the hypothesis leads to no new predictions. Surely it is more likely that they are reacting to what they see as the utter implausibility of the hypothesis.

### j.6 The Notion of Independent Evidence

As we have explained, a standard non-Bayesian account of adhocness asserts that a theory consisting of the combination $t$ & $a$ is only replaced with $t$ & $a'$ in an acceptable, scientific fashion when $a'$ is successfully tested by evidence independent of that which refuted the first theory. This thesis is often associated with another, rather similar view, namely, that no theory is acceptable unless it is supported by evidence independent of that which prompted its initial proposal, whether that evidence also refuted a predecessor or not. We have shown that these views are neither reasonable nor compatible with scientific practice and, moreover, that they fail to live up to the standards of objectivity to which they aspire. (Howson, 1984, addresses a number of other objections.) One problem with the non-Bayesian criterion of adhocness, which we have not needed to exploit in our criticisms, is that the notion of 'independence' with regard to evidence is left vague and intuitive. Moreover, it seems difficult to give it a satisfactory meaning, except in the context of Bayesian induction.

There is an established notion of probabilistic independence, described earlier (Chapter 3, section **e**), which, however, is unable to supply a suitable interpretation. For suppose theory $h$ was advanced in response to a refutation by $e'$ and that $h$ both explains the old $e'$ and makes the novel prediction $e''$. It is the general opinion, certainly shared by Popperians, and a consequence of Bayes's Theorem, that $e''$ confirms $h$, provided it is sufficiently improbable relative to background information. As discussed earlier in this chapter (section **e**), such confirmation is available, in particular, when

$P(e'' \mid h \,\&\, e') > P(e'' \mid e')$. But this inequality is quite compatible with $e''$ and $e'$ *not* being independent in the probabilistic sense.

Another possible way to interpret the independence notion is in terms of logical independence, so that $e'$ and $e''$ would be said to be independent just in case neither entails the other. This would mean that if the two bits of evidence were trivially distinct in, say, relating to different times or slightly different places, then they would be independent in the sense employed in the adhocness criterion. But then practically no theory would be ad hoc. Take Medawar's peculiar theory about the Eskimo's cosy style of life, which was propounded in response to some surprising IQ measurements. Presumably, one could infer from the theory that tests applied during the following week to the same group of Eskimos would produce similarly high IQs. But although this prediction is logically independent of the earlier results, its success would not significantly improve the standing of Medawar's theory. Mere logical independence from the old results is clearly insufficient to ensure evidential support.

Intuitively, new evidence supports a theory only when it is substantially different from known results, not just trivially different in the logical sense described, and it is this intuition which, it seems to us, underlies the standard adhocness criterion. The idea that 'different' or 'varied' evidence gives more support to a hypothesis than a similar volume of homogeneous evidence is an old and widely held one. As Hempel put it, "the confirmation of a hypothesis depends not only on the quantity of the favorable evidence available, but also on its variety: the greater the variety, the stronger the resulting support" (1966, p. 34). So, for example, the report of the rate at which a stone falls to the ground from a given height on a Tuesday is similar to that relating to the stone's fall on a Thursday; it is very different, however, from a report of the trajectory of a planet or of how a given fluid rises in a particular capillary tube. But although the notions of similarity and diversity amongst evidence seem intuitively clear, it is not easy to give them a precise analysis, except, in our view, in probabilistic terms, in the context of Bayesian induction.

The similar instances in the above list have the characteristic that when one of them is known, any other would thereby be anticipated with relatively high probability. This recalls

Bacon's characterisation of similarity in the context of inductive evidence. He spoke of observations "with a promiscuous resemblance one to another, insomuch that if you know one you know all" and was probably the first to point out that it would be superfluous to cite more than a small representative sample of such observations in evidence (*see* Urbach, 1987, pp. 160–64). This idea of similarity between items of evidence is expressed naturally in probabilistic terms by saying that $e_1$ and $e_2$ are similar provided $P(e_2 \mid e_1)$ is higher than $P(e_2)$; and one might add that the more the first probability exceeds the second, the greater the similarity. This means that $e_2$ would provide less support if $e_1$ had already been cited as evidence than if it was cited by itself.

On the other hand, knowing that one of a pair of dissimilar instances has occurred gives little or no guidance as to whether the other will occur. For example, unless Newton's, or some comparable, theory had already been firmly established, a knowledge of the rate of fall of a given object on some specific occasion would not significantly affect one's confidence that the planet Venus, say, would appear in a particular position in the sky on a designated day. Different pieces of evidence may also have a mutually discrediting effect. An example of this might be the observations of the same constant acceleration of heavy bodies dropped at sea level and the unequal rates of fall of objects released on different mountain tops. Both sets of observations would confirm Newton's laws, but in circumstances where those laws are not already well established, the first set might suggest that all objects falling freely (whether on top of a mountain or not) do so with the same acceleration. In other words, with different instances, say $e_3$ and $e_1$, $P(e_3 \mid e_1)$ is either close to or less than $P(e_3)$. Of course, $e_3$ merely differing from $e_1$ in this sense does not imply that it supports any hypothesis significantly; whether it does or not depends on its probability. The notion of similarity, as we have characterised it, is reflexive, as it should be; that is, if $e_2$ is (dis)similar to $e_1$, then $e_1$ is (dis)similar to $e_2$ (this follows directly from Bayes's Theorem).

To summarize, the non-Bayesian appraisal of hypotheses based on the notion of adhocness is ungrounded in epistemology, has highly counterintuitive consequences, and relies on a concept of independence among items of evidence which seems unanalysable except in Bayesian terms. In brief, it is not a success.

## k INFINITELY MANY THEORIES COMPATIBLE WITH THE DATA

### k.1 The Problem

Galileo carried out numerous experiments on freely falling bodies and on bodies rolling down inclined planes in which he examined how long they took to descend various distances. These experiments led him to formulate the well-known law to the effect that $s = ut + \frac{1}{2}gt^2$, where $s$ is the distance fallen by a freely falling body in time $t$, $u$ is its initial downward velocity, and $g$ is a constant. Jeffreys (1961, p. 3) pointed out that Galileo might also have advanced the following as his law:

$$s = ut + \frac{1}{2}gt^2 + f(T)(T - T_1)(T - T_2) \ldots (T - T_n),$$

where $T$ represents the date of the experiment, which could for example be recorded as the number of minutes that have elapsed since the start of the year AD 1600; $T_1, T_2, \ldots, T_n$ are the specific dates on which Galileo performed his experiments; and $f$ can represent any function of $T$. Thus Jeffreys's modification stands for an infinite number of alternatives to Galileo's theory. Although all these theoretical alternatives contradict one another and make different predictions about future experiments, the interesting feature of Jeffrey's unorthodox laws of free fall is that they all imply Galileo's experimental data.

This is a particular problem for those non-probabilistic theories of scientific method which hold that the scientific value of a theory is determined just by $P(e \mid h)$ and, in some versions, by $P(e)$. These philosophical approaches, of which Popper's is one example and maximum-likelihood estimation (Chapter 12, section **d.3**) another, would have to regard the standard law of free fall and those peculiar alternatives described by Jeffreys as equally good scientific theories relative to the evidence available to Galileo, although this is a judgment with which no scientist would agree.

The same point emerges from a well-known example due to Nelson Goodman (1954; for an amusing and illuminating discussion, see Jeffrey, 1983, pp. 187–90). Goodman noted that the evidence of very many and varied green emeralds would normally suggest that all emeralds are green. But he pointed out that that evidence bears the same relation to 'All emeralds are green' as it does to a type of hypothesis he formulated as 'All emeralds are grue'. According to Goodman's definition, something is grue if it was either observed before time *t* and was green, or was not observed before *t* and is blue. If *t* denotes some time after the emeralds described in the evidence were observed, then both the green- and the grue-hypotheses imply that the emeralds observed so far should be green. However, on the assumption that there are unobserved emeralds, the hypotheses are incompatible, differing in their predictions about the colours of emeralds looked at after the critical time. As with Jeffreys's variants of Galileo's theory, the grue-hypothesis represents an infinite number of alternatives to the more natural hypothesis, for *t* can assume any value, provided it is later than now.

Our examples illustrate a general problem for methodology: that a theory which explains (in the sense of implying or associating a certain probability with) some data is merely one out of an infinite set of rival theories, each of which does exactly the same. The existence of this infinite set of possible explanations, it will be remembered, spelled ruin for any attempt at a positive solution to the problem of induction (*see* Chapter 1). The problem with which we are concerned here arises because, in practice, scientists discriminate between possible explanations and typically pick out just one, or at any rate relatively few, as meriting serious attention. An account of scientific method ought to explain how and why they do this.

### k.2 The Bayesian Approach to the Problem

This has not proved easy. For the Bayesian, the nature of the problem, at least, is straightforward. Moreover, Bayesian theory does not imply that every hypothesis similarly related to the data is of equal merit. Suppose one were comparing two theories in the light of the same evidence. Their relative posterior probabilities are given by

$$\frac{P(h_1 \mid e)}{P(h_2 \mid e)} = \frac{P(e \mid h_1)P(h_1)}{P(e \mid h_2)P(h_2)}.$$

If both theories imply the evidence, then $P(e \mid h_1) = P(e \mid h_2) = 1$. And if, in addition, $P(h_1 \mid e)$ exceeds $P(h_2 \mid e)$, then it follows that $P(h_1)$ is larger than $P(h_2)$. More generally, if two theories which explain the data equally well nevertheless have different posterior probabilities, then they must have had different priors too. So theories such as the contrived alternatives to Galileo's law and Goodman's grue-variants must, for some reason, have lower prior probabilities. Indeed, this is clearly reflected in most people finding such hypotheses quite unbelievable. The problem then is to discover the criteria and rationales by which theories assume particular prior probabilities.

Sometimes there is a clear reason why a theory is judged improbable. For instance, suppose the theory concerned a succession of events in the development of a human society; it might, for example, assert that the elasticity of demand for herring remains constant or that the surnames of all future British prime ministers and American presidents will start with the letter Z. These theories, which of course could be true, are however monstrously improbable. And the reason for this is that the events they describe are influenced by numerous independent processes whose separate outcomes are improbable. The probability that all these processes will turn out to favour the hypotheses in question is therefore the product of many small probabilities, and so itself is very small indeed (Urbach, 1987b). The question of how the probabilities of the causal factors are estimated, of course, remains. This could be answered by reference to other probabilities, in which case the question is just pushed one stage back, or else by some different process that does not depend on probabilistic reasoning. For instance, the simplicity of a hypothesis has been thought to have an influence on its initial probability. This and other possible determinants of initial probabilities are discussed in Chapter 15.

It is worth mentioning here that the equation given above,

relating the posterior probabilities of two theories to their prior probabilities, explains an important feature of inductive reasoning. The scientist often prefers a theory which explains the data imperfectly, in that $P(e \mid h_1) < 1$, to an alternative, $h_2$, which predicts them with complete accuracy. Thus, even Galileo's data were not in precise conformity with his theory; nevertheless, he did not consider any more-complicated function of $u$ and $t$ to be a better theory of free fall than his own, even though it could have embraced the evidence he possessed more perfectly. According to the above equation, this is because the better explanatory power of the rival hypotheses was offset by their inferior prior probabilities (*see* Jeffreys, 1961, p. 4).

## I CONCLUSION

Charles Darwin (1868, vol. 1, p. 8) said that "In scientific investigations it is permitted to invent any hypothesis, and if it explains various large and independent classes of facts it rises to the rank of a well-grounded theory". This is, perhaps, an exaggeration, for not any hypothesis would do; the hypothesis must not be refuted or substantially disconfirmed, nor should it be intrinsically too implausible. With these provisos, Bayesianism, we suggest, is just such a well-grounded hypothesis as Darwin referred to. As we showed in Chapter 5, it arises from natural and intuitively reasonable attitudes to risk and uncertainty. It is neither refuted nor undermined by any of the phenomena of scientific reasoning. On the contrary, as we have seen, it explains a wide variety of them. So far, we have concentrated chiefly on deterministic theories. We shall see in the next and following chapters that the Bayesian approach is no less successful when dealing with statistical reasoning.

## EXERCISES

1. Show that a hypothesis that is refuted by some evidence must have zero probability relative to that evidence. Show too that no new evidence can ever raise the probability of a refuted hypothesis above zero.

2. Why is it best to ask someone hard questions in order to discover his or her competence?

3. If $e$ confirms $h$, then $\sim e$ disconfirms $h$. True or false?

4. If $e$ confirms $h$, then $h$ cannot confirm $e$. True or false?

5. Show that it is possible for one theory to confirm a second, for the second to confirm a third, and yet for the first to *dis*confirm the third. Try to find a simple, schematic example in which this pattern of confirmation and disconfirmation occurs. Consider whether the phenomenon is compatible with the Bayesian and Popperian methodologies.

6. If one theory disconfirms a second, and the second disconfirms a third, is it possible for the first to confirm the third?

7. Show that it is possible for a theory to be confirmed by two pieces of evidence separately, but disconfirmed or even refuted by their conjunction. Look for a simple example, and then consider whether the phenomenon is compatible with the Bayesian and Popperian methodologies.

8. If $e$ confirms $h$, and $h$ logically implies $h'$, must $e$ also confirm $h'$? If you think not, present a counter-example. What answer should a Bayesian (respectively, a Popperian) give to the question? (The claim that under the circumstances described $e$ does confirm $h'$ is known as the Special Consequence Condition and is discussed, for example, by Hesse, 1974, p. 143.)

9. $h$ and $h'$ are incompatible hypotheses with non-zero prior probabilities. Show that the probability of $h$ must increase when $h'$ is refuted and that this increase is greater the larger the probability of $h'$ before the refutation (Polya, 1954, vol. 2, p. 124).

10. You pick ticket 185375 in a lottery; your friend draws ticket 654972. The probability of this conjunction of events is tiny, yet you would not be particularly surprised. On the other hand, suppose having made your draw secretly from the lottery, you ask your friend to tell you the first six-figure number that comes to mind, and he gives you

the number of your ticket. This coincidence is also very improbable, but this time your surprise would be great. What is the difference in the two situations that could account for the different psychological reactions? (Schlesinger 1991, pp. 95–113; Urbach, 1992a)

**11.** In June 1988, J. Benveniste published some remarkable experimental results in the journal *Nature*. He steadily diluted an aqueous solution of a certain antibody, checking its biological activity at successive stages. He reported that at some dilutions the activity fell off, that on further dilution it was restored, and that further dilutions produced alternations between activity and inactivity. Particularly remarkable was the fact that the solutions became so dilute that they can hardly have contained even a single antibody molecule. Benveniste ventured the explanation that "antibody molecules once embodied in water leave their internal marks on its molecular structure", a conclusion which is in line with the principles of homeopathic medicine.

The editor of *Nature,* in the same issue of the journal as carried Benveniste's report, described the results as "unbelievable", recommending that they be treated with the utmost caution. On the other hand, he admitted that he had welcomed some other extremely unexpected and surprising results which had recently led to the suggestion of a 'fifth force' between material objects (p. 787). What accounts for the different responses to the two sets of unlikely and surprising results?

To conclude the story, it turned out that Benveniste's laboratory procedures were badly flawed and that when conducted more carefully, his results could not be reproduced (Maddox, et al., 1988).

**12.** Show that Bayesian confirmation respects the Equivalence condition.

**13.** Mackie's solution to the Ravens Paradox is Bayesian in spirit. His argument, in essence, is this (Mackie, 1963, p. 267):

> [I]f we take it as known that the ratio of ravens to non-ravens is $x$ to $1-x$, and that of black things to non-black things $y$ to $1-y$, where $x < y < \frac{1}{2}$, then the

probabilities of each of the four sorts of observation-report $b_1$ ('This is a black raven'), $b_2$ ('This is a non-black raven'), $b_3$ ('This is a black non-raven'), and $b_4$ ('This is a non-black non-raven'), in relation (i) to our background knowledge *k* alone, and (ii) to the conjunction of *k* with *h* [the hypothesis that all ravens are black], are set out below:

| | $b_1$ | $b_2$ | $b_3$ | $b_4$ |
|---|---|---|---|---|
| (i) | $xy$ | $x(1 - y)$ | $y(1 - x)$ | $(1 - x)(1 - y)$ |
| (ii) | $x$ | 0 | $y - x$ | $1 - y$ |

Mackie then applies Bayes's Theorem and argues that this shows that reports of the form $b_1$ confirm *h* much more than those of the form $b_4$, that a $b_3$ report disconfirms *h*, and that the Ravens Paradox is thereby resolved.

Try to reconstruct Mackie's reasoning and evaluate his solution to the paradox. What presuppositions do Mackie's calculations make? Are these justified? Does a black non-raven support *h* on Mackie's reasoning? (For a valuable discussion, *see* Horwich, 1982, Chapter 3.)

**14.** Popper's "third requirement . . . for the Growth of Knowledge" was that a theory should have some of its new independent predictions verified, explaining his reasons thus:

> The first reason why our third requirement is so important is this. We know that *If we had an independently testable theory which was, moreover, true, then it would provide us with successful predictions* (and *only* with successful ones). Successful predictions—though they are not, of course, *sufficient* conditions for the truth of a theory—are therefore at least necessary conditions for the truth of an independently testable theory. In this sense—and only in this sense—our third requirement may even be said to be 'necessary', if we seriously accept truth as a regulative idea. (Popper, 1963, p. 246)

Assess this sequence of claims as a defence of the second aspect of the adhocness criterion.

**15.** Comment on the following defence by Popper against Duhem's criticism.

> As to Duhem's famous criticism of crucial experiments, he only shows that crucial experiments can never *prove*

> or establish a theory; but he nowhere shows that crucial experiments cannot *refute* a theory. Admittedly, Duhem is right when he says that we can test only huge and complex theoretical systems rather than isolated hypotheses; but if we test two such systems which differ in one hypothesis only, and if we can design experiments which refute the first system while leaving the second very well corroborated, then we may be on reasonably safe ground if we attribute the failure of the first system to that hypothesis in which it differs from the other. (Popper, 1960, p. 132)

**16.** Show that the posterior probabilities of Prout's hypothesis and of the auxiliary hypothesis considered in section **h.3** of this chapter depend only on the relative values of $P(e \mid {\sim}t \,\&\, a)$, $P(e \mid t \,\&\, {\sim}a)$, and $P(e \mid {\sim}t \,\&\, {\sim}a)$, and not on their absolute values.

**17.** Prove the results stated in the first paragraph of section **i** in this chapter.

**18.** Descartes proved from the laws of light reflection and refraction and other theoretical assumptions that in appropriate circumstances there should be two rainbows, one at an elevation of 41.5°, the other at 51.5°. However, the corresponding measured angles were 45° and 56°, respectively. Descartes remarked that this discrepancy "shows how little faith we must have in observations which are not accompanied by true reason" (Descartes, 1637, p. 342). Comment on the cogency or otherwise of Descartes' response.

**19.** "And therefore it was a good answer that was made by one who when they showed him hanging in a temple a picture of those who had paid their vows as having escaped shipwreck, and would have him say whether he did not now acknowledge the power of the gods,—'Aye,' asked he again, 'but where are they painted that were drowned after their vows?'" (Bacon, 1620, Book I, xlvi) What is the point of the question at the end of this anecdote?

**20.** Show that similarity and dissimilarity between data, as characterised in section **j.6** of this chapter, are reflexive.

# PART III

# *Classical Inference in Statistics*

We showed in Chapter 7 how numerous aspects of scientific reasoning can be illuminated by reference to Bayes's Theorem. We confined the discussion there mainly to deterministic theories. As already explained, however, scientific theories are often not deterministic but are statistical or probabilistic in character. The evaluation of such hypotheses brings no special problems of principle to a Bayesian analysis, the difference between the cases of deterministic and statistical hypotheses being reflected in the term $P(e \mid h)$ which appears in Bayes's Theorem. In the former case, when $h$ entails $e$, this term equals 1. When $h$ is statistical, $P(e \mid h)$ takes a value equal to the statistical probability which $h$ confers on $e$, this being an application of the so-called Principal Principle, which is discussed in Chapter 13, section **d.** Inductive reasoning about deterministic and probabilistic hypotheses is then explained in a uniform fashion in the Bayesian approach, the former merely constituting a special case of the latter.

No such uniform treatment is afforded, however, by the leading non-Bayesian approaches, which generally offer distinct methods of inference for deterministic and statistical hypotheses. We have followed the pattern thus established by dealing separately with statistical and deterministic hypotheses. This plan is justified, since the main rival to the Bayesian approach, classical, or 'frequentist', statistical inference, is a sophisticated and extremely widely applied body of doctrine whose challenge to Bayesianism requires an answer. Moreover, the strength of the Bayesian approach as a unified and successful account of scientific reasoning can best be appreciated when contrasted with what we hope to show are the inadequacies of its main rivals.

In this part of the book, we shall review the main facets of classical statistical inference. These can be divided roughly into two. First, the theory of significance tests, which purports to inform us when we ought to reject a statistical hypothesis or regard it as false. The two rival versions of significance testing were put forward by Fisher, whose contribution is assessed in Chapter 8, and by Neyman in collaboration with Pearson; the Neyman-Pearson account is presented and criticized in Chapter 9. Secondly, estimation theory is an attempt to arrive at a positive conclusion about the value of a parameter. This also has two aspects: point estimation and interval estimation. Chapter 10 takes up the subject of the classical theory of estimation.

# CHAPTER 8
# *Fisher's Theory*

## a FALSIFICATIONISM IN STATISTICS

Theories seriously entertained by scientists at one time are often later rejected when reviewed in the light of new experimental evidence. The most straightforward form for such rejections is that of a logical refutation, and provided one is prepared to concede certainty to the refuting data, such refutations may be regarded as scientific modes of inference which require no concession to Bayesian principles. Indeed, some philosophers, keen to avoid a subjective probabilistic assessment of hypotheses, maintain that logical refutations are the only significant type of inference in science. However, as we explained earlier, a large part of modern science is concerned with statistical hypotheses, and these are generally not refutable in this way. As an example of a simple statistical hypothesis, take the theory that a particular penny has an even chance of landing heads and tails, with separate tosses being independent (the penny is then said to be a 'fair' coin). This theory cannot be refuted by observing the outcomes of trials in which the penny is tossed; no proportion of heads in any sequence, however large, is precluded by the theory. Nevertheless, scientists do not regard statistical theories as necessarily unscientific, nor have they dispensed with procedures for rejecting them in the face of what they take to be unfavourable evidence. What principles apply here and how can they be justified?

One answer, in the falsificationist spirit, that is occasionally canvassed claims that although statistical theories are not strictly falsifiable, they are falsifiable in an extended sense of the term: as Cournot (1843, p. 155) expressed the idea, events which are sufficiently improbable "are rightly regarded as physically impossible". Popper had the same notion: scientists, he said, should make "a methodological decision to regard highly improbable events as ruled out—as prohibited" (1959a, p. 191). And he talked of hypotheses as having been "practically falsified" if they attached sufficiently low probabilities to events that actually occurred. Watkins, endorsing the idea, called it a "non-arbitrary way of reinterpreting probabilistic hypotheses so as to render them falsifiable" (1984, p. 244).

Popper defended his position with a surprisingly weak argument. He claimed that extremely improbable events that did happen "would not be physical effects, because, on account of their immense improbability, *they are not reproducible at will*" (1959a, p. 203). This unreproducibility of very improbable events, Popper reasoned, means that a physicist "would never be able to decide what really happened in this case, and whether he may not have made an observational mistake". Popper appears to be operating here with a rather eccentric definition of 'physical effect', which would exclude most natural phenomena from that category, for most natural phenomena cannot be humanly controlled and so are not reproducible at will. More importantly, Popper's claim that unreproducible effects cannot be properly checked is clearly mistaken; improbable and unreproducible events—for example, the sequence of heads and tails produced when a coin is tossed ten thousand times—are not necessarily so fleeting as to prevent a close examination.

The Cournot-Popper view overlooks the fact that very improbable events occur all the time. Indeed, it would be difficult to name a probability so small that no event of some smaller probability had not already taken place or is not taking place right now: events of miniscule probability are ubiquitous. Even a probability of $10^{-10^{12}}$, which Watkins considered to be "vanishingly small" and to amount to an impossibility (1984, p. 244), is nothing of the sort. The probability of the precise distribution of genes in the five billion members of contemporary humanity is incomparably smaller than this, relative to Mendel's laws of inheritance, as is the probability that the atoms in the jug of water on this table have a particular spatial distribution at a given time.

Popper attempted to give pragmatic effect to his thesis by propounding a rule that would tell us in particular cases how small a probability should be in order to be classed as a

and invalid. Its irrelevance was highlighted by Fisher, who argued that scientists are not concerned with how rarely they might falsely reject a hypothesis in a series of experiments. As he put it, "To a practical man . . . who rejects a hypothesis, it is, of course, a matter of indifference with what probability he might be led to accept the hypothesis falsely, for in his case he is not accepting it" (Fisher, 1956, p. 42). For this reason, Fisher rightly dismissed the Neyman-Pearson defence as "absurdly academic". Fisher failed, however, to point out that the defence is built on a fallacy. As was mentioned before, the fact that an event has some probability is compatible with any number of occurrences of that event in actual sequences of trials.

It is often argued (mistakenly, we think) that an area where it would not be absurd to act as if some uncertain hypothesis were true is industrial quality control. Even some vigorous opponents of the Neyman-Pearson method, such as A. W. F. Edwards (1972, p. 176), have conceded this. The argument is the following. Suppose an industrialist would lose money if he marketed a production run that included more than a certain percentage of defective elements. And suppose production runs were each sampled with a view to testing whether they were of the loss-making type or not. In such cases there could be no graduated response, it is claimed, since the production run would either be marketed or not. The argument then continues: if the industrialist followed Neyman-Pearson principles, he could comfort himself with the thought that since he will perform the same test and apply the same decision-rule many times, in the long run only about 5 per cent of the batches he will market will be defective, and that may be a failure rate he can sustain.

*But this argument is fallacious,* for just because the industrialist is confined to one of two actions does not imply that he can only entertain one of two beliefs about the success of those actions. The situation is compatible with his attaching probabilities to the various hypotheses and then deciding whether to market the batch or withhold it by balancing those probabilities against the utilities of the possible outcomes of his actions. (The precise way in which this is done is the subject of what is known as Decision Theory.) The approach to decision making just outlined is, surely, closer to both actual and reasonable calculations. For if an industrialist really accepted some hypothesis about the constitution of a batch as definitely true, he would not only be prepared to offer it for sale, he would also accept any bet whatsoever on the truth of the hypothesis, which he clearly would not. We conclude, therefore, that industrial quality control is not a case that either should or does exemplify Neyman-Pearson thinking.

### c.2 The Neyman-Pearson Theory as an Account of Inductive Support

A further reason why it is so difficult to interpret acceptance and rejection as the kind of behaviour one should expect from scientists is that reasonable people do not view theories in such black-and-white terms. As the quotations given in Chapter 1 illustrate, scientific hypotheses are viewed, typically, in varying shades of grey, depending on the weight of evidence in their favour. And it is at least partly the strength of such evidence which dictates how much a scientist will risk on actions whose effectiveness is calculated on the basis of the hypothesis.

This fact is acknowledged by the many classical statisticians who have attempted to interpret their methods of inference in such a way that the impact of evidence on a theory would be a matter of degree. The theory of significance tests appears, on the surface, to have room for such an interpretation. In particular, it is commonly assumed that a theory rejected in a statistical test is the more substantially discredited the smaller the size of the test; and if it is accepted, then its merit in the light of the evidence is sometimes measured by the power of the test.

For instance, Kendall and Stuart have suggested that the results of a Neyman-Pearson test can indicate whether or not a hypothesis is supported by the evidence:

> If the reader cannot overcome his philosophical dislike of these admittedly inapposite expressions, he will perhaps agree to regard them as code words, "reject" standing for "decide that the observations are unfavourable to" and "accept" for the opposite. (Kendall and Stuart, 1979, p. 177)

And at the outset of their exposition of the theory of significance tests, Kendall and Stuart stated that its aim is to answer the question: "which sets of observations are we to regard as favouring, and which as disfavouring, a given hypothesis?" (Kendall and Stuart, 1979, p. 177).

Cramér (1946, p. 421) more explicitly interpreted significance tests as providing a measure of empirical support: "we shall denote a value [of a test-statistic] exceeding the 5% limit but not the 1% limit as *almost significant,* a value between the 1% and 0.1% limits as *significant,* and a value exceeding the 0.1% limit as *highly significant.*" Cramér added that such terminology is "purely conventional"; nevertheless, it is highly suggestive and seems to have been introduced specifically because of the feeling that a significance test should provide an index of how strongly evidence tells either for or against the hypothesis under test. Thus, on the following page of his book, Cramér remarked that since a particular $\chi^2$ value, calculated from some of Mendel's famous experiments on pea plants, differed by only a small amount from the expected value relative to the null hypothesis he was considering, "the agreement must be regarded as good" (p. 422). In another example, when the hypothesis would only be rejected if the level of significance were around 0.9, Cramér said that "the agreement is very good" (p. 423).

Kendall and Stuart go along with this idea that the size of the test which would just lead to a hypothesis being rejected measures the support which it gives the hypothesis. In an example of theirs, where a $\chi^2$-statistic evaluated from a particular experiment would reject a hypothesis if the size of the test exceeded 0.37, the agreement between hypothesis and data was said to be "very satisfactory". However, when the critical test size was only 0.27, Kendall and Stuart (1979, p. 458) described the result as being "still very satisfactory", though they considered it to be "rather more critical of the hypothesis than the other test was". (Oddly enough, these two results were calculated by different methods of grouping the very same data, and there is no indication as to which of the two methods is intended to take precedence. The reader is referred back to section **e** of Chapter 8 for a discussion of this peculiar aspect of the chi-square test.) In general, then, these authors take the view that *the smaller the size of the test which would just result in the rejection of the hypothesis, the more telling is the evidence against that hypothesis.*

As far as we can tell, however, there are no grounds for this thesis. In order to establish it, there must first be an appropriate concept of empirical support, which could then be shown to have the aforementioned connection with the size of significance tests. But no such concept has been formulated. Indeed, there are compelling reasons to think that the concepts of size and power are quite unpromising foundations for a theory of empirical support. First of all, the results of a significance test, either of the Fisher or Neyman-Pearson variety, are often in flat contradiction to the conclusions which an impartial scientist or ordinary observer would draw. Secondly, we shall find that judgments based on significance tests are importantly different from decisions about inductive support, for the former depend on what our intuitions in the matter, such as they are, affirm to be extraneous considerations. The rest of this section details these objections.

### c.3 A Well-Supported Hypothesis Rejected in a Significance Test

Consider the first objection, which can be explained by referring again to the tulip-bulb example. Let us continue to regard $h_1$, the hypothesis that the consignment contained 40 per cent red-flowering bulbs, as the null hypothesis. The following table gives the minimum number of red-flowered plants which would have to appear in a random sample of size $n$ in order for that hypothesis to be rejected at the 5 per cent level.

**TABLE VIII**

| *Sample size* n | *The minimum number of red tulips needed to reject* $h_1$ *at the 5% level, expressed as a proportion of* n | *Power of the test against* $h_2$ |
|---|---|---|
| 10 | 0.70 | 0.37 |
| 20 | 0.60 | 0.50 |
| 50 | 0.50 | 0.93 |
| 100 | 0.480 | 0.99 |
| 1000 | 0.426 | 1.0 |
| 10,000 | 0.4080 | 1.0 |
| 100,000 | 0.4026 | 1.0 |

It will be noticed that as $n$ increases, the critical proportion of red tulips that would reject $h_1$ at the 0.05 level approaches more closely to 40 per cent, that is, to the proportion which $h_1$ asserts is contained in the consignment. Bearing in mind that the only alternative to $h_1$ that is admitted in this simple example is that the consignment contains red tulips in the proportion of 60 per cent, an unprejudiced consideration of these data would, it seems to us, lead to the conclusion that as $n$ increases, these so-called critical values *support* $h_1$ more and more. When the sample size is 1000, the results of the experiment, far from being unfavourable (in Kendall and Stuart's phrase) to $h_1$, appear strongly to favour it. Yet the theory of significance tests requires one to reject that hypothesis, an injunction that is markedly counter-intuitive.

The example envisages that $h_1$ is 'tested against' $h_2$, that is, $h_2$ is regarded as the only alternative to $h_1$. In these circumstances it is a simple matter to calculate the power of the test we are employing, and this information is included in Table VIII. The thesis implicit in the current approach, that a hypothesis may be rejected with increasing confidence or reasonableness as the power of the test increases, is not borne out in the example, which signals the reverse trend. (This objection was developed by Lindley, 1957.)

### c.4 A Subjective Element in Neyman-Pearson Testing: The Choice of Null Hypothesis

Another reason why significance tests cannot produce an account of inductive support is that the results of such tests hang on decisions which would normally be regarded as quite unconnected with the evidential force of experimental data. The first kind of decision concerns the selection of the null hypothesis, and it has a crucial bearing on which hypothesis is finally accepted and which rejected, and hence on which is supported and which undermined by the evidence. Thus in the tulip-bulb example, if an experiment produced 50 red tulips in a random sample of 100, then, with $h_1$ (40 per cent red) as the null hypothesis, it would be rejected at the 5 per cent level and $h_2$ (60 per cent red) would be accepted. If, on the other hand, $h_2$ were the null hypothesis, then the opposite judgment would be delivered, that is, $h_2$ should be rejected and $h_1$ accepted! But as we observed earlier, the null hypothesis is just the one which according to the scientist's personal scale of values would lead to the more undesirable practical consequences, if it were mistakenly assumed to be true. When the scientist is unable to distinguish the hypotheses by this practical yardstick, the null hypothesis may be chosen arbitrarily. But our ordinary understanding of empirical support, such as it is, accords no role to such factors.

### c.5 A Further Subjective Element: Determining the Outcome Space

In any kind of significance test, a hypothesis is assessed according to the probability of the observed outcome relative to the probabilities of other possible outcomes. In other words, whether an observation favours a hypothesis or not depends upon what observations might have been made if the trial had not turned out as it did. This means that the significance of a result of, say, 1 head and 3 tails in an ordinary coin-tossing trial would depend on how the experimental apparatus dealt with other possible outcomes. The apparatus might, for example, be rigged up to record only two results: either the occurrence of 1 head and 3 tails, or else its non-occurrence. In fact, the apparatus might have been designed to report the non-occurring results in many different ways. Each would demand a separate analysis, with no guarantee of always delivering the same significance-test inference.

The outcome space of a trial, and hence the decision whether to reject a hypothesis or not, is also governed by the so-called *stopping rule,* the rule that dictates when the trial should terminate. Thus, consider again the test of the fair-coin hypothesis with which we illustrated Fisher's theory. We assumed there a stopping rule with the instruction to conclude the trial after 20 flips of the coin. This implies an outcome space of $2^{20}$ sequences, each comprising 20 elements, where each element corresponds either to heads or tails.

Suppose, however, the rule had been to stop the experiment after 6 heads had appeared. The outcome space would now differ from that just described, and many of the results that were possible when the number of tosses of the coin was pre-set at 20 would not any longer be possible. If, as before, one ignored the order in which heads and tails appear in a trial and simply noted their number, then the results one could record are (6,0), (6,1), (6,2), . . . , etc., whereas before they were (20,0), (19,1), . . . , (0,20)—see Table I, Chapter 8. This change, the reader may be surprised to learn, has a profound effect on the conclusion of a significance test.

This can be appreciated by calculating the probability of each of the possible results under the new stopping rule, relative to the null hypothesis, namely, that the coin is fair. The calculation proceeds by considering that the result (6,*i*) is obtained when (5,*i*) appears in any order and is then followed by a head. So, on the assumption of the null hypothesis, the probability of (6,*i*) is ${}^{i+5}C_5\ (\frac{1}{2})^5\ (\frac{1}{2})^i \times \frac{1}{2}$, yielding the following distribution:

**TABLE IX**

| *Outcome (H,T)* | *Probability* | *Outcome (H,T)* | *Probability* |
|---|---|---|---|
| 6,0 | 0.0156 | 6,11 | 0.0333 |
| 6,1 | 0.0469 | 6,12 | 0.0236 |
| 6,2 | 0.0820 | 6,13 | 0.0163 |
| 6,3 | 0.1094 | 6,14 | 0.0111 |
| 6,4 | 0.1230 | 6,15 | 0.0074 |
| 6,5 | 0.1230 | 6,16 | 0.0048 |
| 6,6 | 0.1128 | 6,17 | 0.0031 |
| 6,7 | 0.0967 | 6,18 | 0.0020 |
| 6,8 | 0.0786 | 6,19 | 0.0013 |
| 6,9 | 0.0611 | 6,20 | 0.0008 |
| 6,10 | 0.0458 | 6,21 | 0.0005 |
| | | | etc. |

Suppose 6 heads and 14 tails were obtained in a trial conducted according to the present stopping rule. The outcomes which could have occurred and which are at least as improbable as the actual one are (6,14), (6,15), . . . , etc. Since the total probability of these is only 0.0319, which is less than the critical value of 0.05, the result of the trial is significant and so, according to Fisher, the null hypothesis should be rejected. It will be recalled, with concern we hope, that when the stopping rule restricted the size of a sample to 20, this result was not significant, according to Fisher. (An example similar to this is given by Lindley and Phillips, 1976.) The same contradictory conclusions would be permitted by a Neyman-Pearson treatment if the null hypothesis were tested against some alternative hypothesis, for example, one asserting that the probability of tails in a toss of the coin is $p'$, where $p'$ is greater than 0.5.

We have considered just two stopping rules that could have operated to produce some particular result, but there is no limit to the number of possible stopping rules with that property, and not all of them make the decision to stop the trial depend on the outcome, which some statisticians think is not quite legitimate. For instance, suppose that after every toss of the coin, the experimenter had pulled a playing card at random from an ordinary pack with the intention of calling off the coin trial as soon as the Queen of Spades appeared. This stopping rule induces a quite different outcome space from the two we have considered, which is bound to lead to different conclusions in certain cases.

Some stopping rules demand a more complex analysis than that given above. For example, if the experimenter had decided to end the trial as soon as lunch was ready, one would have to evaluate, for every stage of the trial, the probability of the scientist being interrupted by the summons to lunch. If this could not be done, then the experiment would have been useless. The particular experimental result could even be the product of many, conflicting stopping rules. For a scientist might have acted on one such rule, while his collaborator had privately decided to conduct the trial in accordance with another, and yet the outcome of the trial might have happened to satisfy both rules. The outcome space in such a case might be hard to determine, for you would need to discover what each experimenter would have done if the actual outcome had not accorded with his own stopping rule. Would he have tried to overrule his colleague or given in quietly? And if the former, would he have succeeded? Intuitively, such questions do not bear on the testing process, and in practice they are rarely if ever dealt with, yet for the classical statistician, they are vital.

A significance-test inference, therefore, depends not only on the outcome that a trial produced, but also on the outcomes that it could have produced but did not. And the latter are determined in part by certain private intentions of the experimenters, embodying their stopping rule. It seems to us that this fact precludes a significance test delivering any kind of judgment about empirical support, unless our intuitions in the matter of such support were sharply revised. For scientists would not normally regard such personal intentions as proper influences on the support which data give to a hypothesis. (The same point is made by Savage, 1962, p. 18.) In Chapter 14, we shall show that as far as the Bayesian theory of inductive reasoning is concerned, the outcome space and the scientist's intentions that help to create it are irrelevant.

### c.6 Justifying the Stopping Rule

This Bayesian position has been challenged. For example, Whitehead argues that, as the classical, or 'frequentist', position implies, the stopping rule ought to figure in any inductive analysis. He presents his argument in the form of an amusing analogy.

> A football match is played between *Frequentists United* and *Bayesian Wanderers*. I am pleased to learn that *Frequentists United* win 1–0. However, my pleasure is somewhat dulled when I learn that the rules of football were changed specially for the occasion, and that the *Frequentists'* captain was allowed to decide when the match should end. I feel this gave the *Frequentists* an advantage, and my opinion of their superiority is reduced. Some comfort will come from realizing that Bayesian supporters see no need to adjust their opinions, and feel just as beaten as under the normal rules! (Whitehead, 1993, pp. 1412–13)

There is, to be sure, a similarity between a hypothesis test and a competitive game of skill, such as football, in that both represent attempts to evaluate a set of alternatives. A game, however, is an imperfect means of identifying the more skilful of a pair of teams, since it is also designed with an eye to entertainment. And the result of a football match, being reported simply as the number of goals scored by each side, overlooks all sorts of other information relevant to the question. Whitehead believes that the fact that the game was stopped at the behest of the *Frequentists* is one such piece of relevant information.

We agree. The shared intuition is that their single goal might well have been a fluke, that the *Frequentists* perhaps felt played out and perceived that the *Bayesians* were still full of vigour and potential. Such considerations would enter a Bayesian analysis, which has to look at all the moderately plausible explanations of the 1–0 score and at the evidence bearing on each. An analysis along these lines would be complex, no doubt further information, such as how long the game took, would be sought, but in principle the analysis could be done. On the other hand, the problem does not seem susceptible to a classical, or frequentist, treatment—where is the random sample? how is the outcome space determined? what significance test could be performed?

Whitehead's error, in our view, is to assume that for the Bayesian, the stopping rule *never* imparts relevant information. This is not so; indeed, we give another example of an inductively relevant stopping rule in Chapter 14. The Bayesian objects to the classical thesis that because of its role in creating the outcome space, the stopping rule is a necessary part of *every* statistical analysis. To demonstrate that this is false requires only a single counter-example (which we have given); to demonstrate that it is true (as Whitehead wishes) requires more than just one case where the stopping rule was relevant; it requires a general argument showing that the stopping rule must always be taken into account in a statistical inference.

Such an argument has in fact been attempted by Gillies. He claims that "to those who adopt falsificationism (or a testing methodology)", it "seems natural, and only to be expected" that the stopping rule should in general affect a theory's empirical support. Gillies rests his case on the claim that "[w]herever possible the experimental method should be applied, and this consists in designing and carrying out a *repeatable* experiment, . . . whose result might refute $h$ [the null hypothesis]" (Gillies, 1990, p. 94).

It is easy to see why it can do no harm and might well do good to repeat an experiment, for a repetition may produce more data by which to reduce uncertainty about the hypotheses under consideration. But why should an experiment be repeat*able?* Would our confidence in the age of the Turin Shroud, based on certain well-tried tests, be any different if the whole of the cloth had been consumed in the testing process, thus precluding further tests? We do not believe so. And surely many very useful and informative experiments are not repeatable, for example, pre-election opinion polls and certain astronomical observations.

In one sense, no experiment is repeatable, for it could never be redone in *exactly* the way it was done before. Indefinitely many factors alter between one performance of an experiment and another. Of course, not all such changes are relevant or significant. For instance, the person who tossed the coin 20 times might have been wearing yellow shoes or sported a middle parting; but those facts are clearly of no significance, and if you called for the experiment to be repeated, you would issue no instructions as to the experimenter's footwear or hairstyle. On the other hand, whether or not the coin had a piece of chewing gum stuck to one side or a strong breeze was blowing would be highly relevant. The question then is whether the stopping rule falls into the first category of irrelevant factors or into the second of relevant ones. Gillies (p. 94) simply *assumes* the latter. Thus he argues, with reference to the coin trial described above, that "[t]he test of $h$ in this case consists of the whole carefully designed experimental procedure", suggesting thereby that a specification of the whole procedure should include a reference to the stopping rule. But Gillies neither states this explicitly nor provides any reason why it should be so. So his attempt to establish a general inductive role for the stopping rule fails.

## d TESTING COMPOSITE HYPOTHESES

We have, hitherto, presented only a truncated version of the Neyman-Pearson method, restricting it to the case of two alternative hypotheses, one of which is assumed to be true and the other false. Since such cases are atypical (according to Hays, 1969, p. 275, in psychological research they are "almost nonexistent"), Neyman and Pearson extended and substantially modified their approach to enable it to embrace instances where a larger number of alternatives are admitted. For example, they provided a method, which we shall now describe, whereby the hypothesis that some population parameter $\theta$ has a particular value, say $\theta_o$ (call this $h_1$), may be "tested against" the composite hypothesis $\theta > \theta_o$ ($h_2$).

Imagine a trial in which a random sample is drawn from the population and appropriately measured. The old approach can still be used to establish a critical region corresponding to any designated probability of committing a type I error. However, since $h_2$ is equivalent to a disjunction of hypotheses, the calculation of the probability of a type II error is not so straightforward; the probability depends on which element of $h_2$ is true, and this is unknown. $h_2$ itself assigns no probability to events in the outcome space; hence, there is no determinable probability of a type II error in tests of composite hypotheses. This does not imply that such errors may not occur, of course. It does mean, however, that the criterion of low size and high power cannot be applied to the more complicated situation.

The response to this has been to advance a new criterion, which may be explained as follows. Consider, first, the separate hypotheses disjoined in $h_2$ and calculate, on the basis of some fixed significance level, the probability of a type II error for each. The critical region that minimizes this probability for every component of $h_2$ is said to be a *uniformly most powerful* (or UMP) test. In other words, an UMP test is a test of maximum power for the given significance level, whatever hypothesis encompassed by $h_2$ happens to be true. This is a strong recommendation within the Neyman-Pearson scheme.

A weakness of UMP tests, however, is that relative to some elements of $h_2$, including perhaps the true one, the power may be very low. Indeed, it could happen that the power is smaller than the significance level, which would mean that there was a greater chance of rejecting the null hypothesis when it is true than when it is false. If this were the case, there would always be a test which, in Neyman-Pearson terms, is better. Moreover, this test need not even involve sampling the population; one could simply sample a pack of cards. For example, if the pack consisted of 5 red and 95 black cards and the decision whether to accept or reject the hypothesis were determined respectively by whether a black or a red card was drawn in a random selection from the pack, the chances of rejecting the hypothesis would be the same whether it was true or false; in this case it would be 0.05. The idea that one could perform a satisfactory test simply by consulting a pack of cards, without even examining the population referred to in the hypotheses being tested, is rightly regarded by classical statisticians as absurd. They hold this view even though randomized trials (see the end of section **a,** above) appeal to the very similar idea that a randomly selected card, or some other extraneous random event, can influence a significance-test inference. Be that as it may, Neyman and Pearson imposed the further restriction on UMP tests that they should be unbiased.

A test of $h_1$ against $h_2$ is said to be *unbiased* if its power is at least as great as its significance level, in other words, if there is not a greater chance of rejecting $h_1$ when it is true than when it is false. When $h_2$ is composite, the test is unbiased if this condition applies to every element of $h_2$. Neyman and Pearson then suggested that one ought to confine UMP tests to those that are also unbiased, or to UMPU tests. These ideas can be illustrated by means of a simple example. Consider a population that is normally distributed, with a known standard deviation but an unknown mean, which we can label $\theta$, and a test of the null hypothesis $\theta = \theta_0$ against the composite hypothesis $\theta > \theta_0$. Let the test be based on the mean, $\bar{x}$, of a random sample drawn from the population. The following represents the probability distributions of such samples relative to $h_1$ and to some hypothesis, $h_i$, that is included in $h_2$.

**AN UMPU TEST**

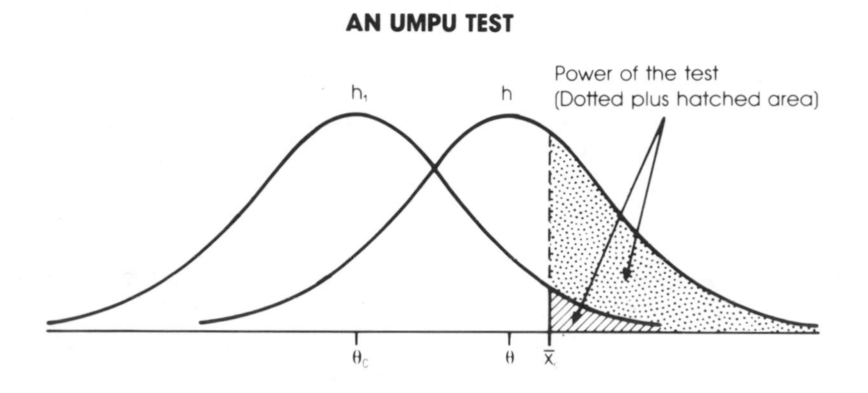


Consider a critical region for rejecting $h_1$ consisting of points to the right of the critical value, $\bar{x}_c$. The area of the hatched portion is proportional to the significance level. The area under the $h_i$ curve to the right of $\bar{x}_c$ represents the probability of rejecting $h_1$ when it is false, which is the power of the test. Clearly, if the standard deviation remained unchanged, the closer $\theta_i$ was to $\theta_0$, the smaller the power. But however close $\theta_i$ approached $\theta_0$, the power could never fall below the level of significance. In other words, the critical region is an UMPU test for $h_1$ against $h_2$. However, although the test is uniformly most powerful and unbiased, its actual power is unknown and, as already pointed out, it might be only infinitesimally different from the significance level. In other words, although the test is unbiased, it may be only just unbiased.

Suppose, now, the case is one in which $h_1 : \theta = \theta_0$ is tested against $h_2 : \theta \neq \theta_0$, it again being assumed that the standard deviation is known. A critical region located in one tail of the $h_1$ curve, as before, would not now constitute an unbiased test, as can be seen by considering a hypothesis $h_i$, in $h_2$, that is centred around the other tail, a situation depicted in the next diagram.

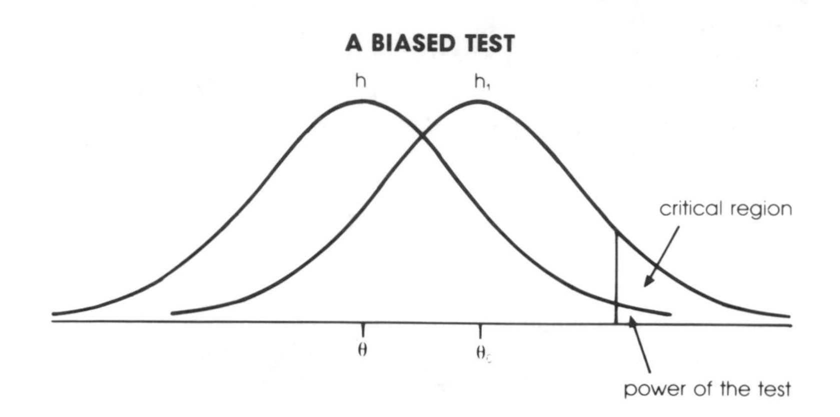


An UMPU test can, however, be constructed for this case by dividing the critical region equally between the two tails. This can be appreciated diagrammatically and can be rigorously shown. An UMPU test thus reinstates the two-tailed test implied by Fisher's theory but for which Fisher gave no adequate rationale.

In order to deal with composite hypotheses, the Neyman-Pearson method needs to depart somewhat from the original ideal. Thus, although the principle of maximizing power for a given level of significance is preserved, one cannot discover what that power is, and, for all one knows, in any particular case the power may be very low. This seems to be a disadvantage, considering that the power of a test was, as Hays (1969, p. 270) put it, intended to reflect "the ability of a decision-rule to detect from evidence that the true situation differs from a hypothetical one". UMPU tests represent a more significant dilution of the original ideal, for they do not have maximum power. It is also worth mentioning that whatever interest UMP and UMPU tests may have is necessarily rather academic, since there are so few circumstances in which they exist. Far more serious, however, than any objection that might be mounted along these lines is the fact that the modifications introduced to deal with composite hypotheses are equally afflicted by the various difficulties that we have already shown to discredit the Neyman-Pearson method in the simple case.

## e CONCLUSION

The significance test in its various forms was created with a view to supplanting Bayes's Theorem as an instrument of inference, in the hope that statistical inductive reasoning could thereby be placed on an entirely objective and rational footing. We have argued that neither of these ideals is actually served and, hence, that significance tests provide a false basis for a theory of scientific inference.

It is instructive to note and ironic that while classical statisticians complain about the subjective aspect of Bayesian methodology—indeed, it is their chief complaint—the methods they advocate cannot operate except with their own, hefty subjective input. This is admitted by one of the founders of the classical approach, Egon Pearson:

> Of necessity, as it seemed to us [him and Neyman], we left in our mathematical model a gap for the exercise of a more intuitive process of personal judgement in such matters—to use our terminology—as the choice of the most likely class of admissible hypotheses, the appropriate significance level, the magnitude of worthwhile effects and the balance of utilities. (Pearson, 1966, p. 277)

Pearson might also have mentioned the subjectivity introduced by the dependence of significance tests on the stopping rule. Bearing in mind the classical objection to the element of subjectivity in the Bayesian scheme, one is put in mind of those people who while straining at a gnat would happily swallow a camel.

## EXERCISES

1. You are invited to comment critically on the following extract from Elston and Johnson's (1987, pp. 130–31) exposition of the significance test. (The *p-value* of some data, relative to a given null hypothesis, is the significance level of the test which would just lead to that hypothesis being rejected.)

   > It is a common but arbitrary convention to consider any value of *p* greater than 0.05 as "not significant". The idea behind this is that one should not place too much faith in a result that, by chance alone, would be expected to occur with a probability greater than 1 in

20. Other conventional phrases that are sometimes used are:

$0.01 < p < 0.05$: "significant"

$0.001 < p < 0.01$: "highly significant"

$p < 0.001$: "very highly significant".

This convention is quite arbitrary . . . : it is far more informative to quote the actual *p*-values. . . . Notice carefully the definition of *p:* the probability of observing what we actually did observe, or *anything more extreme,* if the null hypothesis is true. By "anything more extreme", we mean any result that would alert us even more (than the result we observed) to the possibility that our research hypothesis, and not the null hypothesis, is true. [The term 'research hypothesis' is used here to denote the alternative to the null hypothesis in a Neyman-Pearson test.]

**2.** Show that if data have a low *p*-value relative to a null hypothesis, then that hypothesis is not necessarily improbable.

**3.** Discuss the following claim: "Generally speaking conclusions from a large trial are more reliable than conclusions from a small one. Among the reasons are considerations of statistical power and of errors of the second kind" (Anderson, 1990, p. 147).

How are the errors of the second kind (type II errors) affected by the size of a trial? Can the intuition that the larger the sample, the more reliable the conclusion be understood in terms of type I and II errors? What would be the Bayesian explanation?

**4.** "Tests of significance do provide some assurance, in the form of significant results, that chance has been ruled out as an explanation" (Henkel, 1976, p. 87). Is this claim justified?

**5.** Henkel (1976, pp. 52–53) considers the result 6 tails, 14 heads obtained from tossing a coin, in relation to a null hypothesis which states that the probability of heads is $p = 0.5$. You are invited to comment critically on the following remarks which Henkel makes on how the result should be interpreted:

The probability of a result as extreme as 14 heads is . . . 0.1154. Since the level of significance chosen for our test was the 0.05 level, we would reject the null hypothesis if the probability of our result was less than 0.05 . . . Our result, 0.1154, is greater than 0.05 and thus we would accept it as resulting from chance factors when $p$ is 0.5. In other words, we fail to reject the null hypothesis. Put yet another way, this result (14 heads) is consistent with the hypothesis that $p = 0.5$, though it *does not prove* that $p$ is 0.5.

**6.** 9 heads and 31 tails appeared when a particular coin was tossed. Why is this information insufficient to allow any significance test to be performed?

**7.** "It is well known that, under repeated application of a significance test on accumulating data, a true null hypothesis will eventually be rejected with probability 1" (Jennison and Turnbull, 1990, p. 307). Why is it theoretically impossible to perform repeated significance tests on accumulating data?

**8.** Suppose that after every toss of a particular coin, the experimenter had pulled a playing card at random from an ordinary pack with the intention of calling off the coin trial as soon as the Queen of Spades appeared. What would be the outcome space for this trial?

**9.** You are invited to discuss the following table given in Bourke, Daly, and McGilvray, 1985, p. 71.

**TABLE X** Equivalent descriptions of a statistically significant result

| |
|---|
| Reject the null hypothesis. |
| Accept the alternative hypothesis. |
| There is strong evidence to doubt the null hypothesis. |
| The chance of the result being spurious is small. |
| $p < 5\%$ or $p < 0.05$.* |
| The observed result is not compatible with the null hypothesis. |
| Sampling variation is not sufficient to explain the observed result. |
| Result is unlikely to be due to chance. |

*$p$ here is the $p$ - value defined in Question 1.

**10.** "We may, on the basis of a test statistic, reject $H_0$ when it is in fact true (or at least plausible). This is called a type I test error . . ." (Strike, 1991, p. 163). Is the term 'plausible' appropriate in this context?

**11.** Comment on the following:

> The actual significance level achieved in a [clinical] trial . . . is best regarded as a measure of the conflict between the evidence and the null hypothesis, small values indicating that the conflict is great and that the alternative that the new treatment is superior is better supported. Large values indicate little or no conflict. For example, a significance level of 15% indicates that the apparent evidence in favour of the new treatment could have occurred purely by chance, when the treatments were equivalent, with probability 0.15. An event with such a high probability is quite believable, and so the evidence is inconclusive. (Whitehead, 1983, p. 111)

**12.** Note the following:

> [T]he chi-square test is criticized [by Howson and Urbach] because if 600 throws of a die show 100 'sixes', 'fives', 'fours' and 'threes' but 123 'twos' and 77 'ones' the resulting chi-square value of 10.58 is not significant on 5 degrees of freedom. But of course this is not a real die with 5 degrees of freedom but a philosopher's die (with a bias against ones but not in favour of sixes!) in which 4 degrees of freedom have been rigged to contribute to the overall psychological effect but not to the statistic. On the one honest degree of freedom, 10.58 is highly significant.

This is Senn's (1991, p. 1162) criticism of a part of the first edition of this book. The present edition deliberately retains the earlier discussion of the chi-square test unaltered. Is this justified?

# CHAPTER 10

# The Classical Theory of Estimation

## a INTRODUCTION

Thus far we have considered inferences, based on significance tests, in which a decision is taken as to whether or not "some predesignated value [of a parameter] is acceptable in the light of the observations" (Kendall and Stuart, 1979, p. 175). But many classical statisticians find such conclusions insufficient for scientific purposes, which, they believe, require "principles upon which observational data may be used to estimate, or throw light upon the values of theoretical quantities, not known numerically" (Fisher, 1956, p. 140). Or, as Kendall and Stuart put it, "[w]e require to determine, with the aid of observations, a number which can be taken to be the value of [some unknown parameter] $\theta$, or a range of numbers which can be taken to include that value" (1979, p. 1). In other words, something more precise is wanted than a fallible acceptance that $\theta$ exceeds some value or is unequal to a specific number. The Classical Theory of Estimation, created very largely by Fisher and Neyman, was designed to meet this need.

Scientists frequently estimate a physical quantity and in the process come to regard a certain number or range of numbers as a good approximation to the true value. For instance, the boiling point of a liquid would be estimated by averaging several carefully made thermometric readings. And the average height of people in a large population would be estimated by the mean of an appropriate sample. It is natural to adopt a more-or-less tentative attitude towards estimates arrived at in these ways and to view their accuracy as improving with the size and representativeness of the samples and with the perceived reliability of the measuring devices.

All this is explicable in Bayesian terms. Before the experiment, the scientist typically has only a rough idea of a parameter's value, and so his beliefs would be described by a rather diffuse prior probability distribution. That is, a relatively wide range of possible values would be roughly similar in probability, and no single value or (for the scientist's purposes) sufficiently narrow band of values would stand out as being very probable. But evidence from a well-designed experiment may induce a posterior distribution of probability that is more concentrated around a particular value, the relation between the posterior and prior distributions being expressed in Bayes's Theorem. In most cases, the larger and more representative the sample, the more the posterior probability crowds around some particular value; hence, the greater is the chance that the true parameter value lies in that neighborhood. (A more detailed account of the principles governing this process will be found in Chapter 14.)

Classical statisticians, on the other hand, have looked for principles of estimation which do not assume the unknown parameter to be a random variable and which do not rely on Bayes's Theorem. The resulting procedures of classical estimation have attained a high level of sophistication and now exert an enormous influence on statistical workers and on scientists concerned with statistical hypotheses. In our view, however, which we shall seek to substantiate, this influence is undeserved. We shall, as in our consideration of significance tests, enquire whether the classical principles of estimation are reasonable, and we shall also investigate whether they manage to avoid the kind of subjectivity which their founders and many practitioners found objectionable in the Bayesian system.

Classical estimation theory has two branches, known as *point estimation* and *interval estimation*. Point estimation aims to select a specific number as the so-called best estimate of a parameter; it is contrasted in the literature with interval estimation, a method of locating the parameter within a region and associating a certain degree of 'confidence' with the conclusion that is drawn. These two approaches employ different techniques, and we shall follow standard expositions by dealing with them separately.

## b POINT ESTIMATION

Point estimation is intended to solve the problem posed "when we are interested in some numerical characteristic of an unknown distribution (such as the mean or variance in the case of a distribution on the line) and we wish to calculate, from an observation, a number which, we infer, is an approximation to the numerical characteristic in question" (Silvey, 1970, p. 18). One could, of course, pick any arbitrary number and offer it as an estimate of a physical quantity, but this would not be a reliable technique in general, and in any particular case one would have no idea how accurate it was. Statisticians, on the other hand, would like "an estimate based on the sample [which is] generally better than a sheer guess" (Lindgren, 1976, p. 253). Of course, the Classical Theory does not suppose that estimates arrived at from samples are infallible. On the other hand, it denies that one can meaningfully assign a real, or objective, probability to the proposition that a parameter's true value equals or is close to its estimated value; and classical statisticians repudiate the idea that scientists should rely on subjective probability appraisals of theories.

The classical technique for estimating any population parameter, such as its mean, is the following. First, a random sample of predetermined size is drawn from the population and each element of the sample measured. Suppose $\mathbf{x} = x_1, \ldots, x_n$ denotes the measurements thus derived in a sample of size $n$. An estimating statistic, $t$, is selected, this taking the form of a calculable function $t = f(\mathbf{x})$. Finally, the inference is drawn that $t_o$, the value of $t$ derived from a particular experiment, is the best estimate of the unknown parameter. As classical statisticians deny that such an estimate can be qualified by the probability of its being accurate, they have invented methods by which to decide its quality indirectly. An estimate is said to be 'good' if the method which produced it is 'good'. So-called good methods of estimation, or good estimators, are ones that satisfy certain desiderata, the most frequently mentioned restrictions on methods of estimation being that they should be sufficient, unbiased, consistent, and efficient. We shall explain these concepts in turn and assess their supposed relevance to estimation.

### b.1 Sufficient Estimators

We introduced the notion of sufficiency earlier, in the context of Fisherian tests of significance. It will be recalled that sufficiency in a statistic is usually interpreted as signifying that the statistic contains all the relevant information about a particular parameter. This is a plausible interpretation, for a statistic $t$ is defined to be *sufficient* for $\theta$ when the probability $P(\mathbf{x} \mid t)$ is the same for all $\theta$, where $\mathbf{x} = x_1, \ldots, x_n$ is any element of the outcome space. In other words, once the value of $t$ has been fixed in an experiment, more specific details of the outcome will not depend on $\theta$. On these grounds, sufficiency is standardly incorporated as a criterion for selecting estimators for point estimation. Accordingly, classical statisticians endorse the sample mean, which unlike the sample median or range, say, is sufficient for estimating the population mean. (*Sample median*: if $n$ quantitative observations are arranged in increasing order, the sample median is the value taken by the observation half way along the series—this is the $(n + \frac{1}{2})$th observation if $n$ is odd, and the $(\frac{n}{2} + 1)$th observation if $n$ is even. *Sample range:* the difference between the highest and lowest scores amongst the observations.)

Bayesians also regard sufficient statistics as containing all the relevant information concerning a parameter. But while this must just be incorporated as an assumption into classical point-estimation theory, the Bayesian account has a proof for it. The Bayesian interpretation of the classical condition is that the statistic $t = f(\mathbf{x})$ is sufficient for the parameter $\theta$ just in case $P(\mathbf{x} \mid t \,\&\, \theta) = P(\mathbf{x} \mid t)$. This condition would, of course, not be regarded as meaningful in the classical scheme, for it relies on probability terms conditioned upon $\theta$ and thus treats the parameter as a random variable, which, classically speaking, is impermissible, since it is not itself the outcome of a repeatable experiment over which a physical probability distribution exists. However, the condition expresses in Bayesian terms the same idea which classical statisticians hoped the condition of sufficiency would capture in classical terms. Both conditions assert that $t$ is sufficient for $\theta$ just in case $\mathbf{x}$ is independent of $\theta$, once $t$ is given; in the former case, this independence is a functional independence, in the latter, it is probabilistic.

From the Bayesian point of view, sufficient statistics

sacrifice no relevant information, because the conclusion of a Bayesian inference, describing a posterior probability, is exactly the same whether one uses the raw data or a sufficient statistic based on those data. In other words, $P(\theta \mid t) = P(\theta \mid \mathbf{x})$, and, in fact, Bayesians often use this equation as the defining characteristic of sufficency. The proof is as follows: according to Bayes's Theorem, for all values of $\theta$, $\mathbf{x}$, and $t$

$$P(\mathbf{x} \mid \theta \,\&\, t) = \frac{P(\theta \mid \mathbf{x} \,\&\, t)P(\mathbf{x} \mid t)}{P(\theta \mid t)}.$$

In the continuous case, the corresponding form of the theorem employing densities must be substituted, the following argument proceeding mutatis mutandis. Since $\mathbf{x}$ uniquely determines $t$, so that $\mathbf{x} \,\&\, t$ is logically equivalent to $\mathbf{x}$, it follows that $P(\theta \mid \mathbf{x} \,\&\, t) = P(\theta \mid \mathbf{x})$. Combining this with Bayes's Theorem gives

$$P(\mathbf{x} \mid \theta \,\&\, t) = \frac{P(\theta \mid \mathbf{x})P(\mathbf{x} \mid t)}{P(\theta \mid t)}.$$

This implies that $P(\theta \mid \mathbf{x}) = P(\theta \mid t)$, just in case the condition for sufficiency, $P(\mathbf{x} \mid t \,\&\, \theta) = P(\mathbf{x} \mid \mathrm{t})$, holds. Therefore, sufficient statistics contain all the relevant information about $\theta$.

This argument proceeds from a theory of evidential or inductive relevance (that is, the Bayesian theory) to a demonstration that, as intuition affirms, only sufficient statistics contain all the information that is relevant for estimating a parameter. No analogous argument has been constructed from the Classical Theory of point estimation. Defenders of that theory seem forced to argue in reverse order to the Bayesian; that is, they must start from the intuition that only sufficient statistics include all the relevant information, and then incorporate sufficiency as a criterion for estimators. However, they have provided no evidence that the intuition is well-founded nor any sound explanation for the source of the intuition. In our opinion the source is none other than Bayes's Theorem, applied unconsciously.

**The Total Evidence Requirement.** There is a compelling intuition that in estimating a parameter, all the relevant evidence should be used; hence the interest in sufficient statistics. This intuition, formulated as a desideratum, is known as the Total Evidence Requirement (Carnap, 1947). Despite its natural persuasiveness, the requirement is often claimed to be inexplicable in terms of other inductive principles, the widely held view being that it can be incorporated into methodology only if asserted as a separate postulate. This is certainly the case when the methodology concerned is classical; it is not so with Bayesian methodology, as we explain.

Suppose you have two bits of information, $a$ and $b$, with which to evaluate an unknown quantity $\theta$. There are three posterior distributions to be considered: $P(\theta \mid a)$, $P(\theta \mid b)$, and $P(\theta \mid a \,\&\, b)$. If these probabilities differ, which should you, as a Bayesian, take as your current belief state? The answer is simple. The Bayesian has no choice, for $P(\theta \mid a)$ is your distribution of beliefs were you to learn $a$ and nothing else; $P(\theta \mid a \,\&\, b)$ is your distribution of beliefs if you came to know $a \,\&\, b$ and nothing else. But you have *not* learned $a$ and nothing else; you have learned $a \,\&\, b$ and nothing else. Therefore, your current belief state should be described not by $P(\theta \mid a)$, nor by $P(\theta \mid b)$, but by $P(\theta \mid a \,\&\, b)$. In other words, Bayesian reasoners are obliged to base their current beliefs on all the available relevant evidence; and since sufficient statistics include all the relevant evidence, it would be a mistake to calculate posterior distributions using non-sufficient statistics.

Classical statistics, in sharp contrast, must simply incorporate the Total Evidence Requirement and the claim concerning sufficient statistics into their canon, without justification or explanation. In our opinion, both derive from Bayes's Theorem.

### b.2 Unbiased Estimators

The notion of an unbiased estimator (which is different from that of an unbiased test discussed in the last chapter) is defined in terms of the expectation, or expected value, of a random variable. The latter, as already mentioned, is given by the expression: $E(x) = \Sigma x_i P(x_i)$, the sum (or, in the continuous case, the integral) being taken over all the values that $x$ can assume. As indicated earlier (Chapter 3, section **d**), the expectation of the random variable $x$ is also called the mean of the probability (or density) distribution of $x$. When $x$ is symmetrically distributed (according to the normal law, for example), the mean or expected value of $x$ is also the geometrical centre of the distribution.

A statistic $t$ is defined to be an *unbiased estimator* of a parameter if its expectation equals the parameter's true value. If $t$ is unbiased in this sense, then, other things being equal, the value of $t$ derived from a particular experiment is said to be a good estimate of the parameter.

Many estimators that would be favoured on intuitive grounds are, in fact, unbiased. For instance, the proportion of red counters in a sample is an unbiased estimator of the corresponding proportion in the urn from which they were randomly drawn. Similarly, a sample mean is an unbiased estimator of the corresponding population mean. However, although theory and what seems reasonable practice meet in many cases, we must consider whether this might be merely fortuitous or whether unbiasedness is necessary or generally desirable for estimators, as classical statisticians maintain.

The term that was chosen for the criterion we are discussing strongly intimates fair-mindedness and lack of prejudice: the suggestion clearly is that bias is bad in an estimator. So important is the criterion in the classical context that biased estimators are standardly 'corrected' and general methods have been invented for making such 'corrections for bias'. Thus Kendall and Stuart inform us that while intuition may indicate the sample variance as the appropriate estimator for the variance of the parent population, "that intuition is not a very reliable guide in such matters" (1979, p. 4). The alleged difficulty is that the sample variance is a biased estimator of the corresponding population variance. On the other hand, the 'corrected' statistic obtained by multiplying the sample variance by the factor $\frac{n}{n-1}$ is unbiased "and for this reason it is usually preferred [as an estimator] to the sample variance" (Kendall and Stuart, 1979, p. 5). But these bold claims on behalf of unbiased estimators are not easily sustained.

Consider how the classical statistician recommends we proceed. In estimating a parameter, we should select an unbiased estimator $t$. Then, if a particular experiment gives the result $t = t_o$, we are invited, other things being equal, to infer that the true or approximate value of $\theta$ is $t_o$, or, as Kendall and Stuart have said, $t_0$ "can be taken to be the value of $\theta$".

But there is no necessary connection between an estimator's unbiasedness and the propriety of reposing confidence in its specific estimates; any particular, unbiased estimate may, for all we know, be very inaccurate, and we are certainly not entitled to infer anything about its probability of being close to the truth. It is therefore not surprising that statisticians often defend the unbiasedness criterion in rather muted terms. For example, Barnett said that "within the classical approach unbiasedness is often introduced as a *practical* requirement to limit the class of estimators within which an optimum one is being sought" (1973, p. 120; our italics). Even Kendall and Stuart, who have written so confidently of the need to correct estimators if they are biased, conceded that the prominent place accorded to unbiasedness in classical estimation theory has no epistemic underpinning. Astonishingly, they admitted that

> There is *nothing except convenience* to exalt the arithmetic mean [i.e., the expectation] above other measures of location as a criterion of bias. We might *equally well* have chosen the median of the distribution of [an estimator] $t$ or its mode as determining the "unbiased" estimator. The mean value is used, as always, for its mathematical convenience. (Kendall and Stuart, 1979, p. 4; our italics)

The *mode* of a probability density distribution is the point, if it exists, of maximum density; and the *median,* as already said (Chapter 3, section **d**), is the point at which a vertical line bisects the area under the distribution. Kendall and Stuart's point is that one might just as well have defined an estimator as unbiased when its median or its mode, instead of its mean, is equal to the unknown population parameter. But each of these alternatives corresponds to a different criterion for estimation and is compatible with a different class of estimators; there can be no guarantee whatever that, in particular cases, estimates derived from estimators that are 'unbiased' in these different senses would be identical or even similar. So, according to classical statisticians themselves, an unbiased estimate is best only in a limited sense; it is not the best because it surpasses all others in accuracy, or probability, or rational credibility, but simply because it, allegedly, involves a minimum of tiresome mathematical computation.

Kendall and Stuart (1979, p. 4) confirmed that the bias-criterion is devoid of epistemic force when they warned readers that "the term 'unbiased' should not be allowed to convey overtones of a non-technical nature". But such overtones are inevitably conveyed, and the originators of the proposal that unbiasedness should serve as a standard for good estimation no doubt selected this suggestive name deliberately. Unfortunately, expositions of the classical point of view usually ignore the purely technical meaning of bias. If more neutral terminology had been adopted, this oversight would surely be encountered less frequently, and the criterion of bias would lose what plausibility it possesses and be relinquished as a standard against which our intuitions are judged correct or incorrect. The same might be said of the criterion to be discussed next.

### b.3 Consistent Estimators

The conditions of bias and sufficiency do not, by themselves, determine a unique estimator, hence the need for a number of further criteria. Amongst these is the criterion of *consistency*. An estimator is said to be consistent when its probability distribution shows a diminishing scatter about the true value as the sample size increases. More precisely, a statistic $t$ derived from a sample of size $n$ is a consistent estimator for $\theta$ if, for any arbitrary positive number $\epsilon$, $P(|t - \theta| \leq \epsilon)$ tends to 1, as $n$ tends to infinity. This condition is sometimes referred to as $t$ tending probabilistically to $\theta$. A consistent estimator is said by Kendall and Stuart to exhibit "increasing accuracy", which property they describe as "evidently a very desirable one" (1979, p. 3).

According to Fisher, consistency is the "fundamental criterion of estimation" (1956, p. 141). Indeed, he believed that non-consistent estimating statistics "should be regarded as outside the pale of decent usage" (1970, p. 11). And Neyman (1952, p. 188) agreed "perfectly" in this judgment, adding that "[w]hen one intends to estimate a parameter, . . . it is definitely not profitable to use an inconsistent estimate".

Fisher defended his emphatic preference for consistency with the following obscure and, to our mind, unpersuasive argument:

> . . . as the samples are made larger without limit, the [estimating] statistic will usually tend to some fixed value characteristic of the population, and, therefore, expressible in terms of the parameters of the population. If, therefore, such a statistic is to be used to estimate these parameters, there is only one parametric function to which it can properly be equated. If it be equated, to some other parametric function, we shall be using a statistic which even from an infinite sample does not give a correct value. . . . (1970, p. 11)

This argument was meant to establish that consistent estimators and no others should be employed—but it fails to defend that strong position. Fisher's claim is that because consistent estimators converge to a particular population parameter, they can 'properly be equated' with that parameter alone and not with any other. But he does not claim that this equation can properly be made *only when* the estimator is consistent. So Fisher's argument could not show the necessity of the consistency condition. Moreover, Fisher is surely wrong to claim that an estimate and the quantity being evaluated can ever be properly equated, for, as is granted on all sides, even in the most favourable circumstances, an estimate may be mistaken; indeed, as is often conceded, it is almost certain to be (section **b.4,** below).

Carefully formulated arguments for the consistency requirement do not seem to exist. Its necessity is normally just asserted as more-or-less obvious, and the plausibility of the assertion is, no doubt, due in part to a name which intimates unassailable virtue. However, consistent estimators are sometimes recommended on the grounds that they become more 'accurate' as the sample size increases, it being presumed (*see* section **b.4,** below) that an estimator's accuracy is connected with the spread of its probability distribution about the true parameter value. Any such argument would be like commending the use of a dirty measuring instrument on the grounds that if it were cleaner, its measurements would be more accurate and that in the limit, perfectly clean apparatus would give perfect accuracy. But that an estimate would have been better if a larger sample had been drawn or cleaner apparatus employed, implies nothing at all about its goodness in the case at hand, which surely is what one is, and ought to be, interested in.

We may illustrate this point with an example in which an 'inconsistent' method of estimation yields a perfectly satisfactory and confidence-inspiring estimate. Let the goal of the estimation be the mean of some population and imagine a scientist eccentrically selecting $\bar{x} + (n-100)\bar{x}^2$ as the estimating statistic, where, as before, $\bar{x}$ and $n$ are the sample mean and sample size, respectively. Clearly this odd statistic is not consistent (in the statistical sense), for it diverges ever more sharply from the population mean as the sample is enlarged. Nevertheless, for the special case where $n = 100$, the statistic is just the familiar sample mean, which on intuitive grounds gives a perfectly satisfactory estimate. This example illustrates the essential weakness of the classical principle that an estimate must be evaluated relative to the method by which it was derived.

A corollary of this principle is that an estimate's worth depends on who derived it. For suppose statistician $A$ employed the sample mean to estimate a population mean, while $B$ used some non-consistent and/or biased function of the sample mean; and imagine that they each arrived at identical estimates from the same sample. According to classical ideas, since these identical estimates have different pedigrees, they must be differently evaluated: one would be 'good', the other 'bad'! This, of course, contradicts the difficult-to-gainsay assumption that logically equivalent statements are equally 'good', an assumption enshrined in the probabilistic approach. It also violates the now somewhat battered objectivity ideal that supposedly guides the classical approach.

Hence, we see no merit in consistency (in the statistical sense) as a desideratum, even though many intuitively satisfactory estimators are in fact consistent.

### b.4 Efficient Estimators

It is intuitively appealing that the probability distribution of an estimator should have as narrow a spread about the true value of the quantity being estimated as possible. Hence a thermometer reading would command more confidence if made by a sober professional than by an inebriate layman, for the likely spread of results in the former case is smaller than in the latter. These preferences are understandable if referred to a Bayesian framework, as we shall see in Chapter 14.

They are also acknowledged in classical inference. Classical statisticians see the variance of an estimator as a guide to how good it is, and they express this in terms of the relative efficiency criterion. First they define one estimator to be the *more efficient* when its probability distribution about the true value has the smaller variance. In a natural extension of this idea, one may compare the variances of a statistic calculated from samples of different sizes. The variances of many statistics are inversely related to sample size, in which case their efficiency improves with a larger sample. Classical statisticians maintain that provided the other conditions we have mentioned are met, the more efficient an estimator, the better. For instance, Fisher held that the less efficient of two statistics is "definitely inferior . . . in its accuracy" (1970, p. 12), by which he seems to be saying that the more efficient an estimator, the closer must the estimate it gives be to the truth. But he could not really have meant this. For clearly, in the chancy process of estimation, one can never be sure of any degree of accuracy, even with the best estimating statistic.

A more characteristic classical defence is that given by Kendall and Stuart (1979, p. 7), who argued that since "[a]n unbiased consistent estimator with a smaller variance will . . . deviate less, on the average, from the true value than one with a larger variance . . . we may reasonably regard it as better". The idea here seems to be the following: suppose ${e^i}_1$ and ${e^i}_2$ are the estimates delivered by separate estimators on the $i$th trial, and suppose $\theta$ is the parameter's true value. Then if the first estimator is the more efficient, there is a high probability that $| {e^i}_1 - \theta | < | {e^i}_2 - \theta |$. Kendall and Stuart then translate this high probability into an average frequency in a long run of trials. As we have remarked before, this translation goes beyond logic, though it is quite plausible. Even overlooking this difficulty, the justification of the relative efficiency criterion in terms of average behaviour in a long run of estimations has nothing to say about the accuracy of particular estimates. Since estimates often form the basis of practical actions, and since they are usually expensive and troublesome to obtain, it is reasonable to demand an assessment of their accuracy.

## c INTERVAL ESTIMATION

In practice, this demand is evidently met, for estimates are not normally presented as specific numbers, as a point estimate has to be. This no doubt reflects the fact noted by Neyman (1952, p. 159) that in typical instances, "it is more or less hopeless to expect that a point estimate will ever be equal to the true value". Thus, estimates are normally presented as a range of numbers, in the form $\theta = a \pm b$, the suggestion being that it is reasonable to expect this range to contain the true value. A Bayesian would say that $\theta$ probably falls in the range. Neyman attempted to give a classical expression to this idea with his theory of estimation by confidence intervals, which he developed around 1930 and which is now the dominant theory in statistical estimation. We turn next to that theory.

### c.1 Confidence Intervals

We may explain the idea of a confidence interval through the problem we have already referred to, namely, how to estimate the mean height of the people constituting some large population. For simplicity, we shall include the condition that the standard deviation of heights in the population is already known. We shall refer to this known standard deviation as $\sigma$ and to the unknown mean as $\theta$. The data on which the estimate will be based are obtained by measuring the average height of a predetermined number, say, $n$, of people, sampled at random from the population. Clearly the sample mean, $\bar{x}$, can take many possible values, some more probable than others. The distribution representing this situation is approximately normal for large samples (the larger the sample, the closer the approximation), and its own mean is $\theta$; the standard deviation of the distribution (also known as the 'standard error' of the sample mean), denoted by $\sigma_{\bar{x}}$, is a function of the sample size and the population standard deviation, but is independent of the population mean; in fact $\sigma_{\bar{x}} = \sigma n^{-\frac{1}{2}}$.

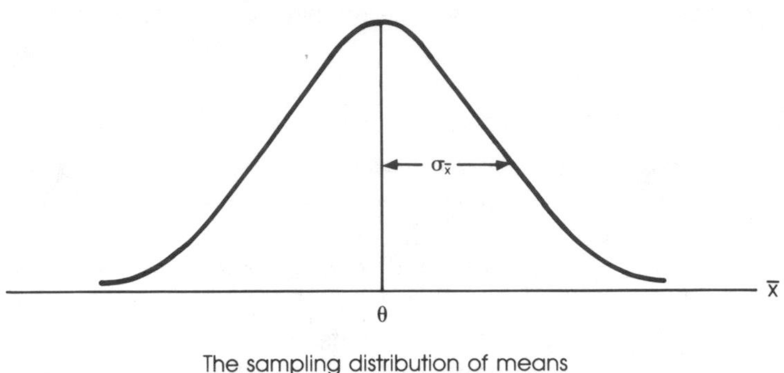


The sampling distribution of means from a population with mean θ and standard deviation σ.

The above distribution is a plot of possible sample means against probability densities, not probabilities; and as explained earlier, this signifies that the probability that $\bar{x}$ lies between any two points is proportional to the area enclosed by those points and the curve. Because the distribution is essentially normal, it follows that with probability 0.95,

$$-1.96\sigma_{\bar{x}} \leq \theta - \bar{x} \leq 1.96\sigma_{\bar{x}}.$$

Rearranging these inequalities gives the result that with probability 0.95,

$$\bar{x} - 1.96\sigma_{\bar{x}} \leq \theta \leq \bar{x} + 1.96\sigma_{\bar{x}}.$$

Suppose that $m$ is the value of $\bar{x}$ that is actually observed in an experimental sample. Then, because we know $\sigma$ and $n$, the terms $m - 1.96\sigma_{\bar{x}}$ and $m + 1.96\sigma_{\bar{x}}$ can be computed; the interval between these two values is called a *95 per cent confidence interval* for $\theta$. Clearly, the mean of a given sample implies other 95 per cent confidence intervals, relating to other regions of the above distribution curve, as well as confidence intervals associated with different probabilities. The probability associated with a confidence interval is known as its *confidence coefficient.*

The probability statements given above are unquestionably correct. As Neyman (1952, p. 209) remarked, they "were *deduced* from the specified assumptions regarding the observable random variables and, therefore, are the result of *deduc-*

*tive reasoning*". What we have said about confidence intervals is also uncontroversial, since it amounts to no more than a definition of that concept. Controversy arises only when confidence intervals are assigned inductive meaning and employed to estimate the value of the unknown parameter. There are two principal proposals for how such estimates are possible: what we shall call the categorical-assertion interpretation and the subjective-confidence interpretation.

### c.2 The Categorical-Assertion Interpretation of Confidence Intervals

This is an interpretation, invented by Neyman, which we have already encountered in the context of the significance test. (*See* Chapter 8, section **c,** and Chapter 9, section **c.1.**) The interpretation is widely supported, for example by Cramér (1946, pp. 512–13), and Kendall and Stuart regard it as "basic to the theory of confidence intervals" (1979, p. 110). Neyman's proposal was that the "practical statistician", in estimating a parameter, should proceed in three steps: first, he should perform the random experiment; secondly, he should calculate the corresponding confidence interval; and, finally, "he must *state*" that the true value of the parameter lies between the two confidence bounds (Neyman, 1937, p. 263; our italics). And while acknowledging that it is never certain that the parameter really lies within the confidence interval, Neyman thought it reasonable for the statistician to state that it does, because (when the confidence coefficient is 0.99) "in the long run he will be correct in about 99 per cent of all cases" (Neyman, 1937, p. 263).

Neyman was careful to observe, however, that when asserting that the parameter is contained in the interval, the statistician is not entitled to conclude or believe that this really is true. Indeed, Neyman held the idea of inductive reasoning to a conclusion or belief to be contradictory, on the grounds that reasoning denotes "the mental process leading to knowledge" and this "can only be deductive". Hence, Neyman claimed that induction could not be a result of reasoning, but was essentially a matter of behaviour; in particular, the proper result of an interval estimate was, he believed, a decision "to behave as if we actually knew" that the true value of the parameter was between the confidence bounds (Neyman, 1941, p. 379).

We argued earlier that no scientist either would or should act as if an uncertain estimate was definitely correct. Typically, theories are evaluated by degree, as the writings and behaviour of scientists amply show. Neyman has argued, on the other hand, that it is sometimes reasonable to act as if one's real beliefs were false or as if what one believed to be false were true. According to Neyman, this is just what we do when we purchase accident insurance before a holiday. In making such a purchase, he said, "we surely act against our firm belief that there will be no accident; otherwise, we would probably stay at home" (1941, p. 380). This seems a perverse analysis of the situation. Reasonable people do not have an absolute conviction that there will be no accident; they believe that there is a larger or smaller chance of an accident, depending upon the type of holiday and on other factors, and they act in accordance with that belief, for example, by proportioning the amount they are prepared to pay for insurance to the degree of risk perceived.

A point in favour of Neyman's categorical-assertion interpretation seems to be that it is objective, since it makes no explicit reference to degrees of confidence, so reminiscent of Bayesian subjective probabilities. But the appearance is deceptive. For suppose some experimentally measured range of numbers constituted a 90 per cent confidence interval for $\theta$, rather than the standardly approved 95 or 99 per cent, one would still be enjoined to assert that $\theta$ is in that range, though with a correspondingly modified justification that now referred to a 90 per cent frequency of making a correct decision. But if the justification had any force at all (we have seen that it does not), it would surely be stronger the lower the ("long run") frequency of error. Hence the categorical assertion that $\theta$ is in some interval must after all be tacitly qualified by an index running from 0 to 100 indicating how 'good' or 'accurate' or 'reliable' it is, and this is hard to distinguish from the index of confidence that is explicit in the subjective-confidence interpretation, with which we deal next.

### c.3 The Subjective-Confidence Interpretation of Confidence Intervals

This much-favoured interpretation claims that we should conclude that the parameter is enclosed in a confidence interval not with complete certainty but with a 'degree of confidence' equal to the confidence coefficient. Gunst and Mason (1980, p. 202), for example, describe a confidence interval as an interval "within which we feel reasonably certain the unknown parameter . . . lies". And Mood (1950, p. 222) held a confidence coefficient to be "a measure of our confidence in the truth of the statement [that the parameter lies within the experimentally computed confidence interval]". Similarly, Harnett (1970, p. 199) maintained that the method of confidence intervals allows us "to assert with a certain degree of confidence that the . . . parameter will fall within . . . [a specified] interval", thus overcoming, in his view, "[o]ne of the major weaknesses of a point estimate [namely] . . . that it does not permit the expression of any degree of uncertainty about the estimate".

The subjective-confidence interpretation has some plausibility. For consider again the task of estimating a population mean $\theta$. We know that in experiments of the type described, $\theta$ is included in any 95 per cent confidence interval with an objective probability of 0.95; and this implies that if one sampled the population repeatedly, calculating 95 per cent confidence intervals in the light of each new sample mean, $\theta$ would be included in those intervals with a relative frequency that would, in the limit, tend to 0.95. It is natural, almost irresistible, to infer from this limit property a probability of 0.95 that the *particular* interval obtained in a *particular* experiment does enclose $\theta$.

Some statisticians endorse that inference. For example, Lewis-Beck (1980, p. 38) examined data on people's annual income and the length of their formal education, from which he calculated a 95 per cent confidence interval for the unknown income of a randomly chosen person with exactly ten years academic training. He interpreted this as showing that "there is a 95% probability that . . . [that person] has an annual income between . . . [the confidence limits]".

But no such probability statement is inferable, for what

could the probability be? It is either an objective, limiting-frequency property—but this is out of the question in the present case, since we are dealing with a singular event—or it is a degree of inductive or rational belief or subjective confidence, but this, we shall argue, is also inadmissible.

The subjective-confidence interpretation seems to be an application of a rule of inference often called the Principal Principle, which is used extensively in Bayesian statistics. The principle states that if the objective, physical probability of a random event (in the sense of its limiting relative-frequency in an infinite sequence of trials) were known to be *r*, and if no other relevant information were available, then the appropriate subjective degree of belief that the event will occur on any particular trial would also be *r*. If the event in question is *a*, the Principal Principle says that $P[a_t \mid P^*(a) = r] = r$, where $a_t$ is a statement describing the occurrence of the event on a particular trial, $P^*(a)$ is its objective probability, and $P$ is a subjective probability function.

For example, the physical probability of getting a number of heads greater than 5 in 20 throws of a fair coin is 0.86 (*see* Table I, Chapter 8). That is, $P^*(K > 5) = 0.86$, where $K$ is the number of heads obtained. According to the Principal Principle, $P[(K > 5)_t \mid P^*(K > 5) = 0.86] = 0.86$, so 0.86 is also the confidence you should place in any particular trial of 20 throws of a fair coin producing a number of heads greater than 5.

Suppose a trial is made and 2 heads are found in a series of 20 throws with a coin that is known to be fair. To infer that we should now be 86 per cent confident that 2 is greater than 5 would be absurd and a misapplication of the Principal Principle. If one could substitute numbers for $K$ in the Principle, it would be hard to see why the substitution should be restricted to the term's first occurrence. But no such substitution is allowed. For the Principal Principle does not assert a general rule for each number $K$ from 0 to 20; the $K$-term is not in fact a number, it is a function which takes different values depending on the outcome of the underlying experiment.

Mistaking this appears to be the fallacy implicit in the subjective-confidence interpretation of confidence intervals. It is true that the objective probability of $\theta$ being enclosed by experimentally determined 95 per cent confidence intervals is 0.95. Let $I_1$ and $I_2$ be variables representing the boundaries of such confidence intervals. By the Principal Principle, $P[(I_1 \leq \theta \leq I_2)_t \mid P^*(I_1 \leq \theta \leq I_2) = 0.95] = 0.95$, which tells us to be 95 per cent confident that any particular performance of the experiment will produce an interval containing $\theta$. Suppose $I_1'$ and $I_2'$ are the values of $I_1$ and $I_2$ obtained from a particular experiment; the subjective-confidence interpretation says that we should now be 95 per cent confident that $I_1' \leq \theta \leq I_2'$.

But as the simple counter-example above shows, this is not a legitimate inference from the Principal Principle. For $I_1$ and $I_2$ are not numbers, but functions of possible experimental outcomes. The principle, therefore, does not license the substitution of numbers for $I_1$ and $I_2$; so the desired inference from experimentally measured intervals to subjective confidences is blocked. (The misinterpretation of the Principal Principle described here was first pointed out in a different context by Howson and Oddie, 1979, and is further discussed in Chapter 15, section **e**, below.)

In reply to this criticism, it might be said that the subjective-confidence interpretation does not depend on the Principal Principle (a Bayesian notion, anyway), that it is justified on some other basis. But so far as we know, no such justification has been presented; moreover, we think none is possible, because, in our view, the interpretation is wrong, for one reason, because it implies that one's confidence in a proposition depends upon factors which manifestly are not and, we think, should not be influential.

### c.4 The Stopping Rule Problem, Again

Confidence intervals arise from probability distributions over spaces of possible outcomes. In the example we have used, the outcome space consisted of the possible means of random samples from the population, all of which are assumed to have the same size, *n*. But unlike the true outcome, the outcome space as a whole is imaginary. It comprises the set of possible results of an experiment conducted in some intended (and feasible) manner. Clearly the same outcome could have resulted from many different experimental intentions; so, before the evidential force of any outcome could be established, one would need to discover from the experimenter what plan motivated the experiment—was the intention to draw a sample of size *n*, or to continue sampling until a simultaneously spun coin produced three heads or until boredom set in, or what? And if the sampling was merely casual and the decision to stop unpremeditated, there would be no definite sample space, and so no confidence interval; hence, no inference could be drawn from the data so gathered. But it seems most implausible and unrealistic that anyone's confidence in an estimate would or should in general be influenced by a knowledge of such facts about the experimenter's state of mind. The subjective-confidence interpretation of confidence-interval statements is in conflict with these, to our mind compelling, intuitions. Until reasons are brought forward that would dispel those intuitions, the interpretation, we believe, should be rejected.

The implausibility of the claim that estimates should be allowed to depend on the unrealised possibilities of a notional outcome space, rather than on the actual outcome alone, as Bayesians recommend, is vividly brought home by Pratt in the following instructive anecdote.

> An engineer draws a random sample of electron tubes and measures the plate voltages under certain conditions with a very accurate voltmeter, accurate enough so that measurement error is negligible compared with the variability of the tubes. A statistician examines the measurements, which look normally distributed and vary from 75 to 99 volts with a mean of 87 and a standard deviation of 4. He makes the ordinary normal analysis, giving a confidence interval for the true mean. Later he visits the engineer's laboratory, and notices that the voltmeter used reads only as far as 100, so the population appears to be 'censored'. This necessitates a new analysis, if the statistician is orthodox. However, the engineer says he has another meter, equally accurate and reading to 1000 volts, which he would have used if any voltage had been over 100. This is a relief to the orthodox statistician, because it means the population was effectively uncensored after all. But the next day the engineer telephones and says, "I just discovered my high-range voltmeter was not working the day I did the experiment you analysed for me". The statistician ascertains that the engineer would not have held up the experiment until the meter was fixed, and informs him that a new analysis will be required. The engineer is astounded. He says, "But the experiment turned out just the same as if the high-range meter had

been working. I obtained the precise voltages of my sample anyway, so I learned exactly what I would have learned if the high-range meter had been available. Next you'll be asking about my oscilloscope". (Pratt, 1962, pp. 314–15)

The engineer is astounded, of course, by the suggestion that the various revelations that so impressed the statistician should disturb his own confidence in the initial estimate. The engineer's response would surely be that of any reasonable scientist. One would require very persuasive reasons to override the natural and traditional attitudes of experienced and successful scientists, reasons which are not forthcoming from the classical theory of confidence intervals.

### c.5 Prior Knowledge

Estimates are frequently made against a background of incomplete knowledge and partial indications about the value of the parameter in question. Suppose, for example, you were interested in discovering the average height of students attending the London School of Economics. Without being able to point to results from carefully conducted studies, but on the basis of common experience and of what you have learned informally about students and the admission criteria of British universities, you no doubt feel pretty sure that the average height of the students is not less than $4\frac{1}{2}$ feet, say, nor above 6 feet. If you had already made an exhaustive survey of the students' heights, had lost all the results, and could recall only that the average was more than 5 feet, your prior distribution of uncertainty would be even more focussed. But as Schlaifer (1959, pp. 665–66), for example, has emphasised, there seems no way of combining informal knowledge such as this with classical estimation methods. So if, in our example, an experimental sample, by chance, produced a 95 per cent confidence interval of $4'0'' \pm 2''$, and if you acted in accordance with classical principles, you would be obliged to repose an equivalent level of confidence in the proposition that the average height of students of this school really is in that interval. But in the light of all you know in addition to the sample information, this clearly would not be a reasonable or credible conclusion.

The classical response to this difficulty could take one of

two forms, neither adequate, we believe. It might claim that classical estimation is applicable only when no relevant information regarding the parameter is available. This, however, would considerably curtail the scope of estimation methods, for there are very few cases, if any, where one can be said to know nothing at all about a parameter's value. And although a little knowledge is certainly a dangerous thing, it would be odd, to say the least, if it condemned its possessor to a state of perpetual ignorance. A second possibility would be to combine in some way informal prior information with formal estimates based on random samples. But it is hard to see how this could be done within the confines of classical methodology. Indeed, if the categorical-assertion interpretation were taken seriously, there would even be difficulties with reconciling two classical estimates if they conflicted, for each separate estimate is supposed, on that interpretation, to be accepted as if correct.

Thus, classical estimation is prone to deliver estimates which prior knowledge would deem unreasonable and improbable, or even definitely false, and it seems to have no facility for revising estimates by taking account of such knowledge. The Bayesian approach meets no comparable difficulty, for it makes explicit provision for prior information in its estimation procedure.

#### c.6 The Multiplicity of Competing Intervals

As we indicated earlier, confidence intervals calculated for any given confidence coefficient are not unique. Indeed, it is easy to see from the diagram of a sampling distribution of means given earlier that indefinitely many regions of that normal distribution cover 95 per cent of its area. Hence, there are infinitely many 95 per cent confidence intervals; and this is generally the case when the sampling distribution is continuous. So, for instance, rather than the usual confidence interval concentrated in the centre of the distribution, one could consider intervals extending further into the tails but omitting a smaller or larger strip in the centre. The multiplicity of equally probable confidence intervals plays havoc with the type of conclusion recommended by Neyman's categorical-assertion interpretation; for since every possible sample value falls into some 95 per cent interval and outside others, unless the permissible confidence intervals were more strictly confined, we would (in Neyman's words) have to "behave as if we actually knew" that the parameter both has some value and fails to have any value!

Defenders of the Neyman version of confidence intervals have tried to meet this difficulty in one of two ways, neither, we believe, satisfactory. The first looks to the length of a confidence interval for a principle of discrimination. Intervals corresponding to a given confidence coefficient may vary in length, and it is frequently held that the shortest interval is best. Thus, when a population mean is estimated in the manner indicated earlier, the centrally symmetrical interval given by $m \pm 1.96\sigma_{\bar{x}}$ is always the preferred 95 per cent interval, on the grounds that it is the shortest interval for that confidence coefficient. Thus Cramér (1946, p. 513) held that it will "[o]bviously . . . be in our interest to find rules which, under given circumstances, yield as *short* confidence intervals as possible", and Hays (1969, p. 290) claimed that there is "[n]aturally . . . an advantage in pinning the population parameter within the narrowest possible range with a given probability".

However, it is hard to see how a preference for the shortest interval could be rationalised, in view of fact that Neyman's only justification for his categorical-assertion rule—that repeated applications of the rule would rarely lead one into error—applies to broad as well as narrow intervals with equal force (though, as we have seen, 'force' is scarcely the right word). Nevertheless, it is standardly held that by restricting inferences to the narrowest confidence intervals, estimates gain precision without sacrificing the associated confidence level. This was, for example, Mood's view (1950, p. 222): in comparing two 95 per cent confidence intervals, he stated that one of them was "inferior" because of its greater length, for "it gives less precise information about the location" of the parameter.

But this is questionable, for when statisticians accept that $\theta$ lies in the range $\{a,b\}$, they necessarily accept that $f(\theta)$ lies in the range $\{f(a), f(b)\}$; and the same applies, mutatis mutandis, under the subjective-confidence interpretation. But while the first interval might be the shortest in a set of intervals for $\theta$, the second might not be the shortest in the corresponding set for $f(\theta)$. If the length of a confidence interval is a measure of its precision, the former interval must then be more precise than the latter. But if $f$ is a 1-1 function, so that $a$ uniquely determines $f(a)$ and vice versa, then the information contained in the confidence intervals $\{a,b\}$ and $\{f(a), f(b)\}$ must be the same; hence that information must be equally precise. And this implies that the minimum-length criterion for the precision of a confidence interval is mistaken.

Another difficulty is this: although the minimum-length criterion fixes a unique confidence interval for a given estimator, it leaves open which statistic should be employed for the estimation. The trouble is that different shortest confidence intervals may often be calculated from a variety of sample statistics. Clearly, the criterion based on the width of intervals needs to be supplemented by a further restriction, and it is usually suggested that the preferred statistic should have a minimum variance.

In justifying this new condition, it is argued that although confidence intervals constructed with a minimum-variance statistic are not necessarily the shortest in any particular case, such statistics have a higher probability than any other of producing the shortest interval. As many statisticians misleadingly express this conclusion, confidence intervals based on the preferred statistic "are shortest on average in large samples" (Kendall and Stuart, 1979, p. 126). We have already criticized both the long-run justification and the short-length criterion; since two wrongs don't make a right, nothing more needs to be said.

Neyman suggested a different criterion for selecting appropriate confidence intervals. He argued, in a manner familiar from his theory of testing, that a given confidence interval should not only have a high probability of containing the correct value but should also be relatively unlikely to include wrong values. This is more precisely formulated as follows: a best confidence interval, $I_0$, should have the property that, for any other interval, $I$, corresponding to the same confidence coefficient $P(\theta' \in I_0 \mid \theta) \leq P(\theta' \in I \mid \theta)$, where $\theta' \in I$ means that $\theta'$ is a member of the set of points constituting the interval; moreover, the inequality must hold whatever the true value of the parameter and for every other value $\theta'$ different from $\theta$ (Neyman, 1937, p. 282). But as Neyman himself showed, there are no so-called best intervals of this kind for most of the cases with which he was originally concerned. In particular, where the probability distribution of the experimental outcomes, governed by the unknown parameter, is continuous, no such best interval exists.

## d PRINCIPLES OF SAMPLING

### d.1 Random Sampling

We have been considering how to estimate population characteristics from information supplied by samples drawn from the population in question. As we have seen, classical point- and interval-estimation methods require a knowledge of the objective probability distribution of the sample data (the sampling distribution); and the chief way of making such knowledge available is by ensuring that the sample is objectively random. What this means is that the sample must be collected in a way that ensures that each element of the population has a known, predetermined, objective chance of being included in the sample. This can be achieved, for example, by placing cards corresponding to each member of the population into a bag, shaking it up thoroughly, and then selecting cards blindfold, one at a time: the names on the sampled cards would represent a random sample from the population. Many other, more sophisticated ways exist of selecting random samples.

What we shall call the *Principle of Random Sampling* asserts that satisfactory estimates can only be obtained from samples that are random in the sense indicated. Our discussion follows Urbach, 1989.

### d.2 Judgment Sampling

The Principle of Random Sampling may be contrasted with another approach, which is motivated by the wish to obtain a *representative sample,* that is, a sample that reproduces in miniature the characteristics which it is desired to estimate. So, for example, if the aim is to measure the average height of a large population or the proportion of its members intending to vote Conservative in a forthcoming election, the ideal sample would have the same mean height or the same percentage of intending Conservative Party voters as the parent population. Selecting a representative sample is not straightforward, however, for if you don't yet know what the population value is, how can you ensure that the population and the sample have the same value?

The standard response is that of course you cannot be sure but that a sample is likely to be more-or-less representative when it resembles the population in respects that are believed to correlate with the characteristic whose value is being sought. So, for example, voting preference is generally agreed to be related to age and socio-economic class: hence, a representative sample would have to recapitulate the population in its age and class structure; quite a number of other factors, such as gender and area of residence, would, no doubt, also have to be taken into account in constructing a representative sample. Samples selected in this manner, with a view to achieving representativeness, are often called *Judgment,* or *Purposive, Samples.*

A kind of judgment sampling that is frequently resorted to in market research and opinion polling is known as *Quota Sampling*. Interviewers are provided with target numbers of people to interview in various categories, such as particular social classes and geographical regions, and are then invited to exercise their own good sense in selecting representative groups within each specified category.

### d.3 Objections to Judgment Sampling

Although it may seem reasonable to base an estimate on a representative sample, judgment sampling is held to be highly unsatisfactory by very many statisticians, particularly those of the classical stripe, who claim that random samples alone are acceptable. Three related objections are often encountered. The first objection is that judgment sampling introduces an undesirable degree of subjectivity into the estimation process. A subjective element does exist, for judgment sampling requires a view to be taken as to the relevant respects in which the sample and the population should be matched. Judgments must be made, for example, on whether a person's social class or the condition of his front garden or the appearance of his cat, and so forth, are related to the characteristics that are desired to be measured. Without making an exhaustive survey of the population, you would be unable to decide categorically on the relevance of the innumerable, potentially relevant factors; there is, therefore, considerable freedom for judgments to vary from one experimenter to another. This is contrasted with random sampling, which requires no individual judgment and is quite impersonal and objective.

The second objection, which is really just an aspect of the first, is that judgment samples are exposed to the danger of being biased, due to the experimenter's ignorance, or through the exercise of conscious or unconscious personal prejudices. Yates (1981, pp. 11–16) has illustrated this possibility with a number of cases in which the experimenters' careful efforts to select representative samples were frustrated by their failure to appreciate and take into account crucial variables. Such cases are often held up as a warning against the bias that can intrude into judgment sampling. Sampling by a random process, on the other hand, cannot be affected by the selector's partiality or lack of knowledge.

But random sampling does not remove bias in every sense of the term, for it may by chance produce samples that are as unrepresentative as any that could result from the poorest judgment sampling. The standard classical response attempts to turn this apparent difficulty into a principal advantage of random over judgment sampling, for it is argued that when the sampling is random, the probabilities of different possible samples can be accurately computed and, using classical estimation methods, systematically incorporated into the inference process. However, judgment sampling—so the third objection goes—does not lend itself to the objective methods of estimation advocated by classical statisticians.

There is a second aspect to the classical response, based on the idea of *stratified random sampling*. This involves partitioning the population into separate groups—strata—and then sampling at random from each. The classically approved estimate of the population parameter is then the weighted average of the corresponding strata estimates, where the weighting coefficients are proportional to the relative sizes of the strata and population. Provided the strata are relatively more homogeneous than the population and significantly different from one another compared with the quantity being measured, estimation through stratified random sampling is more 'efficient' (in the sense defined above) than estimation using ordinary random samples, and so, by classical standards, it is better. Stratified random sampling is clearly a means of avoiding unrepresentative samples, which is generally intuitively advisable.

The classical rationale for stratifying (and implicitly for using relatively representative samples) is questionable, though; indeed, it seems quite wrong. The efficiency of an estimator, it will be recalled, is a measure of its variance. And the more efficient the estimator, the narrower is any confidence interval based on it. So, for instance, a stratified random sample might deliver the 95 per cent confidence interval 4′ 8″ ± 0′ 3″ as an estimate of the average height of pupils in some school, while the corresponding interval derived from an unstratified random sample (which by chance is heavily biased in the direction of the youngest children) might be, say, 3′ 2″ ± 0′ 6″. Classical statisticians seem committed to saying that the first interval estimate is the better one because of its narrower width. But surely this misdiagnoses the fault. It would be more natural to say that the first estimate is probably right and the second almost certainly wrong, which of course cannot be said by a classical statistician.

### d.4 Some Advantages of Judgment Sampling

We have considered certain frequently claimed advantages of random over judgment sampling, in particular, its impartiality; in one respect, however, it is often at a clear disadvantage, namely in practicability. Consider, for example, opinion polls held before an election. These usually need to be conducted quickly. But assembling a random sample, finding the subjects concerned, and persuading them to be interviewed is time consuming, costly, and sometimes impossible, and the election might well be over before the poll has begun. Practical considerations such as these dictate that around 90 per cent of all market research by interview is based on quota samples, according to a recent survey reported by Downham (1988, p. 13), who, however, does not disclose what kind of sample led to this conclusion.

Another point seeming to favour quota sampling is its apparent success. The results of opinion polls, insofar as they can be checked against the outcomes of ensuing elections, are for the most part more-or-less correct, and market research firms are profitable and thrive, their services evidently valued by manufacturers who have the greatest interest in accurately appraising consumers' tastes.

Furthermore, inferences based on non-random samples are often confidently made and believed by others; indeed they seem inevitable in inductive reasoning, where conclusions derived from one sphere often need to be applied to another. For instance, a recent study (Peto, et al., 1988) showed that in two large groups of physicians, only one of which had regularly taken aspirin, the frequencies of heart attacks over a longish period were similar. On the basis of this result, the authors of the study advised against adopting aspirin as a general prophylactic, their implicit assumption being that doctors, in particular the doctors who took part in the study, typified the general population in their cardiac responses to aspirin. Although the rest of the statistical procedures employed in the study were orthodox, this assumption was not checked by means of random samples from the general population, but seems simply to have been adopted on the basis of its high plausibility. (A similar example is used by Smith, 1983, to illustrate the same point.)

In short, judgment sampling in its various forms is convenient, evidently successful, widely practised, and inevitable in scientific reasoning.

Judgment sampling also escapes a strange and disturbing implication of the Principle of Random Sampling. The extreme oddity of the implication is acknowledged even by enthusiastic supporters of the principle, such as Stuart (1962, p. 12), who called it the Paradox of Sampling, though it would be more correctly termed the *Paradox of Random Sampling*. The paradox is that according to the random-sampling principle, while a sample that was drawn by an experimenter applying his or her judgment would be unacceptable for the purposes of estimation, the *very same* sample, had it been generated by a random process, would have been perfectly all right. Stuart admitted that this is a hard pill to swallow, but concluded: "swallow it we must", because only if we do are we able to exploit the correct methods of statistical inference, by which he meant, of course, the classical methods of estimation.

But the burden of the present chapter has been that those

classical methods are anchored in neither logic nor good sense and that they are ill-adapted to the scientific task of estimation. Consequently, they are not a suitable foundation upon which to rest the Principle of Random Sampling, and thus they fail to furnish convincing reasons for swallowing the manifestly indigestible consequences of that principle.

Estimation along Bayesian lines, by contrast to the classical, while not disallowing random samples, does not require them and may be perfectly well applied to appropriate judgment samples, as we shall explain in Chapter 14.

## ■ e CONCLUSION

The Classical Theory of Estimation was designed to meet the need for more informative inferences from experimental data than can be got from tests of significance and to achieve this without reference to Bayes's Theorem and the inevitable element of subjectivity involved in assigning prior probabilities to hypotheses.

Point estimation of parameters issues in precise numerical values. But advocates of classical point estimation are normally quite vague when it comes to explaining the sense in which those values estimate the parameter in question. Although the statement that $t_0$ is a 'good', or the 'best', estimate of $\theta$ sounds quite definite, its meaning is hard to fathom. Kendall and Stuart, who are among the few statisticians who have attended to this question, apparently held that a point estimate is the assertion that the true value and the estimated one are the same, for they claimed that a point estimate is a "number which can be taken to be the value of" the parameter in question. A similar claim is often made for interval estimates, namely, that they must be unequivocally accepted as true, or at least that people should shape their behaviour as if they accepted them as true. But for the reasons we have given, we regard such advice as unrealistic and unreasonable.

Interval estimates, which are expressed in terms of a range of numbers, are thought by most classical statisticians to get round the objection made against point estimates, that being absolutely precise, they are almost certainly false. Interval estimates also associate different ranges of values with their own confidence coefficients, which are often represented as measures of the confidence appropriate to the conclusion; but the idea that they provide such a measure is quite erroneous and appears to be based on a logical fallacy.

We must conclude, then, that confidence intervals simply cannot function as estimates and that they are devoid of inductive significance. We shall see in Chapter 14 that the situation may be remedied, but only if prior probabilities are brought into the picture and the classical approach abandoned.

Although they are thoroughly fallacious, the methods of significance testing and classical estimation are still being advocated in hundreds of books, required texts in thousands of institutions of higher education, where hundreds of thousands of students are obliged to learn them.

## EXERCISES

**1.** "A confidence interval is to be interpreted as follows: if we were to find many such intervals, each from a different sample but in exactly the same fashion, then in the long run about 95% of our intervals would include the true mean and 5% would not. We cannot say that there is a 95% probability that the true mean lies between the two values we obtain from a particular sample, but we can say that we have 95% confidence that it does so" (Elston and Johnson, 1987, p. 115).

Evaluate the three main claims contained in this statement and consider whether it constitutes a satisfactory interpretation of a confidence interval.

**2.** Let $\mu$ be the unknown average level, expressed in appropriate units, of some enzyme in a certain human population and assume that the enzyme levels in the population are normally distributed with a variance of 45. A sample of 10 persons from the population has a mean enzyme level of 22. What conditions must be satisfied by the sample, and the person who drew the sample, in order for a confidence interval to be derived from it? Assume that the conditions are met and calculate a 95 per cent confidence interval for $\mu$. Why must the assumption of normality be an approximation only?

3. Suppose that in a study comparing samples of 100 diabetic and 100 non-diabetic men of a certain age, a difference of 6.0 mm Hg was found between their systolic blood pressures and that the standard error of this difference between sample means was 2.5 mm Hg. Show that the shortest 95 per cent confidence interval for the population difference between the means is from 1.1 to 10.9 mm Hg. Then comment on Gardner and Altman's (1989, pp. 8–9) gloss on this confidence interval:

   > Put simply, this means that there is a 95% chance that the indicated range includes the "population" difference in mean blood pressure levels—that is, the value which would be obtained by including the total populations of diabetics and non-diabetics at which the study is aimed. More exactly, in a statistical sense, the confidence interval means that if a series of identical studies were carried out repeatedly on different samples from the same populations, and a 95% confidence interval for the difference between the sample means calculated in each study, then, in the long run, 95% of these confidence intervals would include the population difference between means.

   Does the second part of the above quotation give a "more exact" analysis of the first part? Are the reference populations adequately characterised?

4. How do the Bayesian and classical concepts of sufficiency differ? What reasons are there for regarding a sufficient statistic as containing all the relevant information?

5. "Let $\hat{\theta}_1$ be any statistic which is not a function of the sufficient statistic $\hat{\theta}$. Then, by the definition of a sufficient statistic, the conditional distribution of $\hat{\theta}_1$, *given* $\hat{\theta}$, does not involve $\theta$. *Hence* $\hat{\theta}_1$ can give us no information about $\theta$ that the sufficient statistic has not already given us" (Mood and Graybill, 1963, p. 168; our italics). Is this a logically valid argument? If not, what kind of extra premiss would be required in order to make it valid?

# PART IV

# *Statistical Inference in Practice*

We have discussed inferential reasoning in fairly abstract terms, and an impression that the subject is far removed from everyday concerns might have been created and then reinforced by our concentration upon examples of little practical application. But the fact is that disagreements over methodology have enormous practical consequences. This is partly because theories dealing with 'big issues' may be variously evaluated by different methodologies, and partly because those methodologies might recommend quite different kinds of investigation.

In the next two chapters we shall illustrate these points by examining two areas where methodology makes a substantial practical difference. The first concerns the investigation of causal connections, particularly those that might exist between a particular drug and the alleviation of symptoms.

Many hundreds of clinical trials to test new drugs are under way at any given time, at immense cost to the drug companies and considerable inconvenience to those who participate in them; their design is largely dictated by the drug regulatory authorities, which more-or-less consistently adopt the classical line with regard to statistical inference. We shall in Chapter 11 enquire whether the restrictions imposed by classically minded methodologists on clinical trials are as beneficial as is claimed.

The second area we shall look at where methodology makes a difference is the study of (possibly causal) relations between physical variables. We may ask, for instance, how the volume occupied by a gas is related to its temperature and pressure, or the boiling point of a fluid to the ambient pressure, or the level of outdoor crime to the intensity of street illumination, or the amount of a good sold to its price. Such questions arise in many contexts, and getting the right answer is often of the greatest practical importance. For example, relationships between the magnitudes of certain earthquake precursor events, such as clusters (or 'swarms') of small earth tremors, and the times and intensities of subsequent earthquakes have been studied and the results used to predict future earthquakes. Striking illustrations of this occurred, in 1975 and 1976, when large areas of China were evacuated in advance of two damaging earthquakes successfully predicted by such means (Rhoades, 1986).

The study of these kinds of relationship is generally known as regression analysis. In Chapter 12, we shall critically review the standard methods of regression analysis which are recommended and employed in this classically dominated area.

# CHAPTER 11

## Causal Hypotheses: Clinical and Agricultural Trials

### a INTRODUCTION: THE PROBLEM

One of the most important areas where modern techniques of statistical inference have been very extensively applied is in testing new drugs and comparing different medical therapies. Fisher was a pioneer in formulating principles for such tests and in applying them to congruent problems in agriculture, where the questions of interest might, for example, be whether a particular fertiliser improves the yield of a certain crop or whether a newly genetically-engineered tomato has improved growth qualities.

These are all questions that concern causal efficacy and, as such, they pose a special difficulty, for to demonstrate, for instance, that a particular therapy causes people to recover from a given ailment, you need to show not only that they recover after receiving the therapy but also that the self same people would not have recovered if they hadn't received the therapy. This appears to suggest that in order to establish a causal link, one ought to examine the results of simultaneously treating and not treating the very same patients under identical circumstances. Since any attempt to put that requirement into effect would meet certain obvious and insurmountable difficulties, how should one proceed?

Well, the next best thing to the impossible ideal of a distinct pair of identical patients would be two patients who are similar in certain crucial respects, in particular, those respects that are relevant to the development of the disease in question. Such causally relevant factors are often known as *prognostic factors*. The therapy could then be applied to one of the patients and withheld from the other and the pertinent changes in their conditions monitored. Experiments of this kind, though normally involving large numbers of patients, not just two, are known as *clinical trials*; the patients that receive the drug or therapy whose effects it is desired to ascertain are normally called the *test group;* those who do not receive it constitute what is known as the comparison or *control group*.

Analogous experiments relating to crops are generally called *agricultural trials*. A simplified agricultural trial for comparing the yield-capacities of two strains of potato, say, might take the following form: a field is divided into pairs of plots, each pair being called a 'block', and one member of each pair is planted with the new strain, while the established variety is grown on the other. The field might look like this:

**TABLE XI** A Possible Distribution of *A*- and *B*-strains of Potato over a Field in an Agricultural Trial

| | *Plot 1* | *Plot 2* |
|---|---|---|
| Block 1 | *A* | *B* |
| Block 2 | *A* | *B* |
| Block 3 | *A* | *B* |
| Block 4 | *A* | *B* |

As with clinical trials, the ideal agricultural trial for comparing the different types of crop would expose all the seeds and the subsequent plants to identical growth-relevant environments. Under these ideal conditions, if one of the crops consistently achieved a greater yield than the other, the effect could reasonably be credited to an intrinsic characteristic of the crop. The inference would be similarly straightforward in a clinical trial in which the patient groups were properly matched on all the prognostic factors, with the possible exception of the treatment itself, for then any difference in average recovery rate or symptom measure could clearly be attributed to the treatment. This sort of inference, where every potential causal factor is laid out and all but one excluded by the experimental information, is a form of what is traditionally called *eliminative induction*.

Unfortunately, the conditions for such an induction are, to put it mildly, difficult to set up. If you wished to arrange for every prognostic factor to be equally represented in the experimental groups of a clinical trial, you would need a comprehensive list of those factors. But since, as Fisher (1947, p. 18) pointed out, there are always innumerably many possible prognostic factors, most of which have not even been thought of and some of which might be mistakenly regarded as causally inactive, it is hard to see how such a list could be compiled.

## b CONTROL AND RANDOMIZATION

To meet this difficulty, Fisher distinguished between factors which are known to influence the course of the disease, or the growth of the crop, and those which are unknown and whose influence is unsuspected. And he claimed that the deleterious effects on the inference process of these two kinds of potentially interfering influences could be neutralised by the techniques of *control* and *randomization,* respectively.

**Control.** A factor is said to have been controlled when it has been deliberately introduced, in equal measure, into both the test and the comparison situations in a trial. So, for example, a clinical trial involving a disease that is known to be sensitive to the patient's age would be conducted on test and control groups with similar age structures. The trial would then be said to have been controlled for age. Medical trials are typically controlled for a number of other factors. Prominent amongst these is the *placebo factor*—the beneficial psychosomatic effect experienced by many patients when they are faced simply with the encouraging paraphernalia of whitecoated, medical attention, whatever the particular nature of that attention. The effect is countered in clinical trials by treating the experimental groups in superficially similar ways, so that the patients are not told and themselves cannot tell whether they are receiving the test treatment or a placebo. When the test treatment is a drug, the placebo would take the form of a substance that looked like the test drug and was administered to the patient under indistinguishable conditions, but which has no known pharmacological effect on the disease.

Similar considerations apply in agricultural trials. Fisher advised that when comparing different crops, they be planted in soil "that appears to be uniform, as judged by the surface and texture of the soil, or by the appearance of a previous crop". Also, the plots on which the varieties are planted should be compact, in view of "the widely verified fact that patches in close proximity are commonly more alike, as judged by the yield of crops, than those which are further apart" (1947, p. 64). And when seeds are sown in plant-pots, Fisher (1947, p. 41) recommended that the soil should be thoroughly mixed before distribution to the pots, the watering of the pots equalized, and precautions taken to ensure that they each receive the same amount of light—and so on, for any other factor known to influence the growth of the plants.

**Randomization.** So much for the known factors. What of the unknown ones, the so-called 'nuisance variables', whose influence on the course of the experiment has not been recognised and which therefore have not been controlled? These, it is often said, should be dealt with by the process of randomization. This requirement means, for example, that in comparing the yields of the two types of potato in an agricultural trial, the plots on which each is grown should be selected at random. And when testing the effectiveness of a drug for some condition in a clinical trial, the randomization requirement demands that whether a particular patient is allocated to the test or the control group should be decided at random.

This is not the place to discuss the thorny question of what randomness really is; we shall, however, consider this question in Chapter 13, section **b.1.ii.** Suffice it to say here that advocates of randomization regard certain repeatable experiments as sources of randomness. Accordingly, they hold that whether a particular seed is planted on the left or the right side of a field is random if it is determined by the throw of a standard coin or die, the draw of a card from a well-shuffled pack, or the decay or otherwise in a given time-interval of a radioactive element. Random-number tables, which are constructed with the help of such random processes, are often recommended as equally effective for applying the randomization rule. However, it would not be good enough, it is alleged, simply to assign experimental units to the various treatments in a way which seems haphazard but which is not objectively random in the sense just specified.

Fisher's randomization procedure is generally regarded as a brilliant and effective solution to the problem of nuisance variables in experimental design. For example, according to Kempthorne (1979, pp. 125–26),

> Only when the treatments in the experiment are applied by the experimenter using the full randomization procedure is the chain of inductive inference sound; it is only under these circumstances that the experimenter can attribute whatever effects he observes to the treatment and to the treatment only. Under these circumstances his conclusions are reliable in the statistical sense.

Similarly, Kendall and Stuart (1983, pp. 120–21) held that although Fisher's contributions to statistical theory were remarkable and wide-ranging, "[n]evertheless, it is probably no exaggeration to say that his advocacy of *randomization* in experiment design was the most important and the most influential of his many achievements in statistics". Kempthorne (1966, p. 17) has claimed that Fisher's ideas "have been taken over by essentially the whole world of experimental scientists". This is an exaggeration—randomization plays little role in physics and chemistry—yet it correctly describes the state of affairs in medical and agricultural investigations, where Fisher's enormous influence is undeniable and where trials that are not properly randomized are frequently written off as seriously deficient or even as entirely useless for the purposes of scientific inference (e.g., by Peto, 1978, pp. 26–27; Gore, 1981; and Altman et al., 1983, p. 1490).

Ensuring that a trial includes the desired random element is not without costs, as we shall shortly explain, but those costs are widely reckoned to be worth incurring on the grounds that by randomizing, the problem of nuisance variables is overcome. Clearly this last claim needs to be demonstrated. It is our belief that it cannot be. In what follows, we shall argue that the standard defences of randomization are unsuccessful and that the problem of nuisance variables is not solved by randomized designs. In Chapter 14 we shall argue that although randomization may sometimes be harmless and even helpful, it is not a sine qua non; we shall maintain (following Urbach, 1985, 1987a, and 1993) that the essential feature of a trial that permits a satisfactory conclusion as to the causal efficacy of a treatment is whether it has been adequately controlled.

Two different accounts of how randomization can help are often suggested, each evidently resting on a different conception of how clinical trial results should be analysed. The first regards classical significance tests as the correct method of analysis, while the second, although advocated by classical statisticians, seems to appeal to a modified form of eliminative induction.

## c SIGNIFICANCE-TEST JUSTIFICATIONS FOR RANDOMIZATION

The first account is due to Fisher, who held that trials such as we are considering should be interpreted through significance tests. Fisher claimed that although we cannot know for sure that the groups are perfectly matched, we can have objective and certain knowledge of the probabilities of different possible mismatchings; and he claimed, and claimed often, that this knowledge is a necessary precondition for a correct test of significance and that randomization furnishes that precondition:

> [T]he full procedure of randomization [is the method] by which the validity of the test of significance may be *guaranteed* against corruption by the causes of disturbance which have not been eliminated [by being controlled]. (Fisher, 1947, p. 19; italics added)

This claim is regularly repeated when randomization is defended, for example by Byar et al., who assert that "[i]t is the process of randomization that generates the significance test, and this process is independent of prognostic factors, *known or unknown*" (Byar et al., 1976, p. 75; italics added). If significance tests and the related procedures of confidence-interval estimation were the only proper ways to analyse clinical trial results, and if randomization were essential to those inference techniques, then it would indeed be a sine qua non, as is often maintained.

### c.1 The Problem of the Reference Population

But despite the widespread agreement that significance tests require randomization, expositions of the standard tests that are employed in the analysis of trials, such as the *t*-test, the chi-square test, and the Wilcoxon Rank Sum test, barely allude to randomization.

We shall relate our discussion here to the example of a clinical trial. The tests we referred to start by considering a certain population of sufferers from the disorder in question, and with a null hypothesis to the effect that people in the population would react similarly to the test- and comparison-treatments or to the test-treatment and a placebo. Suppose that the trial records some quantitative measure of the disease symptoms, giving certain means, $\bar{x}_1$ and $\bar{x}_2$, respectively, in the test and the control groups. The difference $\bar{x}_1 - \bar{x}_2$ is likely to vary from experiment to experiment, and the associated probability distribution has a certain standard deviation, or 'standard error', that is given by

$$SE(\bar{x}_1 - \bar{x}_2) = \sqrt{\frac{\sigma_1^{\,2}}{n_1} + \frac{\sigma_2^{\,2}}{n_2}},$$

where $\sigma_1$ and $\sigma_2$ are the population standard deviations, and $n_1$ and $n_2$ are the two sample sizes. A test-statistic that may be employed in a test of significance is

$$Z = \frac{\bar{x}_1 - \bar{x}_2}{SE(\bar{x}_1 - \bar{x}_2)}.$$

The situation is complicated because the standard deviations of the two populations are usually unknown, in which case they must be estimated from the standard deviations in the corresponding samples. And provided certain further conditions are met, the test-statistic that is then recommended is the $t$-statistic obtained from $Z$ by substituting for $\sigma_1$ and $\sigma_2$ the corresponding sample standard deviations.

The logic of such tests is essentially that of estimating population parameters, or testing differences between such parameters, the validity of which, as we pointed out in Chapter 10, section **d,** requires that the samples be drawn at random from a population.

But which population? According to Bourke, Daly, and McGilvray (1985, pp. 67–68), "the null hypothesis that drug $A$ has the same effect . . . as drug $B$ refers, in a vague sense, to all patients similar to those included in the study". This, however, is too vague to be informative. Pocock's (1983, p. 198) characterisation of the reference population as "all patients with the disease eligible for the trial" sounds less vague but is not really, as it leaves "eligible for the trial" undefined. To be sure, trial eligibility can be, and often is, precisely defined in the trial's design specification. But such a definition cannot adequately characterise a reference population for the purposes of the normal statistical tests; for any reference population would need to include unknown sufferers in faraway places and those presently healthy or yet unborn who will contract the disease in the future, such people being the potential recipients of the treatment, should it prove effective in the trial; but you cannot draw random samples from hypothetical populations full of potential people.

Those tests are, nevertheless, widely applied without the random-sampling condition being met, which seems quite unjustified. Moreover, the role of randomization is unclear. To be sure, if the treatment groups were separate, random samples drawn from some population, those groups could be regarded as automatically randomized, since everybody would have had the same chance of being included in each group. But if the treatment groups cannot be formed in this way, how can ordinary randomization make up the deficiency and validate the statistical test?

An answer to this question has been attempted by Bourke et al. They argued that the standard tests may be properly applied without "the need for convoluted arguments concerning random sampling from larger populations", provided the groups have been constructed by an appropriate randomization process. To illustrate their point, they considered an example in which the random allocation of 25 subjects delivered 12 persons to one group and 13 to the other. These particular groups, they affirm, "can be viewed as one of many possible allocations resulting from the randomization of 25 individuals into two groups of 12 and 13, respectively" (1985, p. 188). They noted the existence of 5,200,300 possible, different outcomes of such a randomization and proposed that the reference population be regarded as this set of possible allocations. The null hypothesis would then assert that in this hypothetical population, the mean difference in treatment-effects in the two groups is zero. They then claimed that a significance test can be applied to detect whether the difference in means in the groups that were in fact drawn is larger than expected by chance.

This approach seems attractive because it defines the reference population unambiguously and ensures that the sample is randomly selected from that population, as required by standard tests. Moreover, these conditions can be met only if the experimental groups were randomized, thus neatly reinstating randomization as an essential ingredient. But the approach does not really work, we submit. First, it is premissed on the apparent fiction of a premeditated plan to accept only groups of 12 and 13, respectively, with the implication that any other combinations that could have arisen in pairs of randomly assembled groups would have been abandoned. Secondly, while the reference population consists of 5 million or so pairs of groups, the sample is just a single pair. But, intuitively, you get very little information about the mean of a large population from a sample consisting of a single member of that population, and this is reflected in the standard $t$-test which, because it requires division by $n-1$, cannot even get off the ground with $n < 2$; so the kind of test Bourke and his colleagues were envisaging is unworkable. Thirdly, as they pointed out, statistical inference in their scheme "relates only to the individuals entered into the study" and may not generalise to any broader category of people. To achieve such a generalisation, they argued, "involves issues relating to the representativeness of the trial group to the general body of patients affected with the particular disease being studied" (1985, p. 188). However, the path from sample to "the general body of patients" is evidently not mediated by significance tests and is left unspecified; but unless that path is mapped out, the clinical trial would have nothing to say on the point that is of most concern.

These difficulties, we suggest, mean that the attempt to rest randomization on the requirements of the $t$-test and other similar tests will not work.

### c.2 Fisher's Argument

When Fisher argued for the need to randomize the experimental units to different treatments in order validly to apply tests of significance, he seems to have had in mind a somewhat different test than we have so far considered.

We may illustrate Fisher's argument through our potato example, in section **a**, which slightly simplifies Fisher's example (1947, pp. 41–42). Suppose that, in reality, the potato varieties $A$ and $B$ have the same genetic growth characteristics and let this be designated the null hypothesis. Imagine, too, that one plot in each block is more fertile than the other and that, this apart, no relevant difference exists between the conditions which the seeds and the resulting plants experience. If pairs of different-variety seeds are allocated at random, one to each of a pair of plots, then, other things being equal, there is a probability of exactly a half that any plant of the first variety will exceed one of the second in yield, even if no intrinsic difference exists. The probability that $r$ out of $n$ pairs show an excess yield for $A$ can then be computed and a simple test of significance, such as we discussed in Chapter 8, applied.

This argument and our subsequent discussion can be easily translated so that it applies to clinical trials, with patients being entered into the trial in matched pairs and the element of the pair that receives the drug being determined at random. Then, if the drug has no effect on the disease (the null hypothesis), and if one of each pair is weaker or has the disease in a more virulent form, the probability that the person assigned the drug does better than the person given the placebo is a half, other things being equal, and the significance test described above can be applied.

### *i Some difficulties with Fisher's argument*

Fisher's justification for randomizing takes it for granted (returning to our agricultural example) that if the two kinds of seed have identical innate growth-characteristics, then, by selecting at random the plot of land on which the seeds are sown, one is *guaranteeing* an objective probability of a half that a difference in yields in either direction will occur. This assumption, however, derives from what seem to be unjustifiable presuppositions. For example, one has to suppose that none of the innumerable environmental variations that emerge after the randomization step introduces a corresponding variation in plant growth, for if one of the plant varieties were selectively subjected to some hidden growth-promoting factor, the probability of getting a difference in yield might not be a half but could have another, unknown value. This difficulty was, of course, perceived by Fisher from the outset. He proposed dealing with it by randomizing at the very last stage in the experimental procedure:

> the random choice of the objects to be treated in different ways would be a complete guarantee of the validity of the test of significance, if these treatments were the last in time of the stages in the physical history of the objects which might affect their experimental reaction. (Fisher, 1947, p. 20)

This condition could only express an ideal, however, for it is hard to imagine how one could identify the last stage in the history of an experiment which *might* affect its outcome, let alone how or what one would randomize at that stage. But this difficulty "causes no practical inconvenience", according to Fisher,

> for subsequent causes of differentiation [subsequent, that is, to the normal randomization step], if under the experimenter's control . . . can either be predetermined before the treatments have been randomized, or, if this has not been done, can be randomized on their own account; and other causes of differentiation will be either (a) consequences of differences already randomized, or (b) natural consequences of the difference in treatment to be tested, of which on the null hypothesis there will be none, by definition, or (c) effects supervening by chance independently from the treatments applied. (Fisher, 1947, pp. 20–21)

Fisher seems to be saying here that influences which might have affected the experimental result and which have not been controlled or dealt with through randomizing do not matter and do not affect the significance test, for they are either produced by differences already randomized and so are automatically distributed at random, or they are chance effects that are independent of the treatment and so are, as it were, subject to a natural randomization.

But this overlooks the possibility of influences which fall into neither category. Take, for example, the potato trial already discussed and suppose that when planted alone, each of the varieties benefits to a similar extent from the attentions of a certain insect, but that if both were growing in the same vicinity, those insects would be preferentially attracted to one of them. Or imagine that the two kinds of potato plant compete unequally for soil nutrients. Effects such as these (of which infinitely many are possible) would favour one of the treatments at the expense of the other, notwithstanding the earlier randomization.

Fisher's defence also fails to take account of disturbing factors that could have operated before the plant seeds were sown. For instance, the different seeds could have been handled by different market gardeners, one of whom had some unwitting effect on subsequent growth; or the sacks in which the seeds were stored may have influenced their future development; and so forth and so on. No later randomization could compensate or undo any unfair advantage that might be imparted by factors such as these. In order to deal with them in the Fisherian way, a separate randomization would need to be devised for each. But because there are infinitely many such possible sources of error, there must be a matching number of randomizations to guarantee the significance test. As this is impossible, randomization cannot provide such a guarantee. Hence the frequently voiced claim for randomization that it "protects against sources of bias that are *unsuspected*" (Snedecor and Cochran, 1967, p. 110) and Fisher's opinion that it "relieves the experimenter from the anxiety of considering and estimating the magnitude of the *innumerable causes* by which his data *may* be disturbed" (Fisher, 1947, p. 43; our italics) are simply wrong.

### *ii A plausible defence*

It might plausibly be argued against these criticisms that many variations in the experimental process, while possible sources of error, are not likely ones. For example, the colour of the market gardener's socks and the size of his shirt collar are conceivable but scarcely credible as influences on plant growth. And in a clinical trial, the shoe sizes of the different experimenters and the shade of undercoat painted on the surgery walls might *possibly* affect a person's medical development but are hardly likely to do so in fact. It seems reasonable to say that such factors can be safely ignored in the experimental design. Indeed, this is what is normally said, or at any rate tacitly held, by advocates of randomization. For instance, Kendall and Stuart acknowledged that in designing a trial, one has to evaluate the relative importance of possible extraneous influences on its outcome:

> A substantial part of the skill of the experimenter lies in his choice of factors to be randomized out of the experiment. If he is careful, he will randomize out all the factors which are suspected to be causally important but which are not actually part of the experimental structure. But every experimenter necessarily neglects some conceivably causal factors; if this were not so, the randomization procedure required would be impossibly complicated. (Kendall and Stuart, 1983, p. 137)

In accordance with these reflections, when examining the effects of various doses of alcohol on a person's reaction time, Kendall and Stuart explicitly omitted the colour of the subjects' eyes from any randomization, since this is "almost certainly negligible" as an influence on the effect being studied. No experimental data were cited to support this claim, presumably because none exists. But even if a careful trial had been conducted on the effects of eye colour on reaction times, certain conceivable influences on *its* outcome would have to have been set aside as negligible. Any demand that an influence be ignored as "almost certainly negligible" only after a properly randomized trial would, therefore, be futile, since it would lead to an infinite regress. Presumably for this reason, Kendall and Stuart (1983, p. 137) concluded that the decision whether a factor should be "randomized out" (their phrase) or neglected *"is essentially a matter of judgement"* (our italics).

So, according to Kendall and Stuart, one cannot ensure that every possibly relevant factor is randomized in a trial. What one should do, according to them, is discriminate, through the exercise of a personal judgment, between those factors that are worth randomizing and those that are not. This seems to us an inevitable consequence of Fisher's position.

It ought, however, to be mentioned in passing that Kendall and Stuart's example of eye colour is not well chosen, for in their experiment, different doses of alcohol are allocated randomly to the subjects, and this means that eye colour and any other characteristic to which the subject is permanently attached is, contrary to their claim, automatically randomized. But this does not detract from Kendall and Stuart's point, for there are many other examples they could have used. For instance, the various doses of alcohol may have been fed to the subjects from different vessels, they will contain unequal volumes of the liquid used to dilute them and different quantities of whatever trace impurities are contained in the alcohol, and so on. None of these factors would have been distributed randomly over the different alcohol doses; but this would not matter since, like eye colour, their disturbing influence would presumably be judged "almost certainly negligible".

Kendall and Stuart's conclusion, which seems to us unavoidable, is that randomization, if it is to achieve the goal set for it, must be restricted to factors that are, in the scientist's judgment, likely to influence the course of the trial.

### *iii Why the plausible defence doesn't work*

This modified principle of randomization neither assists nor reinstates Fisher's case. First, it exposes yet another essentially personal element in what purports to be a purely objective account of scientific inference. Secondly, there now seems no point in randomizing. True, it ensures that presumed or suspected variables that have not been controlled will be randomly distributed, but it does nothing for the unsuspected variables, the nuisance of which it was Fisher's aim to remove. Moreover, the question arises why one should randomize at all, for if it was a good idea to control factors "known" to affect the experimental outcome, then it would appear just as sensible to do the same with factors whose significance is less certain. Indeed, this is just what is almost universally done, though in a roundabout way, for medical researchers are regularly advised always to "check that the groups as randomized do not differ with respect to characteristics as assessed before treatment begins" (Gore, 1981, p. 1959). In other words, before continuing the trial, one should inspect the randomized groups to see whether they display any striking differences; if they do, they should be disbanded and the random allocation started afresh. In practice, this is just what is standardly done, as a glance through virtually any experimental medical journal would reveal. Even Fisher endorsed this practice, as the following conversation reported by Savage shows: "'What would you do,' I had asked, 'if, drawing a Latin square at random for an experiment, you happened to draw a Knut Vik square?'" (These squares are kinds of chequerboard patterns, in which fields are laid out in agricultural trials.) "Sir Ronald said he thought he would draw again and that, ideally, a theory explicity excluding regular squares should be developed" (Savage, 1962a, p. 88).

But Fisher is not entitled simply to discard the products of a random allocation, for this would alter the outcome space of the trial and hence change the sampling distribution. He would have to have specified in advance which configurations were acceptable and which not and recalculated the sampling distribution accordingly. This, so far as we can tell, is never done. But more importantly, if all the factors suspected of playing some causal role in the trial could be controlled, and if, as Kendall and Stuart maintained—correctly, in our view—all other factors may be ignored, then randomization is left without any useful role as far as the significance test is concerned.

## d THE ELIMINATIVE-INDUCTION DEFENCE OF RANDOMIZATION

A second argument for randomized clinical trials claims that in balancing the experimental groups, randomization performs the same or a similar service for the unknown factors as controls perform for the known ones.

This claim is advanced, for example, by Gore (1981, p. 1559), according to whom randomization is "an insurance in the long run against substantial accidental bias between treatment groups", making it clear that "the long run" is meant to cover large trials with more than 200 subjects. And Schwartz et al. say practically the same: "Allocating patients to treatments $A$ and $B$ by randomization produces two groups of patients which are as alike as possible with respect to *all* their [prognostic] characteristics, *both known and unknown"* (1980, p. 7; italics added). Byar et al. echo this claim, though in a slightly weaker form; for they offer no categorical assurance that randomization will balance the groups, but introduce an element of uncertainty about this, speaking instead of a "tendency" for the groups to be balanced: "[r]andomization tends to balance treatment groups in . . . prognostic factors, *whether or not these variables are known"* (1976, pp. 74–80; italics added).

These claims are certainly very credible, as is evidenced by the frequency with which they are voiced (*see,* for example, Ellenberg, 1981, p. 2482; Snedecor and Cochran, 1967, p. 110;

Royall, 1991, p. 54; and Giere, 1979, p. 296). We shall, however, argue that despite their widespread acceptance, strictly speaking, the claims are mistaken, though we shall also argue (in Chapter 14, section **g.1**) that in a modified form, they have a basis within the Bayesian framework.

According to the claims we are considering, all known and unknown prognostic factors are necessarily, or with a necessarily high probability, distributed uniformly across the experimental groups, provided those groups are sufficiently large. In order to examine these claims, let us consider the various possibilities concerning the unrecognised prognostic factors in any experimental situation. First, there is the possibility that there are in fact none; if that were so, the probability that we end up with unmatched groups is clearly zero. Consider next the possibility that in the experimental situation there is a single, unknown factor—call it $X$—carried by some of the patients. In that case, if the groups are randomly formed, they will be substantially unmatched on $X$ with a certain definite probability, $x_n$, say. Clearly, this probability diminishes as $n$, the sample size, increases.

We should keep in mind that the argument under examination makes claims about the unknown prognostic factors. Clearly, if factor $X$ is present, there might, for all we know, be a second factor, $Y$, also carried by some of the patients. In that case, the chance that one of the groups has substantially more $Y$-patients than the other also has a definite value, $y_n$, say, which like $x_n$ diminishes as the sample grows. The probability that the groups are unmatched on at least one of $X$ and $Y$ could be as much as $x_n + y_n - x_n y_n$, if the two factors are independent. And because we are dealing with the unknown, it must be acknowledged that there could be any number of such factors; hence the probability of a substantial imbalance on *some* prognostic factor might, for all we know, be quite large, as Lindley has pointed out (1982a, p. 439). Nor are we ever in a position to calculate how large, since, self-evidently, we do not know what the unknown factors are.

We have so far considered only unknown factors that might, as it were, be attached to some of the patients. What about factors that are independent of the patient but *correlated with the treatment?* We are here thinking of prognostic factors that might be accidentally linked to the treatment in this particular trial and so could not possibly be regarded as an aspect of the treatment itself. For example, suppose the test and control patients were treated in separate surgeries whose different environments, through some hidden process, either promote or discourage recovery; or suppose the drug, despite the manufacturers' best efforts, included some contaminant which influenced recovery; or . . . (one simply needs a rich enough imagination to extend this list indefinitely). For all we know, one or more such factors are active in the experimental situation; and if so, the probability that the groups are imbalanced is, of course, one.

So what can we say with assurance about the objective probability that the experimental groups in a clinical trial differ on unknown prognostic factors? Simply this. Provided the randomization has been carried out properly, the probability of an imbalance on unknown prognostic factors lies somewhere in the range zero to one! In other words, the claim we are evaluating that "[r]andomization tends to balance treatment groups in . . . prognostic factors, whether or not these variables are known" (Byar et al., 1976, pp. 74–80), or that randomized groups "are automatically controlled for ALL other factors, even those no one suspects" (Giere, 1979, p. 296; original emphasis) is not in general true. In some cases it holds, in others not, depending on what unknown factors are operating, which, by definition, is something we do not and cannot know.

To summarise, we considered two closely related claims under this head. First, that in randomized trials exceeding a given size, balanced groups are inevitable. This is perhaps an artificial position, based on too pedantic a reading of the quotations cited earlier, which should no doubt be read as expressing this second view, namely, that the probability that randomized groups will be balanced increases with their sizes, and that with more than 200 subjects in a trial, this probability is close to 1. We showed, however, that just as randomization does not infallibly ensure balanced experimental groups in a clinical trial, so too, it cannot assure any particular level of probability that the groups will be balanced.

Those who advocate randomization on the grounds that it guarantees a high probability that the experimental groups will be balanced with respect to prognostic factors, generally present their case elliptically, for clinical trials are conducted not to learn about balance in groups but in order to discover whether and to what extent a treatment affects some bodily condition. Thus the argument needs to be completed by establishing a link between a high probability of balanced groups and some assessment of treatment efficacy. This task would be straightforward if we could assert, without qualification, that the groups were balanced, for then all possible prognostic factors, other than the treatment, could be ruled out in the kind of eliminative induction mentioned earlier. This perhaps explains the attractiveness of the stronger of the two claims made on behalf of randomization. But as soon as the probability qualification is included, as it surely must be, the conditions of eliminative induction no longer apply, and the corresponding inference is no longer logically valid.

The position we are considering, which is the most frequently heard defence of randomization in clinical trials, is therefore faulty at two points: it is not true that balanced groups are always a high probability; and even if it were true, the cause of randomization would not have been advanced, unless some valid form of inference were available, which it does not seem to be, to lead from there to an assessment of the medical treatment.

In Chapter 14, section **g**, we shall argue that the matter can be resolved in a way that is not only valid but also reflects the main intuitions underlying the arguments here discussed. This resolution becomes available if it is conceded that the probability involved in judging how far the groups are balanced is subjective rather than objective; and if a similar concession is made in regard to the assessment of the experimental treatment. Granted these two, the desired inference from evidence to conclusion can, we shall argue, be validly made via Bayes's Theorem.

## ■ e SEQUENTIAL CLINICAL TRIALS

There are two aspects of the kind of clinical trial we have been considering which have given rise to concern. The first is this. Suppose that the test therapy and the comparison therapy (which may simply be a placebo) are not equally effective and that this fact becomes apparent at the end of a clinical trial through a differential response rate in the experimental groups. This means that the patients in one of the groups will have been treated with an inferior therapy throughout the trial. Clearly, the fewer patients so treated the better, and the question arises whether a particular trial is suited to extracting the required information in the most efficient way.

The second concern arises through the following considerations: the kinds of statistical test that are commonly applied to the results of a trial, such as those we discussed earlier, require that the experimenter fix in advance the number of patients who will enter the trial. And if, for some reason, the trial were discontinued before that predetermined number was reached, the results obtained at that stage would, as Jennison and Turnbull (1990, p. 305) note, have to be reckoned quite uninformative about the treatment under study. Intuitively, however, this is wrong; in many circumstances, quite a lot of information might seem to be contained in the interim results of a trial. Suppose, to take an extreme example, that a sample of 100 had been proposed and that the first 15 patients in the test group completely recover, while a similar number in the control group promptly experience a severe relapse. At this stage, even though the trial as originally envisaged is incomplete, we would nonetheless be inclined to regard the evidence as telling very strongly in favour of the test treatment. And we would be disinclined to continue the trial, because that would entail a further 35 or so patients being deprived of what is apparently the better treatment. As Armitage (1975, p. 27) put it, "a predominant requirement in many trials is that random allocation should cease if it becomes reasonably clear that one treatment has a better therapeutic response than another".

It might be imagined that the last problem could be dealt with by performing a new significance test after each new experimental reading and stopping the trial as soon as a 'significant' result is obtained. But strictly speaking, the suggested sequence of tests is impossible, since significance tests depend essentially on the sample size being fixed and predetermined; in the sequence of quasi-significance tests contemplated here, the sample size is a variable whose actual value cannot be known in advance.

Sequential clinical trials are a type of trial that were designed with a view to overcoming the two problems described. They allow tests of significance to be applied to data even when the sample size is variable; they also permit results to be monitored as they come in, and in many cases they enable a conclusion to be reached with fewer data than would be possible on the fixed-sample design.

The following is a simple sequential trial, which we have taken from Armitage, 1975. Appropriately matched pairs of patients are treated in sequence either with substance *A* or substance *B* (one of which might be a placebo). The treatments are randomized, in the sense that the element of any pair that receives *A,* respectively *B,* is determined at random. We shall assume that *A* and *B* are indistinguishable to the patients and that the medical personnel administering them are also in the dark as to how each patient was treated (the trial is therefore 'blind' and 'double-blind'). For each pair, a record is made of whether, relative to some agreed criterion, the *A*- or the *B*-treated patient did better. The results are assumed to come in sequentially and to be monitored continuously.

Results are recorded for each pair as either *A* doing better than *B,* or *B* doing better than *A;* we label these results simply as **A** and **B.** In the diagrams below, the difference between the number of **A**s and the number of **B**s in a sample is plotted on the vertical axis; the horizontal axis represents the number of comparisons made. The zig-zag line, or 'sample path', starting from the origin, represents a possible outcome of the trial,

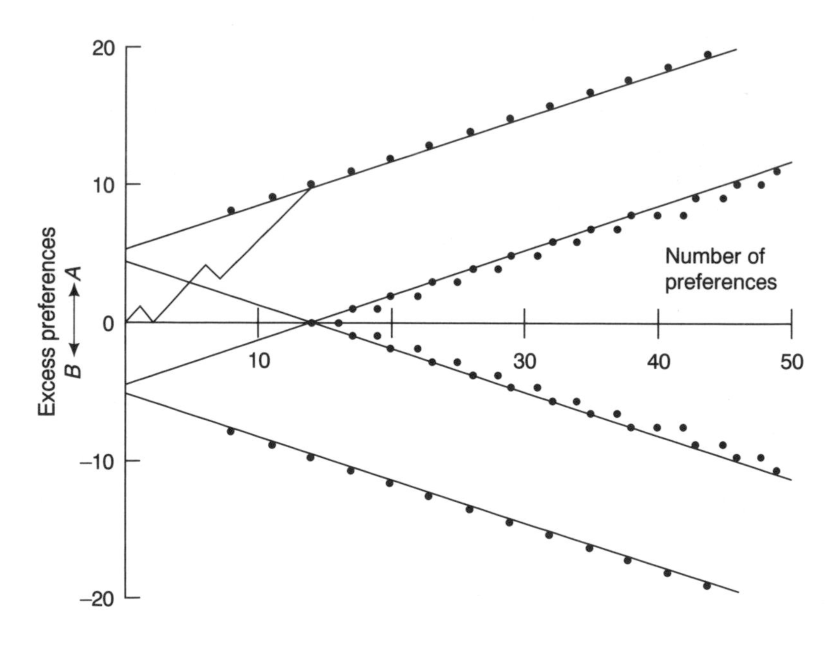


**SEQUENTIAL PLAN a**

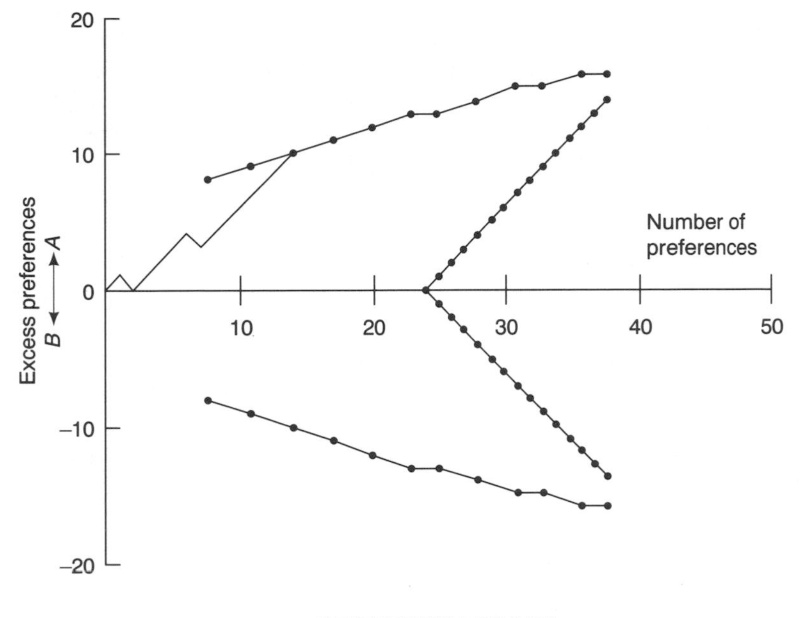


**SEQUENTIAL PLAN b**

corresponding to the 14 comparisons, **ABAAAABAAAAAAA.**

The wedge-shaped system of lines in sequential plan *a* was designed to have the following characteristics: if the drugs are equally effective (the null hypothesis), then the probability of the sample path first crossing one of the outer lines is 0.05. If the sample path does cross one of these lines, sampling is stopped and the null hypothesis rejected. If the sample path first crosses one of the inner lines, the sampling is stopped too, the result declared non-significant, and the null hypothesis accepted. The inner lines are constructed, in this example, so that the test has a power of 0.95 against a particular alternative hypothesis, namely that the true probability of an *A*-treated patient doing better than a *B*-treated one is 0.8.

The sequential test based on plan *a* is, in Armitage's terminology, 'open', in the sense that the trial might never end, for the sample path could be confined indefinitely within one of the two channels formed by the two pairs of parallel lines. Plan *b* represents a 'closed' test of the same null and alternative hypotheses. In the latter test, if the sample path meets one of the outer lines first, the null hypothesis is rejected, and if one of the inner lines, the hypothesis is accepted. This closed test has a similar significance level and power to the earlier, open one, but differs from it in that at most 40 comparisons are required for a decision to be reached.

The above sequential plans were calculated by Armitage on the assumption that as each new datum is registered, a new significance test, with an appropriate significance level, is carried out. (As we pointed out, these are quasi-significance tests, since they presume a fixed sample size when, in fact, the sample is growing at every stage.) The significance level of the tests is set at a value which makes the overall probability of rejecting a true null hypothesis 5 per cent. This is just one way of determining sequential tests, particularly favoured by Armitage (1975), but there are many other possible methods leading to different sequential plans. The sequential tests might also vary in the frequency with which data are monitored; the results might be followed continuously as they accumulate or examined after every new batch of, say, 5 or 10 or 50 patients have been treated (these are 'group sequential trials'). Each such interim-analysis policy corresponds to a different sequential test.

The existence of a multiplicity of tests for analysing trial results carries a strange, though by now familiar implication: whether a particular result is 'significant' or not, whether a hypothesis should be rejected or not, whether or not you would be well advised to take the medicine in the expectation of a cure, depends on how frequently the experimenter looked at the data as they accumulated and which sequential plan he or she was guided by. Thus a given outcome from treating a particular number of patients might be significant if the experimenter had monitored the results continuously, but not significant if he had waited to analyse the results until they were all in, and the conclusion might be different again if another sequential plan had been used. This is surely unacceptable, and it is no wonder that sequential trials have been condemned as "a hoax" (Anscombe, 1963, p. 381). As Meier (1975, p. 525) has put it, "it seems hard indeed to accept the notion that *I* should be influenced in my judgement by how frequently *he* peeked at the data while he was collecting it."

Nevertheless, the astonishing idea that the frequency with which the data have been peeked at provides information on the effectiveness of a medical treatment and should be taken account of when evaluating such a treatment is thoroughly entrenched. Thus the Food and Drug Administration, the drugs licensing authority in the United States, includes in its *Guideline* (1988, p. 64) the requirement that "all interim analyses, formal or informal, by any study participant, sponsor staff member, or data monitoring group should be described in full. . . . The need for statistical adjustment because of such analyses should be addressed. Minutes of meetings of the data monitoring group may be useful (and may be requested by the review division)". But the intentions and calculations made by the various study participants during the course of a trial are nothing more than disturbances in those people's brains. Such brain disturbances seem to us to carry as much information about the causal powers of the drug in question as the goings-on in any other sections of their anatomies: none.

## f PRACTICAL AND ETHICAL CONSIDERATIONS

The principal point at issue in this chapter has been Fisher's thesis, so widely adopted as an article of faith, that clinical and agricultural trials are worthless if they are not randomized. But as we have argued, the alleged absolute need for such trials to be randomized has not been established. In Chapter 14, we shall argue that an intuitively satisfactory approach to the design of trials and the analysis of their results is furnished by Bayes's Theorem, and that when looked at from that vantage point, randomization is seen no longer to be a sine qua non.

It would not, of course, be correct to infer from this that randomization is necessarily harmful, nor that it is never useful—and we would not wish to draw such a conclusion—but removing the absolute requirement of randomization is a significant step which lifts some severe and, in our view, undesirable limitations on acceptable trials. For example, the requirement to randomize excludes, as illegitimate, retrospective trials using so-called historical controls. Historical controls suggest themselves when, for example, one wishes to find out whether a new therapy raises the chance of recovery from a particular disease compared with a well-established treatment. In such cases, since many patients would already have been observed under the old regime, it seems unnecessary and extravagant to submit a new batch to that same treatment. The control group could be formed from past records and the new treatment applied to a fresh set of patients who have been carefully matched with those in the artificially constructed control group (or historical control). But the theory of randomization prohibits this kind of experiment, since patients are not assigned with equal probabilities to the two groups; indeed, subjects finding themselves in either of the groups would have had no chance at all of being chosen for the other.

There is also sometimes an unattractive ethical aspect to randomization, particularly in medical research. A new treatment which is deemed worth the trouble and expense of an investigation has often recommended itself in pilot studies and in informal observations as having a reasonable chance of improving the condition of patients more than established methods. But if there were evidence, however modest, that a patient would suffer less with the new therapy than with the old, it would surely be unethical to expose randomly selected sufferers to the established and apparently inferior treatment. Yet this is just what the theory of randomization insists upon. No such ethical problem arises when patients receiving the new treatment are compared with a matched set who have already been treated under the old regime.

The principle of randomization also represents an implausible departure from normal practice in science. Physicists, for example, do not conduct experiments as Fisher would have them do, yet they are often interested in the same type of question, namely, whether some intervention has a particular causal effect or not. Take a simple experiment to measure the acceleration due to gravity in which a heavy object is dropped close to the earth. The conditions would be controlled by ensuring that the air is still, that the space between the object and the ground is unimpeded, and so on for other factors likely to interfere with the rate at which the object descends. What no scientist would do is divide the earth's surface or any part of it into small plots and then select some of these using random-number tables for the places to perform the experiments. Randomizers might take one of two attitudes to the non-randomizing preference of physical scientists. They could either declare it to be irrational and insist that it be given up or else claim that experiments in physics and chemistry are, in some crucial respect, unlike those in biology and psychology, neither of which seem promising lines of defence.

## g CONCLUSION

We have reviewed the main features of classically inspired designs for clinical and agricultural trials, particularly the alleged requirement for treatments to be randomized over the experimental units. Our conclusion is that neither of the two standard arguments for the randomization principle is effective and that the principle is indefensible as an absolute precondition on trials, even from the classical viewpoint. In Chapter 14, we shall argue in detail that a Bayesian approach gives a more satisfactory treatment of the problem of nuisance variables and furnishes intuitively correct principles of experimental design.

# CHAPTER 12

# *Regression Analysis*

## a INTRODUCTION

We are often interested in how some variable quantity depends on other variable quantities. Such relationships are often guessed at from specific values taken by the variables in experiments; but precisely how those guesses, or inferences, are made and justified is difficult and debatable. The difficulty lies in the fact, which we discussed in Chapter 7, Section **k**, that infinitely many different relationships are compatible with any actual data set. Nevertheless, as we have already seen, despite the immense multiplicity of candidate theories or relationships, scientists are rarely left as bewildered as might be imagined; in fact, they often become quite certain about what the true relationships are and feel able to predict hitherto unobserved values of the variables in question with considerable confidence. How they do this and with what justification is the topic of this chapter (which largely follows Urbach, 1992a).

## b SIMPLE LINEAR REGRESSION

How to infer from specific observed values a general relationship between variable quantities is a problem considered in practically every textbook on statistics, under the heading *Regression*. Usually, the problem is introduced in a restricted form in which just two variables are involved; the methods developed are then extended to deal with many-variable relationships. When it comes to examining underlying principles, which is our aim, the least complicated cases are the best and most revealing; so it is upon these that we shall concentrate.

Statisticians treat a wider problem than that alluded to earlier, by allowing for cases where the variables are related, but, because of a random 'error' term, not in a directly functional way. With just two variables, $x$ and $y$, such a relationship, or regression, may be represented as $y = f(x) + \epsilon$, where $\epsilon$ is a random variable, called the 'error' term. The simplest and most studied regressions—known as simple linear regressions—are a special case, in which $y = \alpha + \beta x + \epsilon$, $\alpha$ and $\beta$ being unspecified constants. In addition, the errors in a simple linear regression are taken to be uncorrelated and to have a zero mean and an unspecified variance, $\sigma^2$, whose value is constant over all $x$. We shall call regressions meeting these conditions linear, dropping the qualification 'simple', for brevity.

Many different systems have been explored as reasonable candidates to satisfy the linear regression hypothesis. The following are typical examples from the textbook literature: the relationship between wheat yield, $y$, and quantity of fertilizer applied, $x$ (Wonnacott and Wonnacott, 1980); the measured boiling point of water, $y$, and barometric pressure, $x$ (Weisberg, 1980); and the level of people's income, $y$, and the extent of their education, $x$ (Lewis-Beck, 1980).

The linear regression hypothesis is depicted below (Mood and Graybill, 1963, p. 330). The vertical axis represents the probability densities of $y$ for given values of $x$. The bell-shaped curves in the $y,P(y \mid x)$-plane illustrate the probability density distributions of $y$ for two particular values of $x$; the diagonal line in the $x,y$-plane is the plot of the mean of $y$ against $x$; and the variance of the density distribution curves which is the same as the error variance, $\sigma^2$, is constant—a condition known as 'homoscedasticity'.

The linear regression hypothesis leaves open the values of the three parameters $\alpha$, $\beta$, and $\sigma$, which need to be estimated from data, such data normally taking the form of particular $x,y$ readings. These readings may be obtained in one of two ways. In the first, particular values of $x$ (e.g., barometric pressures or concentrations of fertilizer) are preselected and then the resulting $y$s (e.g., the boiling point of water at each of the pressure levels or wheat yields at each fertilizer strength) are read off. Alternatively, the $x$s may be selected at random (e.g., they could be the heights of random members of a population) and the corresponding $y$s (e.g., each chosen person's weight) then measured as before. In the examples discussed here, the data will mostly

**THE LINEAR REGRESSION HYPOTHESIS**

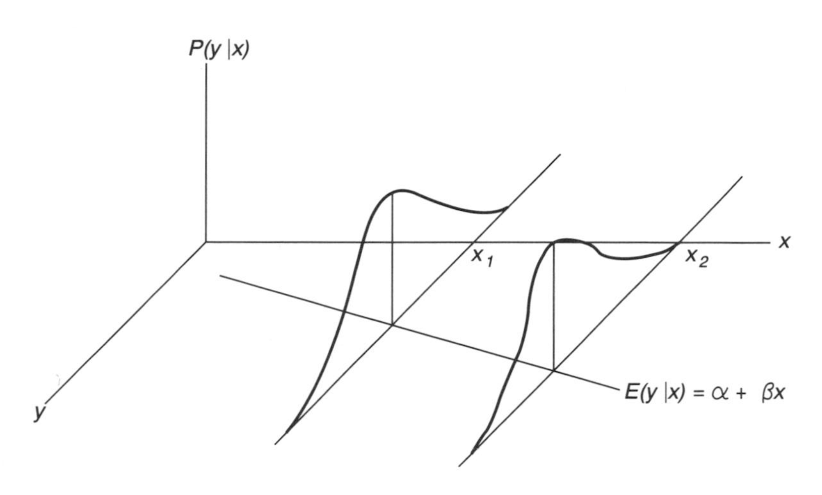


be of the former kind, with the $x$s fixed in advance; none of the criticisms we shall offer will be affected by this restriction.

We are considering the special case where the underlying regression has been assumed to take the simple linear form. For the Bayesian, this means that other possible forms of the regression all have probability zero or are sufficiently close to zero to make no difference. In such special (usually artificial) cases, a Bayesian would continue the analysis by describing a prior, subjective probability distribution over the range of possible values of $\alpha$, $\beta$, and $\sigma$, and would then conditionalise on the evidence to obtain a corresponding posterior distribution (for a detailed exposition, *see,* e.g., Broemeling, 1985). But classical statisticians regard such distributions as irrelevant to science, because of their subjective component. The classical approach tries to put its objectivist ideal into effect by seeking the linear regression equation possessing what is called the 'best fit' to the data, an aim also expressed as a search for the 'best estimates' of the unknown linear-regression parameters. There are also classical procedures for examining the assumption of linear regression, which we shall review in due course.

## c THE METHOD OF LEAST SQUARES

The method of least squares is the standard classical way of estimating the constants in a linear-regression equation and "historically has been accepted almost universally as the best estimation technique for linear regression models" (Gunst and Mason, 1980, p. 66).

The method is this. Suppose there are $n$ $x$, $y$ readings, as depicted in the graph below. The vertical distance of the $i$th point from any straight line is labelled $e_i$ and is termed the 'error' or 'residual' of the point relative to that line. The straight line for which $\Sigma e_i^2$ is minimal is called the *least squares line*. If the least squares line is $y = \hat{\alpha} + \hat{\beta}x$, then $\hat{\alpha}$ and $\hat{\beta}$ are said to be the least squares estimates of the corresponding linear regression coefficients, $\alpha$ and $\beta$, and the line is often said to provide the 'best fit' to the data. The idea is illustrated below.

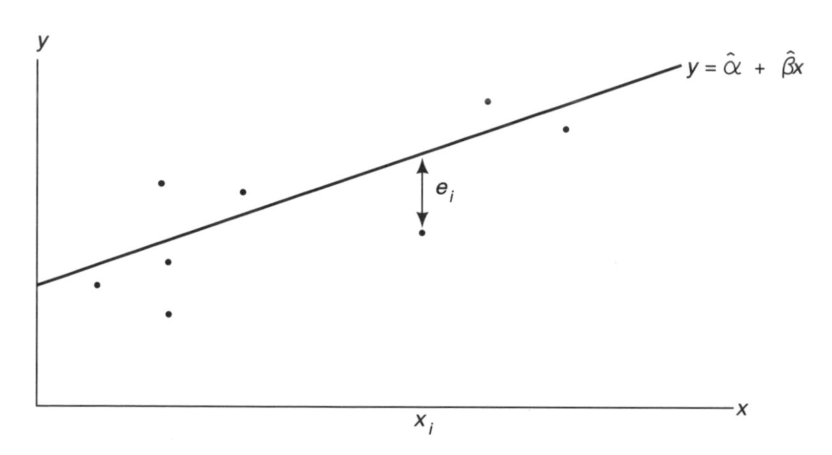


The least squares line, i.e., the line for which $\Sigma e_i^2$ is minimum.

The least squares estimates, $\hat{\alpha}$ and $\hat{\beta}$, are given as follows, where $\bar{y}$ and $\bar{x}$ are the mean values of the $x_i$ and $y_i$ in the sample (the proofs of these formulas will be suggested as exercises):

$$\hat{\beta} = \frac{\Sigma(y_i - \bar{y})(x_i - \bar{x})}{\Sigma(x_i - \bar{x})^2}$$

$$\hat{\alpha} = \bar{y} - \hat{\beta}\,\bar{x}$$

These well-known and widely used formulas, it should be noted, make no assumptions about the distribution of the errors. Note, too, that the least squares principle does not apply to $\sigma^2$, the error variance, which needs to be estimated separately (*see* section **d.3,** below).

## d WHY LEAST SQUARES?

It is said almost universally by classical statisticians that if the regression of $y$ on $x$ is linear, then least squares provides the best estimates of the regression parameters $\alpha$ and $\beta$. The term 'best' in this context intimates some epistemic significance and suggests that the least squares method is for good reason preferable to the infinitely many other conceivable methods of estimation. Many statistics textbooks adopt the least squares method uncritically, but where it is defended, the argument takes three forms. First, the method is said to be intuitively correct, and secondly and thirdly, to be justified by the Gauss-Markov theorem and by the Maximum Likelihood Principle. We shall consider these lines of defence in turn.

### d.1 Intuition as a Justification

The least squares method is often recommended for its "intuitive plausibility" (Belsley et al., 1980, p. 1). However, the intuitions cited do not point unequivocally to least squares as the right method; at best they rule out just some of the alternative methods. For example, the idea that one should minimize the absolute value of the sum of the errors, $|\Sigma e_i|$, is often excluded because, in certain circumstances, it leads to intuitively unsatisfactory regression lines. For instance, in the case illustrated below, the criterion admits two lines for the same data, one of which *(b)* "is intuitively . . . a very bad one" (Wonnacott and Wonnacott, p. 16).

The trouble here is that the criterion of minimum aggregate error can be met by balancing large positive errors against large negative ones, whereas for an intuitively good fit

Two lines both satisfying the condition that $|\Sigma e_i|$ is minimized.

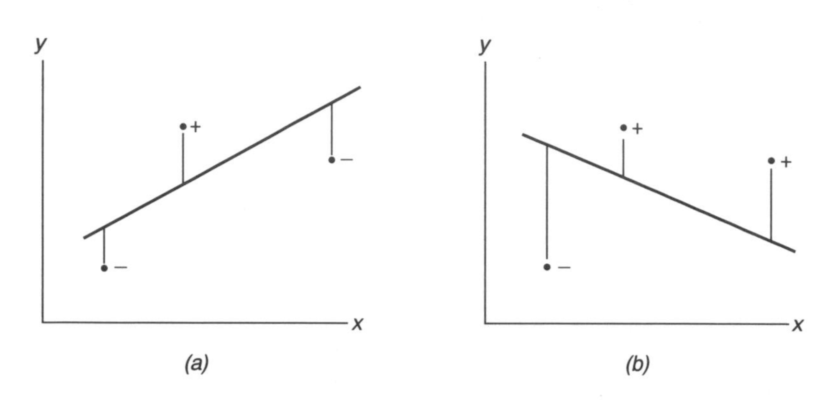


of data to a line, all the errors should be as small as possible. This difficulty would be partly overcome by another suggested criterion, namely, that of minimizing the sum of absolute deviations, $\Sigma|e_i|$, but this too is often regarded as unsatisfactory (for example, by Wonnacott and Wonnacott, who signal their disapproval by referring to it with the acronym MAD). But this highlights a weakness of arguments for least squares based on people's intuitions, namely, that people tend to disagree over the intuitions themselves. Thus, unlike the Wonnacotts, Brook and Arnold (1985, p. 9) find the MAD method perfectly sane, indeed, "a sensible approach which works well", their sole reservation being the practical one that "the actual mathematics is difficult when the distributions of estimates are sought". The reason why it is regarded as important for such distributions to be accessible to calculation is that they figure in the derivation of confidence intervals.

Brook and Arnold, on the other hand, see in the least squares method the intuitively unsatisfactory feature that particularly large deviations, since they must be squared, "may have an unduly large influence on the fitted curve" (1985, p. 9). They refer with approval to a suggestion for mitigating this supposed drawback, according to which one should minimize the sum of a lower power of the errors than their squares. That is, one should minimize $\Sigma|e_i|^p$, with $p$ somewhere between 1 and 2, and they propose $p = 1.5$ as a "reasonable compromise", though why the mid-point value should accord so well with reason is not explained and seems merely arbitrary, despite its air of Solomonic wisdom. But against this proposal, they point out the practical difficulty that, as with the MAD estimation method, "it is difficult to determine the exact distributions of the resulting estimates", and so—for this purely pragmatic reason—they revert to $p = 2$.

The deviations whose squares are minimized in the least squares method appear in a graph as the vertical distances of points from a line. It might be thought that if a close fit between line and points was desired, this could be secured by applying the least squares principle to the perpendicular distances of the points from the line, or to the horizontal distances, or, indeed, to distances measured along any other angle. Why should the vertical deviations be privileged above all others?

Kendall and Stuart defend the usual least squares procedure, based on vertical distances, on the vague and inadequate grounds that since "we are considering the dependence of $y$ on $x$, *it seems natural* to minimize the sum of squared deviations in the $y$-direction" (1979, p. 278; italics added). But this by itself is insufficient reason, for many procedures that we find natural are wrong; indeed, Kendall and Stuart's famous textbook offers numerous 'corrections' for what they view as misguided intuitions.

Brook and Arnold advance an apparently more substantial reason for concentrating on vertical deviations from the regression line, namely that "when our major concern is predicting $y$ from $x$ the vertical distances are more relevant *because they represent the predicted error*" (1985, p. 10; italics added). For most statisticians the main interest of regression equations does indeed lie in their ability to predict (*see* section **e**). But predictions concern previously unexamined values of the variables; the predicting equation, on the other hand, is built up from values that have already been examined. It is misleading therefore to say that "vertical distances [of data points] are more relevant because they represent the predicted error"; in fact, they do *not* represent the errors in predictions of previously unexamined points. To establish their claim, Brook and Arnold would need to show that a regression equation derived by minimizing in the $y$-direction squared deviations of already-observed points gives better predictions (in some yet-to-be-specified sense of 'better') than those that minimized in other directions. However, no such demonstration has appeared.

The following seems a more promising argument for minimizing the squares of the vertical distances. Suppose the units in which $y$ was measured were changed, by applying a linear transformation, $y' = my + n$, $m$ and $n$ being constants. The transformed regression equation would then be $y' = \alpha' + \beta' x + \epsilon$, where $\alpha' = m\alpha + n$ and $\beta' = m\beta$. Ehrenberg (1975) and others (e.g., Bland, 1987, p. 192) have pointed out that if the best estimate of $\alpha$ is $\hat{\alpha}$, the best estimate of $\alpha'$ should be $m\hat{\alpha} + n$; correspondingly for $\beta'$. If 'best' is defined in terms of a least squares based on vertical distances, then this expectation is satisfied, a fact that advocates rightly regard as a merit of that approach. However, the same argument could be made for a least squares method based on horizontal distances, so it is inconclusive. More seriously, the argument is self-defeating, for unless linear transformations could be shown to be special in some appropriate way, non-linear transformations of $y$ should also lead to correspondingly transformed least squares curves; but this is not in general the case. Indeed, it is practically a universal rule in classical estimation—though maximum likelihood estimation is an exception (*see* section **d.3,** below)—that if $\hat{\alpha}$ is its optimal estimate of $\alpha$, then $f(\hat{\alpha})$ is suboptimal relative to $f(\alpha)$; in other words, classical estimators are generally not 'invariant'.

Weisberg (1980, p. 214) illustrates this rule with some data on the average brain and body weights of different species of animals. The least squares line for the data (expressed logarithmically) is log(br) = 0.93 + 0.75log(bo). An unbiased estimate (or prediction) of the log(brain weight) of a so far unobserved species with a mean body weight of 10 kg can now be easily calculated: it would be 0.93log(10) + 0.75 = 1.68. Weisberg regards this as a satisfactory prediction, for one reason, because it is unbiased. From the prediction, it seems natural to infer that 47.7 kg (the antilogarithm of 1.68) would be a satisfactory prediction for the species' average brain weight. Yet Weisberg describes that conclusion as "naive", because estimates of brain weights derived in this way are "biased and will be, on the average, too small".

But it seems commonsensical and not at all naive to assume that estimates should be invariant under functional transformation; indeed, this is often acknowledged in classical texts when maximum likelihood estimation is under discussion and invariance is offered as one of the properties of that approach which particularly recommend it (e.g., Mood and Graybill, p. 185).

So much for intuitive arguments in favour of least squares. They are at best inconclusive, and most classical statisticians believe a more-telling case can be made on the basis of certain objective statistical properties of least squares estimates, as we shall describe.

### d.2 The Gauss-Markov Justification

The objective properties we are referring to are those of unbiasedness, relative efficiency, and linearity, each of which are possessed by least squares estimates of the linear-regression parameters, $\alpha$ and $\beta$. The unbiasedness criterion is well established in all areas of classical estimation, though, as we have already argued, with questionable credentials. The criterion of relative efficiency as a way of comparing different estimators by their variances is also standard; least squares estimates are efficient against a certain class of estimators, a point we shall return to. Linearity, by contrast, is a novel criterion that was specially introduced into the regression context. It is simply explained: estimates of $\alpha$ and $\beta$, $\hat{\alpha}$ and $\hat{\beta}$, are *linear* when they are weighted linear combinations of the $y_i$, that is, when $\hat{\alpha} = \Sigma a_i y_i$ and $\hat{\beta} = \Sigma b_i y_i$. A comparison with the formulas given above confirms that least squares estimates are linear, with coefficients $b_i = \frac{(x_i - \bar{x})}{c}$, and $a_i = \frac{\sum\left(\frac{x_i^2}{n} - \bar{x}x_i\right)}{c}$, where $c = \Sigma(x_i - \bar{x})^2$.

We now come to the Gauss-Markov theorem, which is the most frequently cited "theoretical justification" (Wonnacott and Wonnacott, p. 17) for the method of least squares. The theorem was first proved by Gauss in 1821, Markov's name becoming attached to it because, in 1912, he published a version of the proof which Neyman believed to be an original theorem (Seal, 1967, p. 6). The theorem states that *within the class of linear, unbiased estimators of $\beta$ and $\alpha$, least squares estimators have the smallest variance.*

We have already explored the grounds for unbiasedness

and relative efficiency. Even those who are unconvinced by our arguments against these criteria would still have to satisfy themselves that linearity was a reasonable requirement before calling on the Gauss-Markov theorem in support of least squares estimation. Often this need seems not to be perceived, for instance by Weisberg (1980, p. 14), who says, on the basis of the Gauss-Markov theorem, that "if one believes the assumptions [of linear regression], and is interested in using linear unbiased estimates, the least squares estimates are the ones to use"; but Weisberg simply takes his readers' interest in linear estimates for granted. Seber (1977, p. 49), similarly, calls them "a reasonable class of estimates", without explaining why he thinks they merit such approbation.

The only justification we have seen for requiring estimators to be linear turns on their supposed "simplicity". "The property of linearity is advantageous", say Daniel and Wood (1980, p. 7), "in its simplicity". Wonnacott and Wonnacott restrict themselves to linear estimates "because they are easy to compute and analyse" (p. 31). But *simplicity and convenience are not epistemological categories; the simplest and quickest route might well lead you astray and often does.*

Mood and Graybill do not appeal to simplicity but refer mysteriously to "some good reasons why we *would* want to restrict ourselves to linear functions of the observations $y_i$ as estimators" (1963, p. 349); but they compound the obscurity of their position by adding that "there are many times we would *not"*, again without explanation, though they probably have in mind cases where some data points show very large deviations from the presumed regression line or in some other way appear unusual (we shall deal separately with this concern in section **f.2**).

The limp and inadequate remarks we have reported seem to constitute the only defences which the linearity criterion has received, which surely suggests that it has no epistemic basis. Mood and Graybill confirm this when they admit that since "it will not be possible to examine the 'goodness' of the [least squares] estimator $\hat{\beta}$, relative to all functions . . . *we shall have to* limit ourselves to a subset of all functions" (p. 349; italics added). As an example of such a subset they cite linear functions of the $y_i$ and then show how, according to the Gauss-Markov theorem, the method of least squares excels within that subset, on account of its minimum variance. But how can we be sure that least squares is not just the best of a bad lot? Mood and Graybill do not say.

The Gauss-Markov justification meets a further difficulty. The claim is that among linear unbiased estimates, least squares estimates possess the desirable feature of minimum variance. Whatever that minimum variance was in particular cases, there would normally be alternative, biased, and/or nonlinear ways of estimating the parameters, which, however, have a smaller variance than least squares has. So to establish least squares as superior to such alternative methods, you would need to show that the benefit of the smaller variance offered by the alternatives was outweighed by the supposed disadvantage of their bias and/or non-linearity. But this has not been shown, and we would judge the prospects for any such demonstration to be dim.

We conclude that the Gauss-Markov theorem provides no basis for the least squares method of estimation. Let us move to the third way that the method is standardly defended.

### d.3 The Maximum-Likelihood Justification

The maximum-likelihood estimate of parameter $\theta$, relative to data $d$, is the value of $\theta$ which maximizes $P(d \mid \theta)$. The principle that asserts that the maximum-likelihood estimate is 'good' or 'best' is incorrect judged from a Bayesian viewpoint, though it does hold when the prior probability distribution over $\theta$ is uniform, and it is more and more closely approximated as the sample size increases. That the general principle is incompatible with Bayes's Theorem is obvious, since the $\theta$ with maximum likelihood might possess a relatively low prior probability, in which case it would not necessarily have the greatest posterior probability. In other words, a Bayesian estimate would balance prior information against information conveyed by the sample, whereas such information, unless strictly 'objective', has no role to play in classical inference.

Another difference between maximum likelihood and Bayesian estimation lies in the character of its estimates, which in the Bayesian case takes the form of a statement of how probable various values of the parameter are in light of the data; but maximum-likelihood estimation gives no clear instruction as to the appropriate psychological attitude that should be adopted towards its particular estimates, except, unhelpfully, that they should be regarded as 'good' because they have some allegedly 'good' properties. (For instance, as already mentioned, maximum-likelihood estimators are invariant under functional transformations; they are also consistent and are functions of sufficient statistics, where such statistics exist.)

The maximum-likelihood method can be used to estimate $\alpha$ and $\beta$ in a linear regression equation only if the form of the error distribution is stipulated, for only then does the regression equation confer a specific probability on the data. Least squares, by contrast, can be applied without that information. It turns out that when the regression errors are normally distributed, the maximum-likelihood estimates of $\alpha$ and $\beta$ are precisely those arrived at by least squares, a fact that supposedly provides "another important theoretical justification of the least squares method" (Wonnacott and Wonnacott, p. 54).

Those who regard this as a justification clearly assume that the maximum-likelihood method is correct, and that it furnishes reasonable estimates; they then argue that since least squares and maximum likelihood coincide in the specific case where the errors are distributed normally, the least squares principle gives reasonable estimates in general. Although we have done so, it is unnecessary to take a view on the merits of maximum-likelihood estimation to appreciate that this is a blatant non sequitur; you might just as well argue that because you get the right answer with 0 and 1, $x^3$ is a good way of estimating $x^2$.

## d.4 Summary

The classical arguments in favour of least squares estimation of regression parameters are quite untenable. That based on intuition is inconclusive, vague, and lacking in epistemological force. The Gauss-Markov justification rests on the linearity criterion, which itself rests on nothing. And the maximum-likelihood defence is simply fallacious.

The two theoretical defences of least squares are standardly cited together as if they were mutually reinforcing or complementary. In fact, the reverse is the case. For while the Gauss-Markov justification takes unbiasedness as a desideratum, the maximum-likelihood principle does not. Indeed, it sometimes delivers biased estimates, as for example, in the case at hand: the maximum-likelihood estimate of $\sigma^2$ is

$\frac{1}{n}\Sigma(y_i - \hat{\alpha} - \hat{\beta}x_i)^2$. But when the regression of $y$ on $x$ is linear, this estimate is biased and needs to be increased by the factor $\frac{n}{(n - 2)}$ to unbias it. (It is this modified estimate, labelled $\hat{\sigma}$, that is always employed in classical expositions.)

The two main 'theoretical' defences of least squares estimation are therefore separately ineffective and mutually destructive.

But all this does not mean that the least squares method is entirely wrong. Its great plausibility does have an explanation, a Bayesian explanation. For it can be shown that when the set of possible regressions is restricted to the linear, and one assumes a normal distribution of errors and a uniform prior distribution of probabilities over parameter values, the least squares and the Bayes solutions coincide, in the sense that the most probable value of the regression parameters and their least squares estimates are identical. When the assumption of a uniform distribution is relaxed, the same result follows, though it is reached asymptotically, as the size of the sample increases (Lindley and El-Sayyad, 1968).

## e PREDICTION

### e.1 Prediction Intervals

Fitting a regression curve to data is often said to have as its main practical purpose the prediction of so far unobserved points. The most commonly attempted predictions are of $y$ values from preselected values of $x$. Making such predictions would be straightforward if one knew the precise regression equation as well as the form of the error distribution. For then the distribution of $y_o$, the $y$ corresponding to some particular $x_o$, would also be known, thus enabling one to calculate the probability with which $y_o$ would fall within any given range. Such a range is sometimes called a *prediction interval.*

The problem is that the regression parameters are usually unknown and have to be estimated from data. To be sure, if those estimates were qualified by probabilities, as they would be after a Bayesian analysis, you could still calculate the probability relative to the data that $y_o$ lay in any specified range. But this option is closed to classical statistics, which proceeds from a denial that theories (parameter values, in the present case) do have probabilities, except in special circumstances that do not prevail in the regression problems we are considering.

### e.2 Prediction by Confidence Intervals

The classical way round this difficulty is to make predictions using confidence intervals, by a method similar to the one already discussed. First, an experiment is imagined in which a fixed set of *x*s, $x_1, \ldots, x_m$, is chosen. A prearranged number, $n$ ($\geq m$), of $x,y$ readings is then made. A linear regression of $y$ on $x$ is assumed and $\hat{\alpha}$ and $\hat{\beta}$ are the resulting least squares estimates of the corresponding regression parameters. A single, new reading $x_o, y_o$ is now taken. It is this $y$, which we have labelled $y_o$, whose value it is desired to predict. On the linear regression assumption, $y_o$ is a random variable given by $y_o = \alpha + \beta x_o + \epsilon$. The following analysis requires the error terms to be normally distributed. A new random variable, $u$, with variance $\sigma_u^2$, is now defined as $u = y_o - \hat{\alpha} - \hat{\beta} x_o$. The random variable, $t$, is then considered, where

$$t = \frac{\sigma u}{\sigma_u \hat{\sigma}} \sqrt{\frac{n-2}{n}}.$$

The ratio $\frac{\sigma}{\sigma_u}$ is independent of $\sigma$, being determined just by $x_1, \ldots, x_m$, $x_o$, and $n$, so $t$ can always be computed from the data. Because $t$ has a known distribution (the $t$-distribution with $n - 2$ degrees of freedom), standard tables can be consulted to find specific values, $t_1$ and $t_2$, say, which enclose, say, 95 per cent of the area under the distribution curve. This means that with probability 0.95, $t_1 \leq t \leq t_2$, from which it follows (Mood and Graybill, 1963, pp. 336–37) that

$$P(\hat{\alpha} + \hat{\beta} x_o - \mathbf{A} t_1 \leq y_o \leq \hat{\alpha} + \hat{\beta} x_o + \mathbf{A} t_2) = 0.95,$$

where $\mathbf{A}$ is a complicated expression involving $n$, $x_o$, $\bar{x}$ (the average of the $x_1$ to $x_m$ in the data set), and $\hat{\sigma}$. The terms $\hat{\alpha}$, $\hat{\beta}$, $\mathbf{A}$, and $y_o$ are all random variables that can assume different values depending on the result of the experiment described earlier. If $\hat{\alpha}'$, $\hat{\beta}'$, and $\mathbf{A}'$ are the values taken by the corresponding variables in a particular trial, the interval $\{\hat{\alpha}' + \hat{\beta}' x_o - \mathbf{A}' t_1, \hat{\alpha}' + \hat{\beta}' x_o + \mathbf{A}' t_2\}$ is a 95 per cent confidence interval for $y_o$. As we noted in our earlier discussion, other confidence intervals, associated with other probabilities, may be described too.

On the classical view, a confidence interval supplies a good and objective prediction of $y_o$, independent of subjective prior probabilities, with the confidence coefficient (95 per cent in this case) measuring how good it is. Support for that view is sometimes seen (e.g., by Mood and Graybill, 1963, p. 337) in the fact that the width of any confidence interval for a prediction increases with $x_o - \bar{x}$. So according to the interpretation we are considering, if you wished a constant degree of confidence for all predictions, you would have to accept wider, that is, less precise or less accurate intervals, the further you were from $\bar{x}$. This, it is suggested, explains the "intuition" that "we can estimate most accurately near the 'centre' of the observed values of $x$" (Kendall and Stuart, 1979, p. 365). This does seem to be the general intuition, *provided we are assured that the regression is linear* and the explanation given is a point in favour of the confidence-interval approach. However, it is insufficient to rescue that approach from the radical criticisms already made. As we explained earlier, the two standard interpretations of confidence intervals—the categorical-assertion interpretation and the subjective-confidence interpretation—are both incorrect. Hence, confidence intervals do not constitute estimates; for the same reasons they cannot properly function as predictions. That means that the declared principal goal of regression analysis—prediction—cannot be achieved by classical means.

### e.3 Making a Further Prediction

Suppose, having 'predicted' $y_o$ for a given $x_o$, its true value is disclosed, thus augmenting the data by an additional point, and suppose you now wished to make a further prediction of $y_o'$ corresponding to $x_o'$. It is natural to base the second prediction on the most up-to-date estimates of the linear regression coefficients, which should therefore be recalculated using the earlier data plus the newly acquired data point.

Such a procedure is in fact recommended by Mood and Graybill, but their recommendation does not arise, as it would for a Bayesian, from a concern that predictions should be based on all the relevant evidence. Instead, they say that if the estimated regression equation were not regularly updated in the light of new evidence, and if it were used repeatedly to make confidence-interval predictions, then the basis of the confidence intervals would be undermined. For those confidence intervals are calculated on the assumption that $\hat{\alpha}$, $\hat{\beta}$, and $\hat{\sigma}$ are variable quantities, arising from the experiment described earlier (section **e.2**) and "if the original estimates are used repeatedly (not allowed to vary) the statement may not be accurate" (Mood and Graybill, 1963, p. 337).

But Mood and Graybill's idea of simply adding the new datum to the old and re-estimating the regression parameters does not appear to solve the problem. For although it ensures variability for the parameter estimates, it fixes the original data points, which are supposed to be variable, and it varies $n$, which should be fixed. This means that the true distribution of $t$, upon which the confidence interval is based, is not the one described above. The question of what the proper $t$-distribution is has not, so far as we are aware, been addressed. But until it is, classical statisticians ought properly to restrict themselves to single predictions or else change their methodology, for this inconvenient, unintuitive, and regularly ignored restriction does not arise for the Bayesian, who is at liberty to revise estimates with steadily accumulating data: indeed, there is an obligation to do so.

## f EXAMINING THE FORM OF A REGRESSION

The classical way of investigating relationships between variables is, as we have described, to assume some general form for the relationship and then allow the data to dictate its particular shape by supplying values for the unspecified parameters. In their starting assumption, statisticians show a marked preference for linear regressions, partly on account of the conceptual simplicity of such regressions and partly because their properties with respect to classical estimation techniques have been so thoroughly explored. But statistical relationships are not necessarily nor even normally linear ("we might expect linear relationships to be the exception rather than the rule"—Weisberg, 1980, p. 126); and any data set can be fitted equally well (however this success is measured) to infinitely many different curves; hence, as Cook (1986, p. 393) has said, "some reassurance [that the model used is "sensible"] is always necessary".

But how is that reassurance secured, and what is its

character? The latter question is barely addressed by classical statisticians, but the former, it is widely agreed, can be dealt with in one or more of three ways. First, certain aspects of possible models, it is said, can be checked using significance tests. There is, for example, a commonly employed test of the hypothesis that the $\beta$-parameter of a linear regression is zero. This employs a $t$-test-statistic with $n - 2$ degrees of freedom: $t = \frac{\hat{\beta}}{s(\hat{\beta})}$, where $s(\hat{\beta})$ is the standard error of $\hat{\beta}$, which can be calculated from the data. Another test is often employed to check the homoscedasticity assumption of the linear regression hypothesis, that is, the assumption that $\sigma$ is independent of $x$. Such tests are of course subject to the strictures we made on significance tests in general, and consequently, in our view, they have no standing as modes of inductive reasoning. But even if the tests were valid, as their advocates maintain, they would need to be supplemented by other methods for evaluating regression models, for the familiar reason that the conclusions that may be drawn from significance tests are too weak for practical purposes. Learning that some precise hypothesis, say that $\beta = 0$, has been rejected at such-and-such significance level is to learn very little, since in most cases it would already have been extremely unlikely, intuitively speaking, that the parameter was *exactly* zero. In any case, most practitioners require more than this meagre, negative information; they wish to know what the true model actually is. The methods to be discussed in the next two subsections are often held to be helpful in this respect. They deal first with the idea that the preferred regression model or models should depend on prior knowledge; and secondly with techniques that subject the data to detailed appraisal, with a view to extracting information about the true model.

### f.1 Prior Knowledge

Hays (1969, p. 551) observed that "when an experimenter wants to look into the question of trend, or form of relationship, he has some prior ideas about what the population relation should be like. These hunches about trend often come directly from theory, or they may come from the extrapolation of established findings into new areas."

Such hunches or prior ideas about the true relationship are often very persuasive. For example, in studying how the breaking load *(b)* varies with the diameter *(d)* of certain fibre segments, Cox (1968, p. 268) affirms that as the diameter vanishes, so should the breaking load; Seber (1977, p. 178) regards this as "obvious". More often, prior beliefs are less strongly held. For example, Wonnacott and Wonnacott, having submitted their data on wheat yields at different concentrations of fertiliser to a linear least-squares analysis, point out that the assumption of linearity is probably false: *"it is likely* that the true relation increases initially, but then bends down eventually as a 'burning point' is approached, and the crop is overdosed" (p. 49; italics added). Similarly, Lewis-Beck points out that although a linear relation between income and number of years of education appears satisfactory judged from the data, *"it seems likely* that relevant variables have been excluded, for factors besides education undoubtedly influence income" (p. 27; italics added). Other statisticians express their uncertainty about the general model in less obviously probabilistic terms, such as "sensible" and "reasonable on theoretical grounds" (Weisberg, 1980, p. 126), and "plausible" (Atkinson, 1985, p. 3); while Brook and Arnold talk of "theoretical clues . . . which *point to* a particular relationship" (p. 12; italics added); and Sprent (1969, p. 120), stretching tentativeness almost to the limit, says that an "experimenter is often in a position . . . to decide that it is reasonable to assume that certain general types of hypothesis . . . may hold, although he is uncertain precisely which".

One further example: Cox argues that if, in the case of the breaking load and diameter of fibres, the two lines $b \propto d$ and $\log b \propto \log d$ fit the data equally well, then the "second would in general be preferable because . . . it permits easier comparison with the theoretical model load $\propto$ (diameter)$^2$" (Cox, 1968, p. 268). This is a very circumspect way of recommending a regression model and would seem to be no recommendation at all unless the "theoretical model" was regarded as likely to be at least approximately true. That this is the implicit assumption seems to be confirmed by Seber's exposition (1977, p. 178) of Cox's view, in which he commends the model slightly less tentatively, saying that it "might be" a "reasonable assumption".

One might expect the natural uncertainty attaching to the general model to be revised, ideally diminished, by the data, so producing an overall conclusion that incorporates both the prior and posterior information. This is how a Bayesian analysis would operate. A Bayesian would interpret the various expressions of uncertainty uniformly as prior probabilities and then use the data to obtain corresponding posterior probabilities, though if the distribution of prior beliefs is at all complicated, the mathematics may become extremely difficult; and such difficulties have by no means all been overcome.

In the classical case, the difficulty is not merely mathematical or technical, but arises from a fundamental flaw. For, in the first place, the prior evaluation of a model in the light of plausible theories and previous data seems not, as a rule, to be objectively quantifiable, as the classical approach would demand; certainly none of the authors we have quoted offers any measure, objective or otherwise, of the strength of their hunches. Secondly, even if the reasonableness of a theory could be objectively measured, classical statistics offers no way of combining such measures with the results of standard inference techniques (significance testing, confidence-interval estimation, and so forth) to achieve an aggregate index of appraisal.

The difficulty of incorporating uncertainty about the model into standard classical inferences is apparent from a commonly encountered discussion concerning predictions. Typical expositions proceed by first assuming a linear relation. The apparatus of confidence intervals is next invoked as a way of qualifying predictions with a degree of confidence. It is then explained that the linearity assumption is often doubtful, though perhaps roughly correct over the relatively narrow experimental range, and hence that one should not expect predictions from a least-squares-fitted equation to be quite accurate (e.g., Weisberg, 1980, p. 126; Seber, 1977, p. 6; Gunst and Mason, 1980, pp. 56–63; Wonnacott and Wonnacott, 1980, p. 49).

But instead of measuring this uncertainty about the model and then amalgamating it with the uncertainty reflected in the confidence interval, so as to give an overall level of confidence in a prediction, classical statisticians merely issue warnings to be "careful not to attempt to predict very far outside the [range covered by the data]" (Gunst and Mason, p. 62) and to "be reluctant to make predictions [except for] . . . new cases with predictor variables not too different from [those] . . . in the construction sample" (Weisberg, p. 215). But these vague admonitions do not signify how such caution should be exercised (should one hold off from predicting altogether, or tremble slightly when hazarding a forecast, or what?).

There is also the question of how close $x_0$ should be to the $x$ values in the data, before the corresponding $y_0$ can be predicted with assurance. This is always given an arbitrary answer. For example, Weisberg, in dealing with the case where $y$ is linearly related to two predictor variables $x$ and $z$, suggests, with no theoretical or epistemological sanction, that a "range of validity for prediction" can be determined by plotting the data values of $x$ and $z$ and drawing "the smallest closed figure that includes all the points" (p. 216). Making such a drawing turns out to be difficult, but Weisberg considers that "the smallest volume ellipsoid containing these points" (p. 216) would be a satisfactory approximation to the closed figure. However, he is uneasy with the proposal since, in the example on which he is working, "there is a substantial area inside the [ellipsoid] . . . with no observed data" (p. 217) and where, presumably, he feels prediction is unsafe. Weisberg's demarcation between regions of safe and risky predictions and his suggested approximation to it have some plausibility, but they seem quite arbitrary from the classical point of view.

Classical statisticians are caught in a cleft stick. If they take account of plausible prior beliefs concerning the regression model, they cannot properly combine those beliefs with the classical techniques of inference. On the other hand, if they use those techniques but eschew prior beliefs, they have no means of selecting, arbitrary stipulation apart, among the infinitely many regression models that could be fitted to the data.

### f.2 Data Analysis

A possible way out of this dilemma is to abandon the imprecise, unsystematic, and subjective appraisals described in the previous section and rely solely on the seemingly more objective methods of 'data analysis', though most statisticians would not go so far, preferring to avail themselves of both techniques. Data analysis, or 'case analysis' as it is often called, is an attempt to discriminate in a non-Bayesian way between possible regression models through a close examination of the individual data points. There are three distinct approaches to data analysis, which we shall consider in turn.

***i* Inspecting scatter plots**

The simplest kind of data analysis involves visually examining ordinary plots of the data ('scatter plots'). The inspection of such plots as a source of information is authoritatively endorsed by Cox (1968, p. 268)—"the choice [of relation] will depend on preliminary plotting and inspection of the data"; it is "strongly recommended" by Kendall and Stuart (1979, p. 292), because "it conveys quickly and simply an idea of the adequacy of the fitted regression lines"; and in practice it is widely employed. Weisberg (p. 3), for instance, motivates a linear regression analysis of data on the boiling points of water at different atmospheric pressures by referring to the "overall impression of the scatter plot . . . that the points generally, but not exactly, fall on a straight line". And Lewis-Beck (p. 15) claims that "visual inspection of [a certain] . . . scatter plot suggests the relationship is essentially linear".

Weisberg (p. 99) regards the data of diagram 1 as showing a pattern that "one might expect to observe if the simple linear regression model were appropriate". On the other hand, diagram 2 "suggests . . . [to him] that the analysis based on simple linear regression is incorrect, and that a smooth curve, perhaps a quadratic polynomial, could be fit to the data . . .". Although the data are different in the two cases, they give the same least-squares lines, as the diagrams below illustrate.

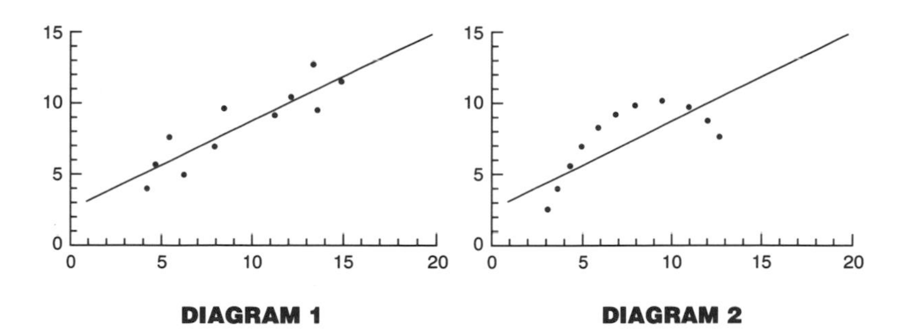


DIAGRAM 1 DIAGRAM 2

Probably most scientists would agree with the various judgments that statisticians make on the basis of visual inspections of scatter diagrams. But how are such judgments arrived at? It could be, indeed, it seems likely, that a close analysis would reveal a Bayesian mechanism, but it is hard to imagine how any of the standard classical techniques could be involved in either explaining or justifying the process, nor do classical statisticians claim they are. In fact, the whole process seems impressionistic, arbitrary, and subjective.

### *ii Outliers*

Another instructive data pattern is illustrated in diagram 3. One of the points stands much further apart from the least squares line than any other and this suggests to Weisberg (p. 99) that "simple linear regression may be correct for most of the data, but one of the cases is too far away from the fitted regression line". Such points are called *outliers*.

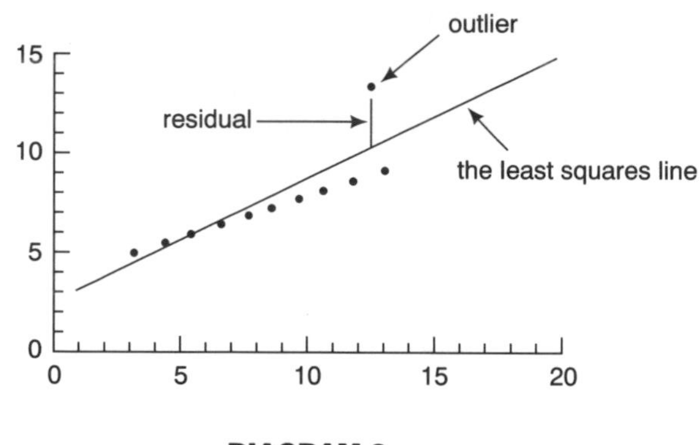


**DIAGRAM 3**

Outliers are defined, rather imprecisely, as "data points [with] . . . residuals that are large relative to the residuals for the remainder of the observations" (Chatterjee and Price, 1977, p. 19). (The residual of a data point is its vertical distance from a fitted line or curve.) As is often noted, a point may be an outlier for three distinct reasons. First, it could be erroneous, in the sense that it results from a recording or transcription error or from an improperly conducted experiment; an outlier could also arise because the assumed regression model is incorrect; on the other hand, the model might be correct and the outlier be simply one of those relatively improbable cases that is almost bound to occur sometimes. Suppose that careful checking has more or less excluded the first possibility, does the outlier throw any light on whether an assumed regression equation is correct or not? This is a question to which classical statisticians have applied a variety of non-Bayesian methods, though, as we shall argue, without any satisfactory result.

Some authors, for example Chatterjee and Hadi, seem not to take seriously the possibility that the least squares line is wrong, when they note that an outlier in their data, if removed, would hardly affect the fitted line. They conclude that "there is little point in agonizing over how deviant it appears" (1986, p. 381). But this is surely a mistake and seems not to have been endorsed by Chatterjee when he was collaborating with Price.

Chatterjee and Price's approach may be illustrated with their own example. Their data consist of 30 $x,y$ readings, the nature of which need not concern us. They applied the linear least-squares technique to the readings, obtaining an upward sloping line. They next examined a number of data plots, first $y$ against $x$, then residuals against $x$, and finally the residuals against $\hat{y}$ ($\hat{y}_i$ is the point on the fitted line corresponding to $x_i$). Four points stand out as having particularly large residuals; moreover, a visual inspection of the various plots makes it "clear [to the authors] that the straight line assumption is not confirmed" (p. 24), though it "looks acceptable" in the middle of the $x$ range. On this informal basis, Chatterjee and Price conclude tentatively that the true line has zero gradient and that $y$ is independent of $x$. They check this conjecture by dropping the four outliers, computing a new least-squares line and then examining the revised plots of residuals. This time they find no discernible pattern; from a casual inspection "the residuals appear to be randomly distributed around [the line] $e$[residual] = 0" (p. 25). Unfortunately, the conclusion from all this is disappointingly weak. It is that the regression with zero gradient "is a satisfactory model for analyzing the . . . data, *after the deletion of the four points*" (p. 25; italics added). But this says nothing about the true relation between $x$ and $y$. Indeed, that this question is still open is acknowledged by the authors, who next ask: "what of the four data points that were deleted?" They repeat the triviality that the points may be "measurement or transcription errors", or else "may provide more valuable information about the . . . relationships between $y$ and $x$". We take the latter to mean that the line derived from the data minus the outliers might be wrong, perhaps wildly so, it should be added (of course, it might be right too). Whether the fitted line is right or wrong, and if wrong what the true line is, are questions that Chatterjee and Price simply leave to further research: "it may be most valuable to try to understand the special circumstances that generated the extreme responses" (p. 25). In other words, their examination of the data from different points of view has been quite uninformative about the true regression and about why some of the points have particularly large residuals relative to the linear least-squares line.

Chatterjee and Price call theirs an "empirical approach" to outliers, since it takes account not just of the sizes of the residuals but also of their patterns of distribution. They disagree with those who use significance tests alone to form judgments from outliers since, in their view, "[i]t is a combination of the actual magnitude and the pattern of residuals that suggests problems" (p. 27). But although significance tests take little account of the pattern of the results, compared with the empirical approach, they seem more in keeping with the objectivist ideals of classical statistics and to offer more definite conclusions about the validity of hypothesized regression equations. Weisberg is a leading exponent of the application of significance tests in this context. His method is to perform a linear least-squares analysis on the data set minus one point (in practice this would be an outlying point) and then to use a significance test to check the resulting regression equation against the removed point.

To see how this is done, suppose the fitted line derived from the data, minus the $i$th point, has parameters $\hat{\alpha}_{-i}$, and $\hat{\beta}_{-i}$. The assumption that this is the true line will be the null hypothesis in what follows. Let $\hat{y}_i$ be the $y$ value corresponding to $x_i$ on the null hypothesis line.

Intuitively, if the data were properly collected and record-

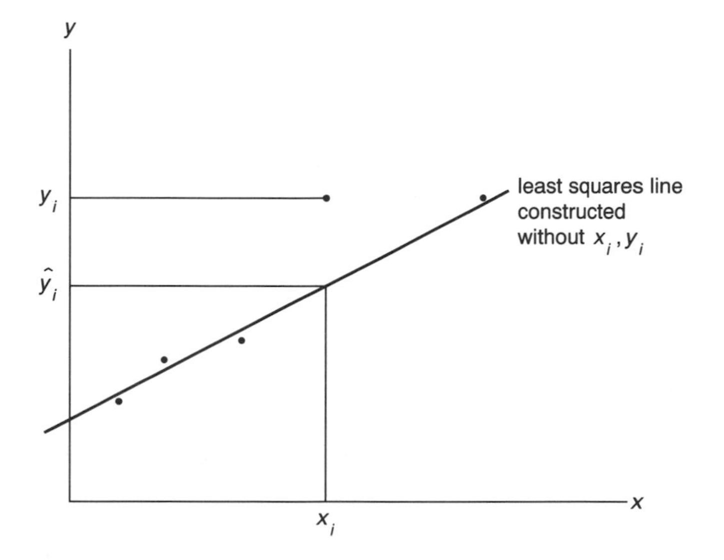


ed, the larger the discrepancy between $\hat{y}_i$ and $y_i$, the less one's confidence that the line drawn is correct. This idea is given classical clothes through a corresponding significance test. Weisberg's test uses as test-statistic a particular function of $y_i - \hat{y}_i$ and other aspects of the data—this has the $t$-distribution with $n-3$ degrees of freedom. If the $t$-value recorded for a data set relative to a particular point $(x_i, y_i)$ in that set exceeds the test's critical value, then the null hypothesis would be rejected at the corresponding level of significance.

But fixing the appropriate critical value is interestingly problematic and shows up the pseudo-objectivity and ineffectiveness of the whole procedure. It is required in significance testing to ensure that the null hypothesis would be falsely rejected with some predesignated probability, usually 0.05 or 0.01. Now Weisberg's test could be conducted in a spectrum of ways. At one end of the spectrum you could decide in advance to perform just one significance test, using the single datum corresponding to a prespecified $x_i$. At the other extreme, the plan could be to check the significance of every data point. As Weisberg points out, if the test was, as a matter of policy, restricted to the datum with the largest $t$-value, one would in effect be following the second procedure.

The probability of a small $t$-value arising in each of these cases can be very different, especially where the data are numerous. In Weisberg's example (p. 116) the probability of $t$ *exceeding 2.00 is 0.05 if a single test is envisaged. But the probability of* $t$ being greater than 2.00 in at least one of 65 tests on 65 data points is approximately 0.96 (since the tests are correlated, it is difficult to work out this probability precisely). To ensure a 0.05 probability of falsely rejecting the null hypothesis in this second case, a very much smaller critical $t$-value would have to be employed than in the first case.

So whether a particular $t$-value is significant is sensitive to how the person performing the significance test planned to select data for testing. But, as we have stressed before, such private plans have no epistemic significance, and without justifiable, public rules to fix the most appropriate plan, the present approach is subject to personal idiosyncracies at odds with its supposed objectivity. Weisberg does in fact propose a rule for choosing a testing plan, namely, that "if the investigator suspects in advance" that a particular case will fail to fit the same straight line as the others, then the larger critical $t$-value, appropriate for a single significance test, should be used (p. 116).

The thinking behind Weisberg's rule seems to be that if one particular point is suspected in advance of being an exception to a general pattern which the other points are expected to show, then only that anticipated exception will be scrutinized as an outlier; hence, only it will be subjected to a significance test. But this would not be a satisfactory argument, for suppose that the anticipated outlier unexpectedly fitted the pattern and that some of the other points were surprisingly prominent outliers. Weisberg's rule implies that these unexpected outliers should be overlooked as possible counter-examples, since the previously formulated policy was not to expose them to a test of significance, a recommendation that is contrary both to practice and reason. The rule is therefore unreasonable, in addition to bringing out clearly the subjective nature of judgments derived from Weisberg's significance tests.

Weisberg introduces and illustrates his rule with data on the average brain (br) and body weights (bo) of different animal species, which show man as the most prominent outlier in a plot of log(br) against log(bo). Weisberg argues that since "interest in Man as a special case is a priori" (p. 130), the outlier test should employ the larger critical value, whereupon the datum on man is significant and the null hypothesis of a log-linear relation is rejected. (Weisberg seems to conclude that such a relation is satisfactory, provided man is excluded as an exception, although the data on several other species are not much less outlying than that for man.) But if "interest in Man" had not been "a priori", the datum would not have been significant and Weisberg would have reached the opposite conclusion.

Weisberg's example, moreover, does not even conform to his own rule: man is not picked out because of an earlier suspicion that he is an exception to the log linear rule, but because of a prior "interest in Man as a special case". No explanation is given for this change, though we conjecture that it is made because, first, from a visual inspection of the data, man is obviously an exception to a log-linear rule; secondly, if the small critical value were employed in Weisberg's 'outlier test', man would *not* be significant and the linear rule would, counterintuitively, have to be accepted; and thirdly, there is no plausible reason to suspect man a priori of being an exception to a log-linear rule—hence, no reason to employ the larger critical value in the significance test. This may perhaps account for Weisberg's changing his rule to suit these circumstances, but clearly it provides no justification.

It seems undeniable that the distribution patterns of data, including the presence of outliers, often tell us a great deal about the regression, but they seem not to speak in any known classical language. It is perhaps too early to say for certain that the medium of communication is Bayesian, since a thorough Bayesian analysis of the outlier notion is still awaited. The form that that analysis will take is perfectly clear, however. As with every other problem concerning the evaluation of a theory in the light of evidence, it will involve the application of Bayes's Theorem. Thus, for a Bayesian, regression analysis is not divided into separate compartments operating distinct techniques, each requiring its own justification; there is just one principle of inference, leading to one kind of conclusion.

#### *iii Influential points*

Another data pattern that statisticians often find instructive is illustrated in diagram 4 (from Weisberg, p. 99).

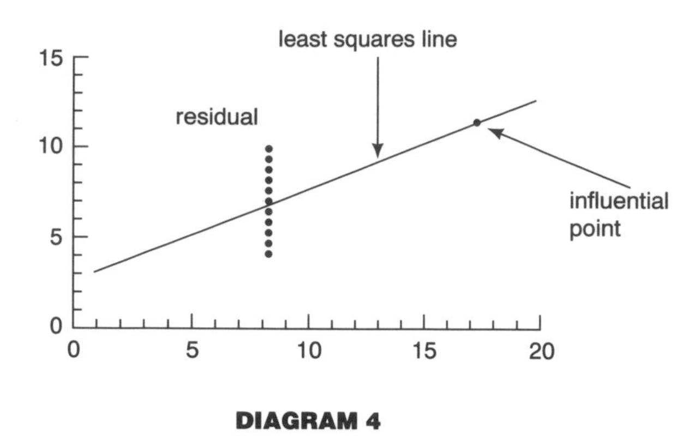


**DIAGRAM 4**

The isolated point on the right has a nil residual, so the considerations of the previous section give no reason to doubt the linear least-squares line. But the point is peculiar in that the estimated line depends very largely on it alone, much more so than on any of the other data. Such points are called 'influential'. Diagram 4 shows an extreme case of an influential data point; it is extreme because without it no least squares solution exists. There are less pronounced kinds of influence which statisticians also regard as important, where particular data points or a small subset of the data "have a disproportionate influence on the estimated [regression] parameters" (Belsley et al., 1980, p. 6).

Many statisticians regard such influential points as deserving special attention, though they rarely explain why; indeed, large textbooks are written on how to measure "influence", without the purpose of the project being adequately examined. All too often the argument proceeds thus: "these [parameter estimates] . . . can be substantially influenced by one observation or a few observations; that is, not all the observations have an equal importance in least squares regression. . . . It is, *therefore,* important for an analyst to be able to identify such observations and assess their effects on various aspects of the analysis" (Chatterjee and Hadi, 1986, p. 379; italics added). Atkinson (1986, p. 398) and Cook (1986, p. 393) give essentially the same argument, which is clearly a non sequitur, without the further and no doubt implicit assumption that conclusions obtained using influential points are correspondingly insecure. Weisberg (p. 100) is one of the few who states this explicitly, when he says that "[w]e must distrust an aggregate analysis that is so heavily dependent upon a single case". Belsley et al. (1980, p. 9) say roughly the same but express it in a purely descriptive mode: "the researcher is likely to be highly suspicious of the estimate [of the slope of a regression line]" that has been obtained using influential data.

But why should we distrust conclusions substantially based on just a few data points? Intuitively, there are two reasons. The first is this: the true regression curve passes through the mean, or expected value, of $y$ at each $x$. In describing our best guess as to that curve, and hence our best estimate of those means, we are guided by the observed points. But we are aware that any of the points could be quite distant from the mean of the corresponding $x$, either because the experiment was badly conducted or because of an error in recording or transcribing, or simply because it is one of those rare and improbable departures from the mean that is almost bound to occur from time to time. Sharp departures of a point from its corresponding mean are relatively improbable, but if such a discrepancy actually arose and the datum was, moreover, influential, the conclusion arrived at would be in error by a corresponding margin. Our intuition is that the more likely a reading is to be discrepant in the indicated sense, and the more influential it is, the less certain would one be about the regression. The second intuitive reason for distrusting conclusions partly derived from influential data applies when they are separated from the rest of the data by a wide gap, as for example in diagram 4; intuitively we feel that in the range where there are few or no data points, we cannot be at all sure of the shape of the regression relation.

A programme of research that was initiated in the early 1970s has aimed to explicate these intuitions in classical (or at any rate, non-Bayesian) terms and to develop rules to guide and justify them. This now-flourishing field is called 'influence methodology', its chief object being to find ways of measuring influence.

The principle governing the measurement of influence is to estimate a regression parameter using all the data, then to re-estimate with a selected data point deleted, noting the difference between the two estimates. This difference is then plugged into an 'influence function' or, as it is sometimes called, a 'regression diagnostic', to produce an index of how influential the deleted point was. To put this programme into effect, one needs first to decide on the parameter to be employed in the influence measure. If the regression has been assumed linear with a single independent variable, one could choose from among the three parameters, $\alpha$, $\beta$, and $\sigma$, as well as from the infinity of possible functional combinations of these. Secondly, a measure of influence is required, that is, some function of the difference noted above and (possibly) other aspects of the data; the choice here is similarly vast. Finally, there has to be some way of judging the reliability of the data point or of the regression assumptions from particular values of an influence measure.

Many influence functions have been proposed and suggestions advanced as to the numerical value such functions should take before the reading concerned is designated 'influential'. But these proposals and suggestions seem highly arbitrary, an impression confirmed by those working in the field. For instance, the author of a function known as Cook's Distance admitted that "[f]or the most part the development of influence methodology for linear regression is based on ad hoc reasoning and this partially accounts for the diversity of recommendations" (Cook, 1986, p. 396).

Arbitrariness and adhocness would not be a feature of this area if a relation could be established between a point's influence as defined by a given measure and the reliability or credibility of a fitted line or curve. No such relation has been discovered. Nevertheless, recommendations on how to interpret levels of influence are sometimes made. For example, Velleman and Welsch (1981) feel that "values [of their preferred function] greater than 1 or 2 seem reasonable to nominate . . . for special attention"; but they do not say what special attention is called for, nor indeed why.

The apparent absence of any objective epistemic contraints when it comes to interpreting influence measures is acknowledged by those most active in this area. For example, Welsch (1986, p. 405), joint author of another influence function, conceded that "[e]ven with a vast arsenal of diagnostics, it is very hard to write down rules that can be used to guide a data analysis. So much is really subjective and subtle". And Velleman (1986, p. 413), who has already been quoted, talks of "the need to combine human judgment with diagnostics", without indicating how this judgment operates, is justified, or coheres with the rest of classical statistics.

A full Bayesian analysis of influential data has yet to be undertaken. It would, we are sure, endorse many of the intuitions that guide non-Bayesian treatments, but unlike these, the path it should take and its epistemic goal are clear; using Bayes's Theorem, it would have to trace the effects of uncertainty in the accuracy of readings and of relatively isolated data points on the posterior probabilities of possible regression models. The technical difficulties facing such a programme are formidable, to be sure (and these difficulties are often referred to by critics wishing to cast doubt over the whole Bayesian enterprise), but the difficulties that any Bayesian analysis of regression data faces arise only from the complexity of the situation and by no means reflect on the adequacy of the methodology. The three-body problem in physics provides an instructive analogy; this is a problem that has so far resisted a complete solution in Newtonian terms and may, for all we know, be intrinsically insoluble, but nobody thinks that Newton's Laws are in the least discredited because of the mathematical difficulties of applying it in this area.

By contrast with a Bayesian approach, the programme of influence methodology based on classical ideas runs into trouble not simply because of intractable mathematics, but, we suggest, because it has no epistemically relevant goal. This explains why the rules and techniques proposed seem arbitrary, unjustified, and ad hoc; it is hard to see how they could be otherwise.

## g CONCLUSION

We have in earlier chapters catalogued various inadequacies of standard classical inferential methods. Those same inadequacies, not surprisingly, surface too when regression problems are at issue. However, some new difficulties are also revealed. For instance, extra criteria for estimation are required, which have led to the plausible, but classically indefensible, least squares principle. There are also extra sources of subjectivity, for instance, in selecting the regression model, both when taking account of 'prior knowledge' and when judging models by informally inspecting scatter plots. They are also present in the process of checking models against outliers and influential data. This subjectivity frequently passes unnoticed. Thus Wonnacott and Wonnacott (1980, p. 15) set themselves firmly against judging the gradient and intercept of a regression line "by eye", on account of its subjectivity—"we need to find a method that is objective"; but without batting an eyelid at their inconstancy, they allow personal judgment and the visual inspection of scatter plots to play a crucial part in determining the overall conclusion.

Unlike the hotchpotch of ad hoc and unjustifiable rules of inference that constitute the classical approach, Bayes's Theorem supplies a single, universally applicable, well-founded inductive rule which answers what Brandt (1986, p. 407) calls the "most important . . . need for integration of this [influence methodology] and many other aspects of [classical] regression and model fitting into a coherent whole".

## EXERCISES

1. Show that the least squares estimates of $\alpha$ and $\beta$ in a simple linear regression have the general form given in section **c,** above. (You need to formulate general expressions for sums of squares of deviations of data points from a line and find conditions for them to have minimum values by differentiating and equating to zero.)

2. Show that least squares estimates are linear, with coefficients $b_i = \frac{(x_i - \bar{x})}{c}$, and $a_i = \frac{\sum\left(\frac{x_i^2}{n} - \bar{x}x_i\right)}{c}$, where $c = \Sigma\,(x_i - \bar{x})^2$.

3. The following example is taken from Daniel (1991). A team of mental health workers in a psychiatric hospital wished to measure the level of response of withdrawn patients to a programme of remotivative therapy. A standardized test was available for this purpose, but because this test was expensive and time consuming, the team developed a new and more-easily administered test. To check the usefulness

of the new test, the team decided to select 11 patients who had scored at each 5-point increment between 50 and 100 on the new test and measure their responses on the standardized test. The results were as follows (the score on the new test being given first): (50, 61); (55, 61); (60, 59); (65, 71); (70, 80); (75, 76); (80, 90); (85, 106); (90, 98); (95, 100); and (100, 114).

Denote the new-test score $x$ and the old-test measurement $y$, and calculate the least squares line for the regression of $y$ on $x$ from these data.

Daniel calls this the "best" line: "The line we have drawn is best in this sense: The sum of the squared vertical deviations of the observed data points $(y_i)$ from the least-squares line is smaller than the sum of the squared vertical deviations of the data points from any other line" (Daniel, 1991, p. 374). Is this an adequate defence of least squares estimation?

**4.** Show that Bayesian and maximum-likelihood estimation give similar results in the special case of a uniform prior distribution.

**5.** Show that there are always infinitely many hypotheses with maximum likelihood relative to any data. How, if at all, does this fact reflect on the merits of the maximum-likelihood method?

**6.** "Outliers are simply data points that diverge noticeably from the general pattern of the data on an $X$–$Y$ plot. . . . An outlier may be telling you something important and the research report should acknowledge its existence. It may be the only really interesting thing that happened in your study. There are *no* statistical grounds for *excluding* outliers although a number of statistical criteria can be used to *identify* such points" (Strike, 1991, p. 213).

What, on this reasoning, is the point of identifying outliers?

# PART V

# *The Bayesian Approach to Statistical Inference*

In this part of the book we shall first consider the nature of the probabilities that statistical hypotheses purport to describe. This will involve a discussion of whether those probabilities represent, as some believe, states of the mind, or whether, as we shall contend, they are physical qualities of the external world. This will be the topic of Chapter 13. We shall then give an account of the manner in which statistical hypotheses may be evaluated by reference to their subjective probabilities, in accordance with Bayes's Theorem. It is not our purpose to provide a manual of Bayesian methods for dealing with every kind of hypothesis and prior structure of beliefs. A number of excellent works of this nature already exist. Our object, which we pursue in Chapter 14, is to show how, in general, the Bayesian approach works for inferences in statistics and how it unifies the various intuitions that scientists have developed in regard to inductive reasoning—intuitions which are only accommodated, if at all, in the classical treatment by means of what we have argued are unconnected and unsatisfactory rules. In the final part of the book we shall respond to the numerous objections with which the Bayesian approach has been assailed.

# CHAPTER 13
# Objective Probability

## a INTRODUCTION

Statistical inference is the experimental evaluation of statistical hypotheses, that is to say, hypotheses ascribing probabilities to types of event, such events usually being characterised by the values of one or more random variables, like the number of heads to appear in $n$ tosses of a coin. We have already discussed the principal classical techniques for evaluating such hypotheses and have argued that they are thoroughly fallacious. Before we argue that the Bayesian approach succeeds where these fail, we must say something about the nature of the probabilities which statistical hypotheses talk about.

So far we have simply assumed that these probabilities characterise certain types of experimental situation without finding it necessary to give any finer analysis of what it means to say that a given experiment has a particular probabilistic structure. Because the probabilities purport to describe quite objective, structural features of experiments, we shall call them objective probabilities to distinguish them from the subjective, degree-of-belief probabilities introduced in Chapter 2. Some discussion of the nature of these objective probabilities is necessary, however, since merely saying that a particular experiment is characterised by an objective probability-distribution over its outcome space gives you no information concerning what, if any, empirical features you should look at in order to evaluate this claim.

We shall conduct, in this chapter, an examination of the extant accounts of objective probability. We shall show that all of them are vulnerable to powerful objections and that for all but one of them these objections appear to be quite decisive. We shall show that the sorts of criticism usually thought to be decisive also against this remaining theory (von Mises's) lose their force when a Bayesian framework for evaluating hypotheses about the values of those probabilities is adopted. That the Bayesian theory offers the only sound foundation for a logic of inductive inference was the burden of chapters 4 to 8; we shall reinforce this conclusion by arguing that only within such a framework do statistical hypotheses acquire any empirical significance at all.

The principal objection to the extant theories of objective probability is that they, or rather the statements within them which attribute particular values to events, do not seem to be either strictly verifiable or falsifiable. Some people, most notably de Finetti (1937 and 1974) and Savage (1954), conclude from this that their empirical content is nil and that all appeals to objective probabilities therefore merely import unwelcome metaphysics into empirical science. De Finetti and Savage have gone on to argue that statistics can function perfectly adequately, moreover, without making any reference at all to objective probability.

We shall discuss this last claim in due course. First let us look at some of the accounts of objective probability currently on offer.

## b VON MISES'S FREQUENCY THEORY

*Frequency theories of probability* define probabilities in terms of properties of a sequence of outcomes of an, in principle, repeatable experiment. The best-known and most-developed frequency theory is undoubtedly von Mises's; and as it is also the one which we shall eventually adopt, we shall consider it in some detail.

We shall start by describing the essential features of the theory, paying particular attention to those which seem to make it vulnerable to apparently fatal objections. We shall conclude, however, that while von Mises's theory cannot, as he presented it, furnish an adequate scientific theory of probability, it can when embedded suitably within the theory of subjective probability.

### b.1 Relative Frequencies in Collectives

The fundamental notion of von Mises's theory is of the *relative frequency of an attribute in a collective,* for it is with such relative frequencies that probabilities will be identified. A useful start will be to elucidate the three terms 'collective', 'attribute', and 'relative frequency'.

A *collective* is an infinite sequence $w = (w_1, w_2, \ldots, w_n, \ldots)$ of 'attributes' drawn from some set *A*. The sequence must also satisfy certain further conditions which we shall mention in due course. The attributes in *A* are intended to define a space of exclusive and exhaustive outcomes of some repeatable experiment, like throwing a die and recording the number uppermost on the die when it comes to rest. The attribute set here is {1,2,3,4,5,6}, though it is always possible that we might want to 'collapse' this into the binary set {odd, even} or to represent it in terms of the values of any other random variable we wish to employ. The set *A* need not be finite, by the way, and it may even be a continuum.

A collective with attribute space *A* is intended to represent a sequence of outcomes defined in *A* which would be obtained if, *per impossibile,* some specified repeatable experiment were to be repeated infinitely often in time or space. Collectives are, of course, ideal objects which we shall never be privileged to witness. They are introduced by von Mises because he believed that certain measurable characteristics of actual sequences of outcomes are best explained by the hypothesis that those sequences are initial segments of collectives. To what extent such explanations are successful will occupy a considerable part of the subsequent discussion.

The measurable characteristics of outcome-sequences von Mises was particularly interested in are the *relative frequencies* with which the attributes which make up the sequence occur. Suppose that we have a finite sequence, of length *n,* of attributes from some set. The *relative frequency* $\frac{n(A)}{n}$ of the attribute *A* in that sequence is the number $n(A)$ of times that *A* occurs in it, divided by *n.* For a collective *w* we can then consider the sequence of relative frequencies $\frac{n(A)}{n}$ for each initial segment of length *n* of *w*. Von Mises lays down two further conditions on collectives, both relating to properties of the sequences $\frac{n(A)}{n}$. The first is what he called the *Axiom of the*

*Existence of Limits,* and which we shall call by the less-cumbersome name of the *Axiom of Convergence.*

> **Axiom of Convergence:** the relative frequency of any attribute $C$ in the first $n$ members of $w$ must tend to a definite limiting value as $n$ tends to infinity.

For von Mises this 'axiom' is the theoretical expression of an apparently well-confirmed empirical fact, which he termed the *Empirical Law of Large Numbers.* According to this, those repeatable experiments which appear to generate their various possible outcomes 'at random', do so in such a way that the relative frequency of each type of outcome seems to converge within an increasingly small interval. For example, a few hundred tosses of a particular coin are found to give a relative frequency of heads whose first decimal place does not change subsequently; some thousands more tosses determine the second decimal place, and so on.

The second condition von Mises imposed on collectives is precisely the condition that the attributes are distributed randomly in $w$. This condition he called the *Axiom of Randomness,* and it employs a characterisation of randomness which is one of von Mises's great and lasting contributions to the theory of probability. Before we state the 'axiom' in the precise form in which von Mises gave it, we shall try to motivate it in the context of a simple example: a sequence of tosses of a coin.

Besides the apparent convergence of the relative frequencies of heads and tails we observe in such a sequence, we also observe that the *order* in which heads occur appears to defy all attempts at systematic prediction. And not only does it defy prediction, but there also seems to be no method of selecting outcomes in advance which yields a higher or lower frequency of heads than the frequency of heads in the sequence as a whole, if the latter is sufficiently long. Were such methods of prediction available, they would clearly form the basis for betting on or against the various possibilities at more favourable odds, and they would therefore constitute a *gambling system.*

To say that some collective-generating device is immune to a gambling system is therefore equivalent to saying that no system of prediction would, if indefinitely pursued, yield different frequencies of occurrence than those in the collective as a whole. This in effect will be von Mises's *Axiom of Randomness.*

To state that 'axiom' precisely it is helpful to introduce the idea of a *place-selection.* This is any effectively specifiable method of selecting indices of members of $w$, such that the decision to select or not the index $i$ is allowed to depend at most on the first $i-1$ attributes $w_1, \ldots, w_{i-1}$ in $w$. Suppose, for example, that the attribute class $A$ of $w$ is $\{0,1\}$; place-selections for $w$ might be any of the following: (i) select every prime index, (ii) select the index of every element of $w$ preceded by a 1; select the index $i$ of every element $w_i$ of $w$ such that $i$ is odd and $w_i$ is preceded by the sequence 010101. We can now state the *Axiom of Randomness.*

> **Axiom of Randomness:** the limit of the relative frequency of each attribute in a collective $w$ is the same in any infinite subsequence of $w$ which is determined by a place-selection.

The notion of an *effectively specifiable* method of selection may seem too vague to be useful, but it can be made precise in the following way. Whether $i$ is to be selected by some given place-selection or not may depend, as we have seen, only on $i$ and the sequence of predecessors of $w_i$ in $w$. A place-selection is therefore formally a *function* $f(s_n, n)$, defined on the possible initial segments of length $n$ of $w$, which takes the values 1 and 0. 1 means 'select the $n + 1$th index of $w'$, and 0 means 'do not'. That the place-selection is effectively executable is therefore the same as saying that the function $f$ should be *effectively computable.*

If the attribute space $A$ is finite or denumerably infinite, we can be even more precise by employing the highly developed theory of computable, or recursive, functions from natural numbers to natural numbers. Kurt Gödel, in his famous proof of the incompleteness of formalised arithmetic, developed a now well-known method of 'coding' finite sequences of symbols as positive integers (Mendelson, 1964, Chapter 3, is one of many texts giving a clear account of this). Using this—or other—method we can code the first $n$ elements $(w_1, \ldots, w_n)$ of $w$ by means of a natural number $v_n$, and define a place-selection to be any computable function $f(n, v_n)$ which takes only the values 0 or 1 (such a definition of a place-selection was first suggested and carried out by the American mathematician Alonzo Church, 1940).

The von Mises–Church characterisation of randomness is not the only formal explication of the informal idea of randomness, nor is it the only one to have been based on the notion of an effective or computable procedure. Kolmogorov (1965), for example, employed a quite-different approach, based on the observation that a computer program for generating any finite sequence of zeros and ones is longer, the more irregular the sequence. An upper bound on the length of the program will, of course, be the length of the sequence itself. Unlike the von Mises–Church approach, this suggests the following *complexity criterion* of randomness for *finite* sequences: a sequence is random if the minimum length of computer program required to compute it exceeds some appropriate positive integer less than the length of the sequence.

There are still other characterisations of randomness, for both finite and infinite sequences (Fine, 1973, and Earman, 1986, contain excellent discussions). These various mathematical theories of randomness are strongly non-equivalent, however, in that no pair determines exactly the same class of random sequence. Indeed, it seems highly doubtful that there is anything like a unique notion of randomness there to be explicated.

But there is nevertheless one notion of randomness which all statisticians employ when they talk about random samples, which is defined purely probabilistically: a sample determined by the values of $n$ random variables $X_1, \ldots, X_n$ is random if the $X_i$ are independent and identically distributed, that is to say, they have the same probability distribution. The von Mises–Church definition of randomness, based on the idea of immunity to gambling systems, turns out to satisfy this condition: as we shall see, *the successive outcomes of a collective turn out to be probabilistically independent with a common distribution.* To this extent, von Mises's Axiom of Randomness is entirely appropriate given the central objective of his theory, which was to provide a unified explanatory account of the scientific uses of probability. We shall discuss the connection between randomness and independence in von Mises's theory shortly; it is now time to give the explicit definition of *probability relative to a collective.*

**b.2 Probabilities in Collectives**

If the two axioms, that of convergence of the relative frequencies of each type of outcome and their invariance under all place-selections which select infinitely many members of $w$, are satisfied, then the limiting relative frequency of an attribute in $w$ is called by von Mises its *probability* relative to $w$. The apparent dependence of the probabilities in von Mises's theory on the particular collective $w$ is misleading, though such a dependence has often been cited in criticism of that theory. It is clear that von Mises intended the probabilities to characterise the experiment $E$ rather than $w$ itself. The particular finite sequence of outcomes we observe is not, as far as we are aware, the only *possible* one which $E$ might have generated, and von Mises certainly did not intend the probabilities to depend on any particular realisation of $E$. He took it as a fact about the world, as everybody does who accepts a frequency interpretation of probability, that all those sequences which might have been generated by the same collective-generating experiment $E$ would possess the same long-run characteristics.

It remains to show that the use of the term 'probability' is formally justified; that is to say, we have to show that von Mises's theory satisfies the axioms of the probability calculus as we presented them in Chapter 2.

First we need to enlarge the descriptive apparatus of the theory. Consider a collective $w$ with attribute space $A$. By allowing arbitrarily many random variables to be defined on $A$, and using the standard logical operations on statements describing their values, we can generate an extensive class of event-descriptions. Let $H$ be the class of descriptions $h$ such that the limiting relative frequency with which $h$ is satisfied exists (it is known, incidentally, that this class is not always closed under even the ordinary truth-functional operations). Let the domain of $P$ be $H$ and let $P(h)$ be the limiting relative frequency with which $h$ is true in $w$.

It is now straightforward to check that each of the axioms 1 to 3 of Chapter 2 is satisfied, and we shall leave the task of showing this to the reader. As for axiom 4, suppose that $h'$ is such that $P(h') > 0$. Let $w'$ be the subsequence of $w$ composed of all the successive attributes instantiating $h'$ in $w$. Since $P(h') > 0$, it follows that $w'$ is infinite. It is fairly straightforward to show that $w'$ is also a collective; that is to say, $w'$

satisfies both the condition of convergence and that of randomness. Define $P(h \mid h')$ to be the limiting relative frequency with which $h$ is satisfied within $w'$. It is easy to show that $P(h \mid h')$ is well-defined if $P(h \,\&\, h')$ is (i.e., the limits exist), and that it is then equal to $\frac{P(h \,\&\, h')}{P(h')}$. For let $n(h \,\&\, h')$ be the number of instances of $h \,\&\, h'$ in the first $n$ members of $w$, and similarly for $n(h')$. Let $n'(h)$ be the number of instances of $h$ in the first $n'$ members of $w'$. Then

$$\lim \frac{n(h \,\&\, h')}{n(h')} = lim \frac{n'(h)}{n'}.$$

But

$$\frac{n(h \,\&\, h')}{n(h')} = \frac{n(h \,\&\, h')}{n} \div \frac{n(h')}{n},$$

and by assumption, the limits of both terms on the right-hand side exist and the second is positive. Hence

$$P(h \mid h') = \frac{P(h \,\&\, h')}{P(h')}.$$

We can note that the principle of countable additivity is not true in general, however. A simple counter-example is provided by Giere (1976, p. 326), as follows. Let the attribute space be denumerably infinite, of the form $(A_1, \ldots, A_n, \ldots)$. Suppose that there is a collective of these attributes and that the $i$th member is $A_i$. Then the limiting relative frequency of $A_i$, for each $i$, is the limit of the sequence $n^{-1}$, as $n$ tends to infinity, which is 0, and the limiting relative frequency with which the disjunction '$A_1$ occurs or $A_2$ occurs or . . .' is satisfied is obviously 1. But each of the disjuncts has probability 0 and a denumerable sum of zeros is 0.

Van Fraassen regards facts such as this, and the fact that closure of the domain of $P$ under countable and even finite disjunction and conjunction is not guaranteed, as providing strong reasons against characterising probability—at any rate as it is used in science—as a limiting relative frequency (1980, pp. 184–85). His argument is that science makes extensive use of probabilities whose domains are not only fields but also sigma fields (Chapter **2C**). Moreover, those probabilities are countably additive, because they are often 'geometric', that is to say, defined for events of the form 'the outcome is in $Q$', where $Q$ is a measurable region of $n$-dimensional Euclidean space, and made proportional to the 'volume', or Lebesgue measure of $Q$ (Lebesgue measure is an extension of the ordinary length, area, and volume measures for intervals in Euclidean space of one, two, three, etc., dimensional space to a wider class of sets, known as the Borel sets, generated from these intervals). Since Lebesgue measure is countably additive, any probability measure $P$ proportional to it and defined on the same domain is also countably additive; for if $Q$ above is the union of a countable family of pairwise disjoint sets $Q_i$, then $P(Q) = k\,\lambda\,(Q) = k\,\lambda\,(\cup\, Q_i) = k\,\Sigma\,\lambda\,(Q_i) = \Sigma\,k\,\lambda\,(Q_i) = \Sigma\,P(Q_i)$, where $\lambda\,(Q_j)$ is the Lebesgue measure of $Q_j$. So where probability is proportional to Lebesgue measure, then it is countably additive.

But this argument does not show that in every type of scientific application probability must be countably additive, only that it is in those contexts where proportionality to Lebesgue measure is deemed necessary. Also, there is the consideration that smooth mathematical theories, like that of countably additive probabilities defined on sigma fields, might owe their acceptance into science precisely to their being smooth, well-understood mathematical theories. Indeed, van Fraassen's arguments rest on taking such theories at their face value, rather than regarding them as justified on largely pragmatic grounds. We see this again in his next argument against a limiting relative-frequency account of probability, in which he shows (1980, p. 185) that limiting relative-frequency probability spaces can never be identified with 'geometrical' probability spaces (a probability space is essentially just the probability function and its associated attribute space). He considers the example of a dart's being thrown 'randomly' at a dart-board. Idealising, suppose it hits at a mathematical point each time and that the throws generate a collective. Let $A$ be the set of points hit in denumerably infinitely many throws. Hence the limiting relative frequency of the event 'the hit is in $A$' is 1. Given the 'randomness' of the throws, it is plausibly assumed that the appropriate probability model is one in which the probability of a hit in any region of the board is proportional to the area of the region. But then the probability of a hit in $A$ is 0, since the Lebesgue measure of a denumerable set is 0. Also, the limiting relative frequency of a hit in the complement of $A$ is 0, while its 'geometrical' probability is clearly 1.

For a conventionalist (as is van Fraassen), examples like this represent a decisive objection to a limiting relative-frequency theory of probability. But a realist should no more feel forced to make a choice between limiting relative frequency and 'geometrical' probabilities than he should between a belief that a piece of matter is a discrete bundle of molecules and the adoption of a continuous function to describe its mass distribution. Continuous mass distributions are a fiction, but a useful one, for they offer a mathematically simple approximation to the true state of affairs. Likewise, 'geometrical' probability distributions are a sufficiently good approximation to appropriate relative frequency distributions to warrant their use as simple mathematical models of them. Consider, for example, a standard 'geometrical' probability distribution, in the kinetic theory of gases. For those subsets of the phase space of a gas which physicists want to consider, the relative frequency, in a collective of gas samples in identical macroscopic states, of phase points belonging to those sets is thought to be closely approximated by their measure —or at any rate, it is thought so by those who take the view that the probabilities involved are objective. It is only fair to add that many people do not take that view; who is right in this, it is not for us to say. What we can conclude, however, is that van Fraassen's objections to relative frequency definitions of probability will deter only those who mistake the mathematical appearances of science for its substance.

### b.3 Independence in Derived Collectives

Consider the two stochastic experiments (i) of tossing a coin once and (ii) of tossing the same coin $n$ times, for some $n>1$. The attribute space $A$ of the first experiment is $\{T,H\}$ and of the second the set $A^n$ of all $n$-termed sequences of $T$'s and $F$'s ($A^n$ is the so-called Cartesian product of $A$). In the orthodox treatment these two experiments are treated as two distinct probability spaces. The first is a triple $(A,\mathbf{F},P)$, where $\mathbf{F}$ would usually be the set of subsets of $\mathbf{F}$, that is to say the set of *events* defined in **F**, and $P$ is a probability function on **F**. The second is a triple $(A^n, \mathbf{F}_n, P_n)$ where $\mathbf{F}_n$ would usually again be the set of all subsets of $A^n$, and $P_n$ is a probability function defined on $\mathbf{F}_n$. Usually $P_n$ is assumed to be related to $P$ as follows: if $P(T) = p$, then $P_n\ [s(r)] = p^r(1-p)^{n-r}$, where $s(r)$ is any sequence in $A^n$ containing $r$ $T$'s. In other words, it is assumed that the probability function in the second space is such that the event $X_i = 1$ has probability $p$, and that the $X_j$ are independent, where for each $i$, $1 \leq i \leq n$, $X_i(s) = 1$ if the $i$th member $s_i$ of the sequence $s$ is $H$, and $X_i\ (s) = 0$ if $s_i$ is $T$.

But that $P_n$ is related to $P$ in this way is an *assumption*, or rather, a hypothesis. Nothing in the orthodox treatment requires that it be so. Yet it is an assumption that must be made to tie the mathematics to its canonical interpretation for applied statistics, $n$-fold random samples. It is in our opinion one of the great strengths of von Mises's theory that it does not require such an additional assumption: that $P_n$ has the required properties is a straightforward consequence of the theory. For suppose that repetitions of some experiment $E$ with attribute space $A$ have the capacity to generate a collective $w$. Now consider the experiment $E_n$ of repeating $E$ $n$ times. If we successively repeat $E_n$, we get a sequence $w_{(n)}$ of members of $A^n$. *But this sequence is also, if we remove the brackets around the* n*-tuples, a sequence of outcomes of* E. Any sequence can be partitioned into a sequence of pairs, of triples, and so forth; for example, we can partition the natural number sequence into segments of length 4:

[(0,1,2,3), (4,5,6,7), (8,9,10,11), . . . ].

The brackets are, so to speak, a conceptual artefact which can be added or removed at will.

By assumption, $w$ is a collective, and it is not difficult to show (we leave it as an exercise) *that so too is* $w_{(n)}$; in other words, each attribute in $A^n$ has a limiting relative frequency in $w_{(n)}$, which is unaltered by any place-selection. This is as it should be. Now suppose that $C$ is some property of outcomes of $E$ (extensionally, $C$ is a subset of $A$). If we define $n$ binomial random variables $X_i$ on the members of $A^n$, such that $X_i$ takes the value 1 on $s$ if $s_i$ is an instance of $C$, then it is straightforward to show that the convergence and randomness properties of $w$ together imply that the $X_i$ are independent with the same probability, relative to $w_{(n)}$, of being equal to 1 as the probability of $C$ relative to $w$. So if $P(C) = p$ in $w$, then $P_n[s(r)] = p^r(1 - p)^{n-r}$ in $w_{(n)}$, where $s(r)$ is a sequence in $A^n$ containing $r$ instances of $C$.

It might be objected that this actually proves too much. After all, a great deal of applied probability theory deals with sequences of events which are not independent, like those forming Markov chains, for example. This type of application seems to be precluded if the account we have just given is correct. It is not difficult to see, however, that this conclusion is incorrect. Suppose, for example, that $E_n$ is the experiment of tossing a coin $n$ times. Let $X_1, \ldots, X_n$ be $n$ random variables defined on the $n$-tuples in $A^n$ such that $X_i = 1$ on a given sequence if the $i$th member of that sequence is a head. Then the $X_i$ are independent (given that $E$ does in fact generate a collective). Let $Y_m$, $1 \leq m \leq n$, be the $n$ random variables which record the number of heads in the first $m$ members in those sequences. Then the $Y_m$ are certainly not independent, though the $X_i$ are. In other words, it is quite possible to construct dependent random variables within this theory in which successive attributes in the collective are independent.

### b.4 Summary of the Main Features of Von Mises's Theory

Before we proceed to the next main topic of discussion, whether von Mises's theory does, as he claimed, furnish the foundation for scientific applications of probability, we shall summarise the main features of that theory.

1. Some—and it is the task of science to discover which—repeatable experiments $E$ with attribute spaces $A$ have the property that, were they to be repeated indefinitely often, they would generate collectives of members of $A$.
2. The limits in $w$ are determined by $E$ itself: if $w_1$ and $w_2$ are both generated by $E$, then the probability of any property in $w_1$ is the same as that in $w_2$.
3. For any given collective of members of $A$, let $H$ be the class of those descriptions $h$ such that $P(h)$ exists in $w$. Then $P$ with domain $H$ satisfies the probability calculus.
4. $H$ is not always closed under truth-functional operations.

5. $P$ is a finitely, but not countably additive, probability function.
6. Within the derived collective $w_{(n)}$, the successive outcomes of $E$ are independent with the same probability of occurrence as in $w$.

**b.5 The Empirical Adequacy of Von Mises's Theory**

In our opinion, von Mises's theory is a very great intellectual achievement which has never received its just recognition. One reason for this was the initial vagueness with which the notion of place-selection was formulated and the attendant suspicion that it could generate inconsistencies. Church's characterisation of place-selections in terms of recursive functions, and related work by Wald, established the consistency of the theory, but only for it to face the apparently insurmountable objection that *it could never be applied.*

This is not, as some have charged, because no collective exists or will exist in any real sense. A collective is admittedly an ideal object, a purely hypothetical extrapolation of actual sequences of repetitions. It certainly does not follow, however, that von Mises's theory is necessarily meaningless or unscientific. Classical mechanics is paradigmatically scientific, yet it tells you what would happen if, for example, two perfectly elastic spheres, travelling with given velocities, were to collide. There are no perfectly elastic spheres; nor are there ideal gases, nor are there point masses, nor are there many other of the entities whose existence appears to be postulated by science and which furnish the basic entities of its theories. Science accommodates them because their postulation leads to scientifically successful theories.

The very serious, because apparently justified, charge against von Mises's theory is precisely that the ideal entities it postulates *cannot* be incorporated into scientifically successful theories: the theory seems to be irremediably metaphysical. On a limiting relative-frequency interpretation of probability statements, *a hypothesis of the form* P(h) = p *makes no empirically verifiable or falsifiable prediction at all,* for it is well known that a statement about the limit of a sequence of trials hypothetically continued to infinity contains by itself absolutely no information about any initial segment of that sequence. In other words, any initial segment of a collective—

and we are, of course, only ever capable of observing initial segments—can be replaced with any arbitrary sequence of the same length without affecting any of the limits in the collective.

Von Mises was well aware of this apparently rather devastating objection to his theory, and he attempted to evade its force by means of a variety of arguments. We shall now consider these in turn and try to judge to what extent he was successful.

#### *i The Fast-Convergence Argument*

This is the weakest of von Mises's responses to the charge of empirical emptiness against his theory. It is that one in practice makes the "tacit assumption" that the convergence to the limit in a collective is fairly quick (von Mises, 1964, p. 108). The obvious defect of this reply is that it is hopelessly vague and does not allow us to decide whether obtaining, for example, 273 heads out of 500 tosses of a coin provides us with any definite information as to whether the coin is fair. Clearly, without some method of making such decisions, the theory of collectives does seem to remain a piece of untestable metaphysics.

#### *ii The Laws of Large Numbers Argument*

Von Mises invokes limit theorems of mathematical statistics, like Bernoulli's, about indefinitely long sequences of indepedent trials, as the means of making the "silent assumption" a little more audible and definite. He asserts that these limit theorems provide information which, in relating properties of initial segments of collectives to the limits of the relative frequencies within them, provide the required information about the speed with which convergence to the limits occur. Thus he claimed that "Poisson's Theorem [a variant of Bernoulli's in which the assumption of constant probability is relaxed] has a practical meaning for real sequences of observations" (1939, pp. 170–71), and he described Bernoulli's Theorem as "a very important theorem of practical statistics" (1939, p. 189).

Von Mises regarded such laws-of-large-numbers results as giving information about the rate at which convergence proceeds within a collective, for reasons which seem to be implicit in the results mentioned in section **b.3.** For suppose that an experiment $E$ generates a collective $w$ with attributes 0 and 1, and that $X$ is a random variable taking the value 1 on 1 and 0 on 0, such that $P(X = 1) = p$. Then $w_{(n)}$, the sequence obtained by partitioning $w$ sequentially into sequences of length $n$, is also a collective; and, moreover, the random variables $X_1, \ldots,$ $X_n$, defined in section **b.3** on these sequences, are independent and such that for each $i$, $i = 1, \ldots, n$, $P_n(X_1 = 1) = p$, where $P_n$ is, as before, the probability relative to $w_{(n)}$. Where $Y = X_1 + \ldots + X_n$, it follows that $Y$ has the binomial distribution with probability parameter $p$. Hence the probability that $Y$ lies within, say, 3 standard deviations of $np$ is, for large $r$, close to 0.99, where the 3 standard-deviation points are $np - 3\sqrt{np(1-p)}$ and $np + 3\sqrt{np(1-p)}$. If $E$ is the experiment of tossing a coin once, and this generates a collective $w$ in which $P(H) = \frac{1}{2}$, then in the derived collective of sequences of 100 tosses of the coin, the probability that the number of heads lies between 35 and 65 is close to 0.99.

But such a result gives no information whatever about how fast convergence actually proceeds within $w$. For the meaning, within von Mises's theory, of the statement

$$P_n(Y \text{ is within 3 s.d.'s of } np) \approx 0.99,$$

where 's.d.' stands for standard deviation, is merely that in the collective $w_{(n)}$, the relative frequency of attributes ($n$-fold sequences of zeros and ones) in which $Y$ is within 3 standard deviations of $np$ tends to some number in a small neighbourhood of 0.99. That statement does *not* assert that if you perform $E$, $n$ times in succession, you will with overwhelmingly large probability get a value of $Y$ within 3 standard deviations of $np$; indeed, such an interpretation of that inequality is, as von Mises himself stressed in another context (that of so-called single-case probabilities, which we discuss in section **c**), totally illegitimate: $P_n$ refers, and refers only, to a limiting relative frequency of attributes within $w_{(n)}$.

It might be objected against this deflating conclusion that you may confidently expect a value of $Y$ lying outside the 3 standard deviation limits not to occur in any sequence of $n$ repetitions of $E$, if the hypothesis, call it $H$, that $E$ generates a collective in which 1 has the probability $p$, is true. For such an outcome would, by virtue of its small limiting relative frequency in $w_{(n)}$, be extremely unlikely to occur this time: there are, if $H$ is true, so very few of them that for this outcome to be among them, an enormous coincidence would have to occur. If we were to observe such an outcome, would we not, therefore, be justified in regarding it as prima facie evidence against $H$?

Whether or not we should be justified, we should certainly not be justified for that reason. First, we have to remember that von Mises's theory entails that if an outcome $C$ of an experiment $E$ has any finite probability, *however small,* then in an infinite sequence of repetitions of $E$, $C$ must occur *infinitely* many times. To say that there are only a "few" occurrences of an outcome with a small probability has, therefore, a distinctly Pickwickian ring. Secondly, we have only to choose $n$ large enough and we will *always* obtain a type of outcome of $E_n$ which occurs with arbitrarily small relative frequency in $w_{(n)}$: for in $w_{(n)}$ the limiting relative frequency of any given value of $Y$ tends very quickly to 0 as $n$ tends to infinity (assuming that $0 < p < 1$). "Enormous coincidences" therefore occur literally everywhere in $w_{(n)}$ for sufficiently large $n$, which means that without further information, no significance can be attached to the occurrence of any one of them. Indeed, phrases like "so very few" and "enormous coincidence" are no more than rhetoric and ought not to conceal from us that all that one is justified in inferring from an occurrence of an event to which $H$ assigns a small probability is that an event has occurred to which $H$ assigns a small probability.

### *iii The Limits-Occur-Elsewhere-in-Science Argument*

Von Mises pointed out, in a further attempt (1939, p. 124) to rebut objections to the apparent empirical emptiness of his limit definition of probability, that the use of limits is ubiquitous in science. Velocity, acceleration, density, work, for example, are all defined in terms of limits of infinite sequences, since the first three are derivatives and the fourth is an integral; and integration and differentiation are defined as limits (or in terms of limits if there are complex quantities involved) of sequences of real numbers. Furthermore, nobody complains that the introduction of these notions renders those parts of science in which they appear metaphysical. The problem of how to relate what you observe in 500 trials of a putatively collective-generating experiment to the postulated frequency limits "is exactly similar to that occurring in all practical applications of theoretical sciences".

But this is not true. The difference, from the point of view of empirical significance, between, say, classical kinematics and dynamics based on velocity and acceleration and von Mises's probability theory is exactly that the former straightforwardly generate testable hypotheses while the latter does not. The Galilean law of free fall, which asserts that the acceleration of a freely falling body is constant, predicts distance fallen after time elapsed, and both these quantities are straightforward observables. Statistical hypotheses incorporating von Mises's probabilities do not, we should be aware by now, deductively predict *any* observable state of affairs at all.

But perhaps von Mises can make valid his analogy with the physical sciences, where, after all, few if any of the principal theories make any prediction directly about observables. These theories are saved from relegation to the status of metaphysics because they are enabled to make such predictions through being coupled with suitable auxiliary hypotheses. Why not here also? Gillies (1973) has proposed that probabilistic hypotheses are rendered empirically significant in just this way. Although Gillies does not define objective probabilities as frequency limits in collectives (he claims that probabilities should simply be regarded as primitive), he suggested a type of auxiliary hypothesis which might serve also in von Mises's theory to render its hypotheses testable.

**Gillies's rule.** Gillies's proposed auxiliary hypothesis says that part of the range of a random variable $X$ which has a sufficiently small probability density, according to some hypothesis $H(X)$, is effectively prohibited to $X$, if $H(X)$ is true, the prohibition being subject to certain conditions being satisfied by the form of the density function (1973, pp. 171–72). Gillies calls this stipulation a Falsifying Rule for Probability Statements; by invoking it, he claims to have made probabilistic hypotheses effectively amenable to the empirical check of possible falsification. His rule, being a rule, might not seem like a factual statement at all and therefore not like an auxiliary hypothesis as normally understood. Gillies claims, nevertheless, that it is analogous to statements like 'The mass of the earth can be ignored in comparison with the Sun's in certain calculations of gravitational attractions'. Be that as it may, the falsifying rule Gillies proposes leads, unless qualified in some way, to inconsistencies (as Redhead, 1974, demonstrates) and also represents just that theory of significance testing which we have already examined and rejected as a legitimate way of evaluating statistical hypotheses. We therefore conclude that as it stands, Gillies's proposed rule is unacceptable.

**The Kolmogorov-Cramér rule.** Kolmogorov proposed, in his classic 1933 work, a weaker, less categorical type of auxiliary assumption. While he did not attempt to define the probability function directly, he stipulated that it is "practically certain" that if the repeatable experiment, characterised by a probability-distribution *P* over its class of outcomes, is repeated "a large number of times", then the relative frequency of any event *A* "will differ very slightly from *P(A)*". Cramér (1946) repeats this account using almost exactly the same words and refers to it as the Frequency Theory of Probability.

Without further information about what the phrases in quotation marks are supposed to mean, however, this stipulation is far too vague to be of any use. 'Almost certain' possesses a quite precise meaning in mathematical probability theory. It means 'with probability 1', and Kolmogorov's and Cramér's theory sounds rather like a paraphrase of the Strong Law of Large Numbers, that with probability 1 the relative frequency of outcomes *A* in a sequence of independent trials with constant probability *p* tends to *p*. But for this to be informative would presuppose that we already know what 'with probability 1' means.

Kolmogorov and Cramér do provide a further empirical criterion, which is, incidentally, independent of the original one. This is that in a single performance of the experiment, we can be practically certain that any event with a probability close to zero will not occur. What they seem to mean by this statement is that you are entitled to regard a hypothesis as refuted by sample data if it ascribes a very small probability to an event instantiated by that sample.

This, however, is just Cournot's rule again. We pointed out

in our earlier discussion of this (Chapter 8, section **a**) that it is very obviously unsound: by tossing a coin enough times one can generate an event—a particular sequence of heads and tails—which has an arbitrarily small probability whatever the true probability may be of heads; according to the rule, then, every hypothesis about the value of that probability is rejected, which contradicts the assumption that one of those values is the true one.

### b.6 Preliminary Conclusion

Von Mises does not seem to have been able to rebut at all successfully the charge that his theory fails the minimal condition for being regarded as an adequate scientific account of random phenomena, namely, that it permit an empirical evaluation of its constituent hypotheses. But there is a fundamental difficulty, in principle, which any empirical theory of random phenomena will face, and that is that *by their very nature random phenomena defy any attempt at categorical prediction.* This feature of its subject matter was actually emphasised in von Mises's theory and seems to lead to the conclusion that there can in principle be no successful empirical theory of random phenomena. Had von Mises's theory, for example, predicted that within some specifiable number of trials the relative frequency of a particular type of outcome would be confined within some proper subinterval of the unit interval, then that would have been a ground actually for rejecting the theory; for it seems to be the case that once in so many hundred, or so many thousand, or so many million times, these bounds are exceeded.

Indeed, this is one fact which the theory of sequences of independent random variables seems in some sense to explain, since it assigns a small but finite probability to such rare deviations. The trouble with any suggestion that departures from large-scale regularity ought to be explicitly accommodated by any adequate theory, however, is again that the suggestion is accompanied by no systematic account of how these probabilistic 'predictions' are to be empirically evaluated. We shall, we hope, be able to provide such an account; before we do this let us first look briefly at the other theories of objective probability and see whether they succeed where von Mises (apparently) fails.

## c POPPER'S PROPENSITY THEORY, AND SINGLE-CASE PROBABILITIES

### c.1 Popper's Propensity Theory

In a series of papers between 1950 and 1970, Popper introduced what he called the "propensity interpretation" of the probability calculus. According to one characteristic statement of this interpretation (1959), certain types of repeatable experiment are endowed with dispositions or *propensities* to produce fixed limiting relative frequencies of their various outcomes, were those experiments to be continued indefinitely under similar conditions. Popper claims that these propensities "are not only as objective as the experimental arrangements but also *physically* real" (1959, p. 69; his italics). But we have heard something very like this before, and not from Popper. Indeed, the following statement antedates Popper's writings:

> the probability of a 'double six' [at dice] is a characteristic property of a given pair of dice (or it may be a property of the whole method of throwing) and is comparable with all its other physical properties. The *theory of probability is only concerned with relationships existing between physical quantities of this kind.* (von Mises, 1939, p. 18; our italics)

That probabilities are physical properties of experiments, manifested in a disposition to generate constant limiting frequencies, is the foundation stone of von Mises's theory; it is no less than "the 'primary phenomenon' (*Urphänomen*) of the theory of probability" (ibid.). Indeed, how could probabilities, characterised as relative frequencies which "would tend to a fixed limit if the observations were indefinitely continued" (von Mises, 1957, p. 15), be anything other than dispositional properties of experimental set-ups? The fact is that this theory of objective probabilities, as dispositions of repeatable experiments to generate convergent relative frequencies when repeated indefinitely, is as much von Mises's theory as it is Popper's and was certainly announced much earlier than Popper's.

Popper nevertheless did proceed beyond von Mises in one respect. In his *Logic of Scientific Discovery* (1959a), Popper argues that since probabilities, characterised as dispositional properties of some experiment, appear to depend only on the experiment, these same probability-values may be attributed to predictions about the outcomes of *particular* occurrences of the performance of the experiment:

> This [dependence of the probabilities on the experimental conditions] allows us to interpret the probability of a *singular* event as a property of the singular event itself, to be measured by a conjectured *potential or virtual* statistical frequency rather than by the *actual* one. (1959a, p. 37; his italics)

These putative probabilities are therefore, according to Popper, to be considered as *objective single-case probabilities.*

Von Mises not only did not make this step of transferring the collective's probabilities to the single case, but he also explicitly denied that it made any sense to attribute probabilities to single events at all. And he was right to do so, as we shall see. But first we must discuss an influential objection of Popper's to von Mises's theory.

### c.2 Jacta Alea Est

Popper's objection employs a simple example. Two throws of a fair die are interpolated into a long sequence of throws of a heavily loaded die. Popper claims (1959a, pp. 31–35) that we should want to say that the probability of a six occurring on each of the two occasions of a throw of the fair die is not equal to the probability of a six as estimated by the long-run relative frequency in the sequence in which those throws actually occur, and he goes on to urge this as a criticism of "the frequentist", who "is forced to introduce a modification to his theory". The modification is to make the probabilities properties of the structural features of the experiment rather than of a particular sequence of events (1959a, p. 35).

However, if "the frequentist" to whom Popper addresses these remarks is von Mises—and von Mises appears to be the target—then the latter is certainly not forced to modify his theory because of Popper's observation. Von Mises would, as we have seen, have agreed with Popper that the probability of a six for the fair die is not the same as the probability of a six for the loaded die, and like Popper's, his justification of this claim would be that were the fair die to be tossed indefinitely often, the relative frequency of six would tend to a limit equal to one sixth. But that would be the extent of his agreement, however;

he would not have agreed that these observations justify, as Popper himself goes on to allege, the attribution of the probability one sixth—or any other value—to the outcome of either of the two throws of the fair die. As we observed, von Mises explicitly denied that such attributions were possible, and his denial is well-founded.

For there is an age-old, and quite insuperable, objection to Popper's attempt to define single-case probabilities in this way. The objection rests on the observation that the outcome of a particular throw of the die, to return to Popper's example, is influenced in practice by a number of factors, *only one of which is the mass distribution of the die itself.* Variations in air density, convection currents, the strength and point of contact of the initial impulsive force, the distance above the point of landing at which the die is cast, and so on, also play a part in determining on which face the die will fall. Suppose that the face on which the die falls is uniquely determined by parameters $q_1, \ldots, q_k$. Suppose also that at the first throw of the fair die, these take values $q_{10}, \ldots, q_{k0}$, and that the outcome of that throw is a four. Consider the experiment $E_0$ consisting in throwing the die in such a way that the $q_i$ take the values $q_{i0}$ (no matter that $E_0$ may be impossible to perform in practice, because of the limits to which precise measurements can be made; it is an issue of principle, not practice, which is in question here). Clearly, the relative frequency of a six in any sequence of repetitions of $E_0$ is 0, whereas the relative frequency of a six in a long sequence of repetitions of the experiment $E$, consisting simply of throwing the fair die in the usual way, is one sixth. Now the first throw of the fair die in Popper's imagined hybrid sequence is an instance both of $E_0$ and of $E$. Hence the single-case probability of a six at the first throw of the die is both 0 and one sixth, and we have a contradiction (assuming, as is assumed, that single-case probability, like probability itself, is a single-valued function).

This sort of objection to Popper's attempt to apply statistical probabilities to single trials is standard and well-known—it is the old 'problem of the choice of reference class'. He seems to have been led to his theory of single-case probabilities by his inference that *since* probabilities are properties of experiments, and *since* the experiment is the same (trivially) at every repetition of it, then the probability must be the same at each repetition. As we see, however, this inference is a non sequitur.

The idea of single-case propensities nevertheless continues to appeal to many people. It was taken up by Giere, who seems to have been the first to point out (1973, p. 473) that Popper's theory founders on the objection that any actual observation of some effect may instantiate many different probability-determining experiments, and whose own response to that objection is to restrict a Realist interpretation of single-case probabilities to those examples, which can be tentatively identified only in the quantum domain, of irreducibly indeterministic phenomena. Giere reserves a purely instrumental, as-if interpretation, of the theory for the macro-world, where in practice the 'propensities' for particular trials to yield their possible outcomes are obtained from the most specific reference class for which statistics exist; "but it must be realised that this is only a convenient way of talking and that the implied physical probabilities really do not exist" (1973, p. 481).

Giere diverges sharply from Popper in denying any necessary link between propensities and long-run relative frequencies. For Giere, propensities are simply measures of "causal tendencies [in single trials] not reducible to relative frequencies whether actual or possible" (1976, p. 327). There are two objections to this account of single-case probabilities. First, the authentic domain of applicability of the theory—irreducibly indeterministic trials—is uncertain (the micro-world might turn out to be deterministic, however unlikely an eventuality that would seem today) and very restricted relative to the standard applications of statistical theory. Our objective is to find a theory of objective probability that will fit in with the practice of statistics, not to truncate the latter because we have found a theory which does not.

The second objection is more fundamental and seems to be unanswerable. Von Mises's theory may seem stubbornly deficient in empirical content, but the present account is, if anything, even worse. For Giere's single-case propensity theory conveys no information of any sort about observable phenomena, not even, we are told, about relative frequencies in the limit. It is not even demonstrable, without the frequency link, that these single-case probabilities are probabilities in the sense of the probability calculus, so their place in a discussion of objective *probabilities* is, in default of further development, based on nothing more than assertion. Nor do the limit theorems of the probability calculus help. How, without further information, is one to give empirical meaning to phrases like 'with probability one' or 'independent, identically distributed random variables'?

Others besides Giere and Popper have attempted to construct a satisfactory theory of single-case propensities. There are too many variants of the Popper-Giere approach to consider in detail, and we believe that none survives the main criticisms that we have brought against Popper's and Giere's own theories, but there is a rather different type of single-case theory, the theory of objective chance, so-called, proposed by Mellor (1971), Levi (1980), Lewis (1981), and Skyrms (1984), which purports to be both empirically significant and internally coherent. Let us now see what this theory says and whether its claims can be sustained.

## d THE THEORY OF OBJECTIVE CHANCE

This theory is most simply presented in the familiar context of the tossed coin. The coin is credited by the advocates of this theory with a characteristic *chance* of yielding heads when tossed, the chance being present whenever the coin is tossed in the prescribed way. The chances of successive outcomes are, moreover, independent. The chance distribution, with its characteristic independence property, is to be thought of as a permanent characteristic of the coin, just as much a function of its physical structure as its resistance to bending forces.

The adverse criticisms levelled against Giere's similarly non-frequentist theory of objective single-case probabilities seem to have been met by the advocates of this theory. In particular, they have an answer to the question why chances should obey the probability calculus. This (rather ingenious) answer is that chances license numerically equivalent fair betting-quotients, and these, by the same sort of reasoning based on the Dutch Book argument which we gave in Chapter 5, satisfy the probability calculus. Hence, so must chances.

We shall discuss the adequacy of this answer shortly. Let us go on to address the question why the chance of heads from toss to toss should be regarded as constant. According to Mellor, it is a propensity of the coin, when it is tossed, to possess a particular chance of landing heads, that chance depending on the mass distribution of the coin; and this remains (very largely) the same from toss to toss. The independence of the successive tosses is likewise a plausible assumption: a head at the $i$th toss does not seem to causally affect the chance of a head at the $(i + 1)$th.

It would appear to follow from these considerations that the subjective probability of $r$ heads out of $n$ tosses of the coin, given that the chance of heads is $p$, is equal to $p^r(1 - p)^{n-r}$. So we can, it seems, simply insert this as the value of the likelihood term $P(e \mid h)$ in Bayes's Theorem, where $P$ is our subjective probability function, $e$ is the sample data, and $h$ the given chance distribution, to compute, in conjunction with our prior degrees of belief $P(h)$ and $P(e)$, a posterior degree of belief in $h$. In other words, we appear to have a simple method of evaluating hypotheses about the values of chances from the data.

The principle, or axiom, that chances license numerically equivalent degrees of belief is called the *Principal Principle* by David Lewis (1981). It is a very powerful principle, for it implies that chances are probabilities in the sense of the probability calculus, and as we saw in the previous paragraph, it also enables us to evaluate the likelihood terms in Bayes's Theorem. From these we can evaluate empirically hypotheses about chance by constructing posterior distributions for them. The theory of objective chance seems to supply the deficiencies of von Mises's as well as of the single-case theorists' accounts.

But this is not really the case. For the Principal Principle is a synthetic statement relating chance to degree of belief which, however, receives no independent justification. According to Lewis, the Principal Principle contains *all* that we know about the structure of chances (1981, p. 276), while Levi contends that a characterisation of chances independently of this principle is neither possible nor necessary (1980, p. 258). The principle itself, though a synthetic one, stands unsupported, and according to its advocates *necessarily* stands unsupported, by any empirical argument. The Kantian synthetic a priori has reappeared in the least likely of places.

Chances, in other words, are simply postulated entities which justify corresponding degrees of belief. That may well be all we know on Earth, but it is certainly not all we need to know. The theory of objective chance attempts to bootstrap itself up by means of the Principal Principle, in a way slightly reminiscent of the arguments of Rational theology for God's existence. Ex nihilo nihil, we respond. But the chance theorists have nevertheless correctly identified the crucial role played by the Principal Principle in relating empirical data to hypotheses about objective probabilities. In the next section we shall actually supply an argument for the validity of the principle, though the objective probabilities it will link to degrees of belief will not be chances, *but limiting relative frequencies in collectives.*

## e A BAYESIAN RECONSTRUCTION OF VON MISES'S THEORY

None of the theories we have looked at so far seems able to furnish an adequate, empirically significant account of objective probabilities. We shall argue that this is not the case, however, and that despite the adverse considerations we have already dwelt on, von Mises's theory *can* furnish such an account. But, we shall argue, it can do so *only when embedded in a theory of subjective probability.* To this extent the relation between objective probabilities and observables is even less direct than that between the theoretical entities of physics and observables. There is, in particular, no deductive relation between the hypothesis that this experiment generates a collective and the outcomes of the first $n$ tosses, for any $n$.

It is a consequence of a well-known result of the mathematical theory of probability, however (Halmos, 1950, p. 213, Theorem B), that if a repeatable experiment generates a von Mises collective with probability $p$ of some attribute, then in the limit as you collect data you will infer that this is indeed the case. Let $\Omega$ be the set of all denumerably infinite sequences of 0's and 1's. $\Omega$ can be regarded as the space of all possible infinitely continued performances of an experiment with the two possible outcomes 0 and 1. Let $H_p$ be the hypothesis which is true relative to any member $\omega$ of $\Omega$ just in case $\omega$ is a von Mises collective with probability $p$ of the attribute 0. The theorem we have referred to implies that there is a non-empty subset $A$ of $\Omega$, such that for each $\omega$ in $A$ the probability that $H_p$ is true, conditional on data $e$ forming an initial segment of $\omega$, tends to 1 if $H_p$ is true on $\omega$ and to 0 if not (in fact, the theorem says that such sequences occur with probability 1).

That theorem still does not tell anything like the whole story, however. One of the most celebrated features of the Bayesian theory is its revealing the striking way in which the posterior probability distributions of hypotheses like $H_p$ reflect characteristic features of the data, and in particular how they are sensitive to observed relative frequencies. The technique used is a Bayes's Theorem computation of $P(H_p \mid e)$, where $e$ is the sample data, the likelihood terms $P(e \mid H_p)$ being evaluated by means of the Principal Principle where it is assumed that you have no other information relevant to $e$. The use of the Principal Principle is justified as follows. Suppose that you are asked to state your degree of belief in a toss of this coin landing heads, conditional upon the information *only* that were the tosses continued indefinitely, the outcomes would constitute a von Mises collective, with probability of heads equal to $p$. And suppose you were to answer by naming some number $p'$ not equal to $p$. Then, according to the definition of degree of belief in Chapter 5, you believe that there would be no advantage to either side of a bet on heads at that toss at odds $p' : 1 - p'$. But that toss was specified *only* as a member of the collective characterised by its limit-value $p$. Hence you have implicitly committed yourself to the assertion that the fair odds on heads occurring at *any* such bet, conditional just on the same information that they are members of a collective with probability parameter $p$, are $p' : 1 - p'$. Suppose that in $n$ such tosses there were $m$ at which a head occurred; then for fixed stake $S$ (which may be positive or negative) the net gain in betting at those odds would be $mS(1 - p') - (n - m)Sp' = nS[(\frac{m}{n}) - p']$. Hence, since by assumption the limit of $\frac{m}{n}$ is $p$, and $p$ differs from $p'$, you can infer that the odds you have stated would lead to a loss (or gain) after some finite time, and one which would continue thereafter. Thus you have in effect contradicted your own assumption that the odds $p' : 1 - p'$ are fair.

Thus we have established a version of the Principal Principle which says that if you have any degree of belief in an outcome $A$'s occuring in a performance of an experiment $E$, then, conditional *only* on the information that $E$ generates a collective in which $P(A) = p$, that degree of belief must on pain of inconsistency equal $p$. The principle, understood in this way, clearly represents a consistent theory of single-case probabilities, and single-case probabilities which are equal to certain objective probabilities. But the single-case probabilities it introduces *are not themselves objective probabilities*. They are subjective probabilities, which considerations of consistency nevertheless dictate must be set equal to the objective probabilities just when all you know about the single case is that it is an instance of the relevant collective.

Now this is in fact all that anybody ever wanted from a theory of single-case probabilities: they were to be equal to objective probabilities in just those conditions. The incoherent doctrine of objective single-case probabilities arose simply because people failed to mark the subtle distinction between the values of a probability being objectively based and the probability itself being an objective probability. Once the distinction is drawn, it becomes clear that single-case probabilities are really subjective probabilities which, relative to the right sort of information, are equal to probabilities in collectives.

Let us return to the computation of $P(H_p \mid e)$. It is the Principal Principle itself which is responsible for the striking manner in which the posterior probability of $H_p$ responds to augmenting $e$. We can see this most easily if we consider the *ratio* of the posterior probabilities of $H_p$, $H_{p'}$, where $p \neq p'$. Let $e_{r,n}$ report that the observed frequency of 1's was $\frac{r}{n}$. The data from which $e_{r,n}$ was abstracted constitute an initial segment of a collective, according to both $H_p$ and $H_{p'}$. It follows, as we noted earlier, that successive outcomes are independent with constant probability $p$ and $p'$, respectively. Therefore, the von Mises probability of the sample outcome $e_{r,n}$ is $p^{r}(1 - p)^{n-r}$ if $H_p$ is true, and $p'^{r}(1 - p')^{n-r}$ if $H_{p'}$ is true. By the Principal Principle applied to these derived collectives, we infer that $P(e_{r,n} \mid H_p) = p^r(1 - p)^{n-r}$, and $P(e_{r,n} \mid H_{p'}) = p'^r(1 - p')^{n-r}$ and

$$\frac{P(H_p \mid e_{r,n})}{P(H_{p'} \mid e_{r,n})} = \frac{p^r(1-p)^{n-r}\,P(H_p)}{p'^r(1-p')^{n-r}P(H_{p'})}$$

Suppose, now, that as $n$ increases, the relative frequency $\frac{r}{n}$

approaches and remains within a very small neighbourhood of $p'$. Then, since the function of $x$ given by $x^{r'}(1 - x)^{n-r}$, $0 < x < 1$, peaks very sharply, for large $n$, in the neighbourhood of $x = \frac{r}{n}$, and is close to 0 elsewhere, the ratio above will tend to 0, independently of the values of the prior probabilities $P(H_p)$ and $P(H_{p'})$, assuming neither is 0 or 1. In other words, the support of $H_p$ by $e_{r,n}$ will eventually be negligible compared with the support of $H_{p'}$.

We shall see in the next chapter how Bayes's Theorem enables us to evaluate many more types of statistical hypotheses in the light of sample data. We can sum up this section by saying that the Bayesian theory allows us the means to evaluate hypotheses about probabilities in collectives on the basis of sample data. Von Mises's theory has, after all, been shown to be a viable scientific theory of objective probability. Subjective probability, in other words, is sufficiently strong to link frequencies in the limit with finite data segments. In the next section we shall investigate the very influential claim that the subjective theory is actually so strong that it even renders the need for a theory of objective probability superfluous.

## f ARE OBJECTIVE PROBABILITIES REDUNDANT?

Before we turn our attention to Bayesian statistical inference, we must briefly consider the very influential thesis, advanced by de Finetti and Savage, that the use of objective probabilities should be rigorously eschewed and that such probabilities are anyway redundant in just those contexts where they appear to possess explanatory value.

According to de Finetti and Savage, hypotheses about objective probabilities lack empirical significance (de Finetti, a strong positivist, actually declared them absolutely meaningless). It is sufficient to show them wrong in this claim actually to construct an empirically significant account of objective probability, and this we believe we have done. The redundancy thesis is not so easily disposed of, though it too is false, as we shall now show.

De Finetti proved the theorem, which he and Savage

interpreted as implying the explanatory redundancy of objective probabilities, in 1937. Before stating this theorem and evaluating the de Finetti–Savage claim, some preliminary explanation is in order. Suppose that $P$ is a subjective probability defined for all finite sequences of 0's and 1's (this is more general than it might sound, for we can always regard 1 as representing the possession of some property $C$ or other, and 0 as signifying its absence). Suppose also that $P$ assigns a probability to every statement of the form 'the $i_1$th member of the sequence is $x_1$ and the $i_2$th member of the sequence is $x_2$ and . . . and the $i_k$th member of the sequence is $x_k$', where each $x_j$ is 1 or 0. Suppose, finally, that the $P$-probability of any such statement is the same whatever the order of the 0's and 1's occurring in it. If $P$ has these properties, then the binomial random variables $X_1, \ldots, X_n, \ldots$, whose values on any sequence are the first, second, . . . , $n$th, . . . members of that sequence, are called *exchangeable* (relative to $P$) by de Finetti.

De Finetti then showed that if the $X_j$ are exchangeable, then the $P$-probability of any sequence of $n$ of them taking the value 1 $r$ times and 0 the remaining $(n - r)$ times, in some given order, is equal to

$$(1) \quad \int_{-\infty}^{\infty} z^r(1 - z)^{n-r}dF(z)$$

where $F(z)$ is a distribution function, uniquely determined by the $P$ values on the $X_i$, of a random variable $Z$ equal to the limit, where it is defined, of the random variables $Y_m = m^{-1}\Sigma X_i$.

But (1) is also the expression you get for the value of the probability of $r$ of the $X_i$ taking the value 1 if you believed (i) that there is a constant unknown probability $p$ of a 1 at any point in the sequence, where $F$ defines your subjective probability-distribution over $p$'s values in the closed unit interval, (ii) that the $X_i$ were independent relative to this unknown probability, and (iii) you evaluate the subjective probability of $r$ ones and $(n - r)$ zeros to be equal to the value that that outcome would have relative to that unknown probability (you use the Principal Principle, in other words).

To see this, let $h$ be the hypothesis that the $X_i$ are independent with constant but unspecified probability; let $h_p$ be the hypothesis that the value of this unknown probability is $p$; and let $e$ be the statement that $r$ ones are observed, in a sample of size $n$. If, for simplicity, we first assume that $p$ is discrete and takes only finitely many values, then, by the probability calculus,

$$\begin{aligned} P(e \mid h) &= \Sigma P(e \mid h_{p_j} \& h) P(h_{p_j} \mid h) \\ &= \Sigma p_j^r (1 - p_j)^{n-r} P(h_{p_j} \mid h). \end{aligned}$$

In the passage from discrete to continuous $p$, the sum is replaced by an integral; $P(h_{p_j} \mid h)$ is replaced by the distribution function $F$; and we obtain an equation formally identical to (1).

Now, it is customary to invoke the hypothesis of constant objective probability and independence (with respect to that probability)—in other words to invoke $h$—to explain those sorts of features of, for example, long sequences of coin tosses which von Mises singled out as requiring some sort of scientific explanation, namely the non-existence of gambling systems and the marked convergence of the relative frequencies. From our point of view the explanatory hypothesis is that a sequence of coin tosses is the initial segment of a von Mises collective (for it is in terms of collectives that we have elected to understand objective probabilities), which implies, as we saw, that the hypotheses of independence and constant probability are satisfied.

But de Finetti's proof implies that the $P$-probability of any feature of the $X_i$ obtained on the condition that $h$ is true is obtainable simply by assuming that the $X_i$ are exchangeable and without, or so it seems and de Finetti claims, introducing objective probabilities at all. Hence their introduction—at any rate, from the point of view of explaining any given property of the $X_i$ which we feel receives an explanation in terms of appeal to the hypothesis of independence plus constant probability from trial to trial—appears to be quite unnecessary.

## g EXCHANGEABILITY AND THE EXISTENCE OF OBJECTIVE PROBABILITY

We shall now argue that de Finetti's thesis is untrue and that although objective probabilities are not explicitly introduced into the domain of $P$, they are nonetheless to a very great extent implicit in the very assumption of exchangeability itself. First, consider two events in the domain of $P$, (i) the (cylinder) sequence $s_1$, 010101 . . . 01, of length $2n$, and (ii) the sequence $s_2$, of the same length $2n$, also ending with 1, also having $n$ zeros and $n$ ones, but differing from $s_1$ in that the arrangement of its ones and zeros is very disorderly. Assume that the $X_i$ are exchangeable, so that $P(s_1) = P(s_2)$. By the probability calculus

$$P(s_1) = P(1 \mid 010101 \ldots 0)P'(010101 \ldots 0)$$

and

$$P(s_2) = P(1 \mid s_3)P(s_3),$$

where $s_3$ is the segment of $s_2$ of length $2n - 1$ preceding the terminal 1. By exchangeability again

$$P(010101 \ldots 0) = P(s_3)$$

and hence

$$P(1 \mid s_3) = P(1 \mid 010101 \ldots 0).$$

But these equations hold for all values of $n$, however large, and $s_3$ is by assumption a highly disorderly sequence in which there are as many zeros as ones. Were we actually to deny that the $X_i$ are independent, there could be no reason for this last equation; we should in these circumstances expect the right-hand side to move to the neighbourhood of 1 and the left-hand side to tend to $\frac{1}{2}$ or thereabouts. We should certainly not expect equality. (This example is to be found in Good, 1969, p. 21.)

This is not a proof that exchangeability implies independence (which is not true), but it is a very powerful reason for supposing that exchangeability assumptions would not be made relative to repeated trials of this type *unless* there was already a belief that the variables were independent. But there is in addition a proof, due to Spielman (1976), which we shall not reproduce here, that if the $X_i$ are exchangeable, for all $n >$ 0, then you must be certain that the relative frequency of ones tends to a limiting value and that whatever character is generated at the $i$th place in the sequence is independent, with respect to that limiting value, of the characters generated both before and after that place (though that you are certain that the relative frequency of ones tends to a limit is a consequence of de Finetti's theorem).

Spielman's important result requires that $P$ be countably,

and not merely finitely, additive, a condition that de Finetti remains strongly opposed to. But as we saw in Chapter 5, it is justified as a consistency constraint no less than the other axioms of the probability calculus, so we feel that Spielman is quite correct to disregard that objection. We shall, therefore, conclude that objective probabilities, far from being redundant, are on the contrary necessary to justify the exchangeability of the $X_i$.

## h CONCLUSION

We have shown, admittedly only in outline so far, how hypotheses about objective probability distributions are to be evaluated against observational data. The only understanding of objective probability which permits empirical evaluations of hypotheses about its magnitude in given cases appears, moreover, to be in terms of long-run relative frequencies. So, despite a very adverse press over the last quarter of a century, von Mises's theory seems, as he himself believed, to offer the only scientific account of objective probability, where by 'scientific' we mean that hypotheses about the magnitudes of such probabilities are in principle capable of empirical evaluation. The task now is actually to evaluate some particular types of hypothesis against some particular types of sample data and see what results are obtained. This will be done in the next chapter.

## EXERCISES

1. Explain why the infinite sequence 01010101 . . . with attribute space $\{0,1\}$ is not a collective.

2. "There are no collectives other than attribute-homogeneous ones, because if the attribute $A$ occurs infinitely often in any sequence $w$, then the infinite subsequence obtained by selecting all the occurrences of $A$ in $w$ will have a limiting relative frequency of 1. Hence, if the limiting relative frequency of $A$ in $w$ is not 1, the axiom of randomness is infringed". Why is this comment mistaken?

**3.** Show that the probability axioms 1–3 are satisfied in von Mises's theory.

**4.** Show that if $P(h) > 0$ in a collective $w$, then $h$ must be instantiated infinitely many times in $w$.

**5.** Show that if $w$ is a collective, then so is $w_{(n)}$.

**6.** Suppose $w$ is a collective with attribute space $A = \{0,1\}$. What are the attributes in $A^n$?

**7.** Let $w$ be as in question 7, and for each $i$, $1 \leq i \leq 1$, let $X_i(s)$ = the $i$th member of $s$, for $s$ in $A^n$. Show that the random variables $X_1, \ldots, X_n$ are independent and identically distributed; i.e., each of them has the same probability of taking the value 1.

**8.** The *limit supremum* of a sequence of real numbers $x_n$ is the limit of the sequence of least upper bounds of the sets $\{x_i : i \geq n\}$, $n \geq 1$. The *limit infimum* is the limit of the corresponding sequence of greatest lower bounds. Let $w$ be an infinite sequence of attributes whose relative frequencies may or may not converge. Let $P_*(A)$ be the limit infimum of the relative frequency of $A$, and $P^*(A)$ be the limit supremum. Show that $P_*$ and $P^*$ satisfy the conditions on lower and upper probabilities listed in Chapter 5, section **b** (Milne, 1992).

**9.** Show that the infinite sequences

0000000000 . . .
1111111111 . . .

both satisfy the axioms of convergence and randomness and are therefore both collectives.

# CHAPTER 14

# *Bayesian Induction: Statistical Hypotheses*

Bayesian induction involves the computation of a posterior probability, or density, distribution from a corresponding prior distribution, by means of Bayes's Theorem. This theorem does not discriminate between deterministic and statistical hypotheses; hence, unlike classical approaches, it affords a unified treatment for every kind of hypothesis. We have already considered, in Chapter 7, the way that Bayes's Theorem deals with a number of issues raised in the evaluation of deterministic hypotheses. In this chapter we shall, in the main, be concerned with the type of hypothesis that is most often discussed in expositions of classical statistical inference, namely, those characterised by the value of a parameter. We have already pointed out the necessity in the classical scheme of restricting possible hypotheses in such ways, and we have criticized the artificiality and arbitrariness of those restrictions. Our approach will be first to operate with the restricted models made standard by their widespread use in classical manuals, so as to allow a comparison. But in order to give a more realistic analysis, we shall also consider how and under what circumstances the restrictions might be lifted.

## a THE PRIOR DISTRIBUTION AND THE QUESTION OF SUBJECTIVITY

In estimating the value of a parameter, the Bayesian approach, unlike the classical, starts out from a prior probability, or density, distribution over a set of possible values of the parameter. The distribution reflects a person's opinions before the results of the experiment are known. As we have made clear, such opinions are subjective, in the sense that they are shaped in part by elusive, idiosyncratic influences and that they may, and often do, vary from person to person. The subjectivity of the premisses of a Bayesian inference might suggest that the conclusion must be similarly idiosyncratic, subjective, and variable. Were this the case, the Bayesian approach would fly in the face of one of the most striking facts about science, namely, its substantially objective character. However, it is not the case. As Edwards, Lindman, and Savage (1963, p. 527) have pointed out, "If observations are precise, in a certain sense, relative to the prior distribution on which they bear, then the form and properties of the prior distribution have negligible influence on the posterior distribution". Hence, from a practical point of view, "the untrammelled subjectivity of opinion about a parameter ceases to apply as soon as much data become available. More generally, two people with widely divergent prior opinions but reasonably open minds will be forced into arbitrarily close agreement about future observations by a sufficient amount of data". We may illustrate these points by reference to two sorts of problem that frequently crop up in statistical literature and practical research.

### a.1 Estimating the Mean of a Normal Population

It will be recalled that we used this customary example to present the ideas of classical interval estimation. The population whose mean is to be estimated is assumed to be normal and to have a known standard deviation, $\sigma$. The assumption of a known standard deviation is a plausible approximation in many cases, for instance, where an instrument is used to measure some physical quantity. The instrument would, as a rule, deliver a spread of results if used repeatedly under similar conditions, and experience shows that this variability often follows a normal curve, with an approximately constant standard deviation. Making measurements with such an instrument would then be practically equivalent to drawing a random sample of observations from a normal population of possible observations, whose mean is the unknown quantity and whose standard deviation has been established from previous calibrations.

Let $\theta$ be a variable that ranges over possible values of the population mean. We shall assume for the present that the prior density distribution over $\theta$ is also normal, having a mean of $\mu_o$ and a standard deviation of $\sigma_o$. In virtually no real case would the prior distribution be strictly normal, for physical considerations would place limits on possible values of a parameter, while normal distributions assign positive probabilities to every range. For instance, the average height of a human population could not be negative, nor could it be five thousand miles. Nevertheless, a normal distribution often provides a mathematically convenient idealisation of sufficient accuracy. (We are assuming a normal prior distribution in order to simplify this illustration of Bayesian inference at work. We shall show later that the assumption may be considerably relaxed without affecting the conclusions substantially.)

Let a random sample of size $n$ be drawn from the population, and let $\bar{x}$ be the mean of that sample. It is now, surprisingly, a simple matter to calculate the posterior distribution of $\theta$, relative to these data. This distribution is also normal (like the prior distribution) and its mean, $\mu_n$, and standard deviation, $\sigma_n$, are given by

$$\mu_n = \frac{n\bar{x}\sigma^2 + \mu_0\sigma_0^{-2}}{n\sigma^{-2} + \sigma_0^{-2}} \text{ and } \frac{1}{\sigma_n^2} = \frac{n}{\sigma^2} + \frac{1}{\sigma_o^2}.$$

These results (which are proved by Lindley, 1965, for example) are illustrated below.

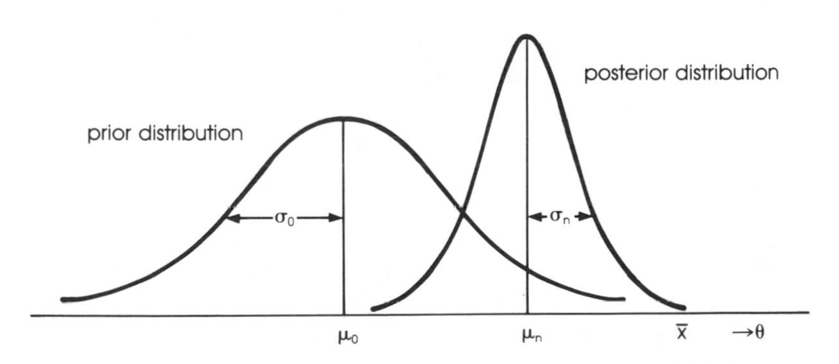


The prior and posterior distributions for the mean, θ, of a population with mean and standard deviation μ and σ, relative to a sample with mean $\bar{x}$.

*The precision of a distribution* is often defined as the reciprocal of its variance. So the second of the equations above tells us that the precision of a posterior distribution increases with the precision, $\sigma^{-2}$, of the population whose mean is being evaluated. By the same token, the precision of a measuring instrument is the reciprocal of the variance of its error distribution. Thus, the more precise such an instrument is, the more precise will be the corresponding posterior distribution of the quantity being measured. Assuming that a precise inference is desirable, this accounts for the appeal of the efficiency criterion that classical statisticians have imposed, for inadequate reasons, on estimators. (*See* Chapter 10, section **b.4.**)

It is easy to see, too, from the above formulas that, as $n$ increases, $\mu_n$ (the mean of the posterior distribution) tends to $\bar{x}$ (the mean of the sample). There is a similar result for the variance, $\sigma_n^2$, of the posterior distribution, namely, that it tends to $\frac{\sigma^2}{n}$, a quantity that is independent of the prior distribution, being fixed just by the population (through its standard deviation, $\sigma$) and the size of the sample. Thus, as the sample is enlarged, the contribution of the prior distribution and, so, of the subjective part of the inference, lessens, eventually dwindling to insignificance. Moreover—and this is the crucial point for explaining the objectivity of the inference—the objective information contained in the sample becomes the dominant factor relatively quickly. Hence, two people proceeding from different normal prior distributions would, with sufficient data, arrive at posterior distributions that were arbitrarily close together and, moreover, the convergence of their opinions would be rather rapid.

We can show how rapid the convergence may be by an example in which one person's normal prior distribution over the mean of some normal population is centred on 10, with a standard deviation of 10, and a second person's is centred on 100, with a standard deviation of 20. This difference represents a very sharp divergence of initial opinion, for the region that each considers almost certainly contains the true mean is regarded by the other as practically certain not to contain it. But even profound disagreements such as these are resolved after relatively few observations. The following table shows the means and standard deviations of the posterior distributions for the two people, relative to random samples of various sizes, each sample having a mean of 50. We shall assume that the population is normal and has a standard deviation of 10.

**TABLE XII** The means and standard deviations of the posterior distributions of 2 people with different prior distributions, relative to a random sample with a mean of 50. The population standard deviation is 10.

| Sample size | Person 1 | | Person 2 | |
|---|---|---|---|---|
| n | $\mu_n$ | $\sigma_n$ | $\mu_n$ | $\sigma_n$ |
| 0 | 10 | 10 | 100 | 20 |
| 1 | 30 | 7.1 | 60 | 8.9 |
| 5 | 43 | 4.1 | 52 | 4.4 |
| 10 | 46 | 3.0 | 51 | 3.1 |
| 20 | 48 | 2.2 | 51 | 2.2 |
| 100 | 50 | 1.0 | 50 | 1.0 |

We have selected a rather extreme example, in which the two people differ substantially in their prior opinions, there being almost no overlap in the initial distributions. The first line in the table above, corresponding to no data, represents this initial situation. Nevertheless, a sample of only 20 brings the two posterior distributions very close, while one of 100 renders them indistinguishable. Not surprisingly, the closer opinions are initially, the less evidence is needed to bring the corresponding posterior opinions within given bounds of similarity. (This is proved, for instance, by Pratt et al., 1965, Chapter 11.) Hence, although a Bayesian analysis of the case under consideration must operate from a largely subjective prior distribution, the most-powerful influence on its conclusion is the experimental evidence. We shall see that the same is true of more general cases.

### a.2 Estimating a Binomial Proportion

Another standard problem in inferential statistics is how to estimate a proportion, for instance, of red counters in an urn, or of Republican sympathisers in the population, or of the physical probability of a coin turning up heads when flipped. Data are collected by sampling (with replacement) the urn or population at random, or by flipping the coin, and then noting for each counter sampled whether it is red or not, and for each person his or her sympathies, and for each toss of the coin whether it landed heads or tails. If, as in these cases, such experiments have just two possible outcomes, with probabilities $p$ and $1 - p$, each of which is constant from trial to trial, then the data-generating process is known as a Bernoulli process, and $p$ and $1 - p$ are called the Bernoulli parameters or binomial proportions. The two outcomes of a Bernoulli process are often labelled 'success' and 'failure'.

The Bayesian way of estimating $p$ commences, of course, with the description of a prior distribution. As a point of departure, we shall assume that this takes the form of a *beta distribution,* for then it becomes very easy to calculate the corresponding posterior probability. A more important advantage is that since beta distributions take on a variety of shapes, depending on the values of two parameters, $u$ and $v$, an appropriate one can often be chosen that best approximates your actual prior distribution of beliefs. And, as we shall see, with sufficient data, posterior probabilities are not much affected by even quite substantial changes in the corresponding priors, so that inaccuracies in determining the prior distribution do not have a significant effect.

**SOME BETA DISTRIBUTIONS**

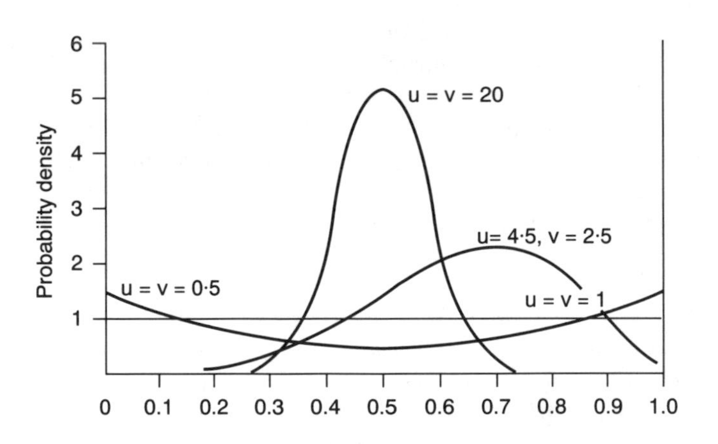

The preceding diagram illustrates some beta distributions for different values of $u$ and $v$ and we see, for instance, that if $u = v = 1$, the distribution is uniform. The mean and variance of a beta distribution are given by

$$\text{mean} = \frac{u}{u+v} \qquad \text{variance} = \frac{\left(\frac{u}{u+v}\right)\left(\frac{v}{u+v}\right)}{u+v+1}.$$

If the prior distribution over the Bernoulli parameters is of the beta form, Bayes's Theorem is particularly easy to apply. Suppose a random sample of $n$ observations obtained from a Bernoulli process shows $s$ successes and $f$ failures. It turns out that the posterior distribution is also of the beta form, with parameters $u' = u + s$ and $v' = v + f$. Hence, the mean of the posterior distribution is $\frac{u+s}{u+v+n}$, which tends to $\frac{s}{n}$, as the number of trials increases to infinity, while the variance of the posterior distribution tends, though more slowly, to zero. (*See,* for example, Pollard, 1985, Chapter 8.) Thus, as with the earlier example, the influence of the prior distribution upon the posterior distribution steadily diminishes with the size of the sample, the rate of diminution being considerable, as simple examples, which the reader may construct, would show.

## b CREDIBLE INTERVALS AND CONFIDENCE INTERVALS

Estimates of parameters are often reported in the form of a range of possible values, e.g., $\theta = \theta^* \pm \epsilon$. Such estimates have an obvious Bayesian interpretation, namely, as a range that, with very high probability, contains the true value of $\theta$. This would be an example of a so-called credible interval. Such intervals may be calculated from a posterior distribution. If $P$ is the probability that $\theta$ lies between $a$ and $b$, then the interval $(a,b)$ is said to be a $100 \times P$ per cent credible interval for $\theta$. Bayesians recommend credible intervals as useful summaries of posterior distributions.

The idea may be illustrated with the first example we gave. As we showed, in the presence of sufficient data, the posterior distribution of the mean of a normal population is symmetrical

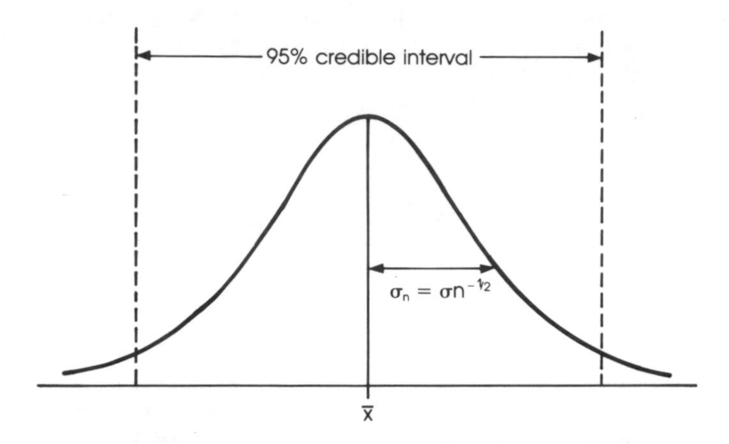


A posterior distribution showing a 95% credible interval for the mean of a normal population with standard deviation $\sigma$

about the sample mean, $\bar{x}$, and has a standard deviation $\sigma_n = \sigma n^{-\frac{1}{2}}$, where $n$ is the sample size, and $\sigma$ the population standard deviation. Since the distribution is normal, the range $\bar{x} \pm 1.96 \times \sigma n^{\frac{1}{2}}$ contains $\theta$ with probability 0.95 (*see* Chapter 3, section **d**), and so is a 95 per cent credible interval.

Of course, by selecting other areas of the distribution, one could construct different credible intervals. For instance, the infinite range of values defined by $\theta > \bar{x} - 1.64 \times \sigma n^{-\frac{1}{2}}$ is also a 95 per cent credible interval. Another would be an interval that extends further into the tails of the posterior distribution but omits a narrow band of values around its centre. There is sometimes a discussion about which of the possible intervals should be 'chosen', which we believe is misconceived (*see,* for example, Lindley, 1965, vol. 2, pp. 24–25), for strictly speaking, one should not *choose* any interval, because a choice implies a commitment in excess of that permitted by the probability of the interval. All 95 per cent credible intervals are on a par.

Credible intervals clearly resemble the confidence intervals by which classical statisticians estimate parameters. Indeed, in the particular case before us, the 95 per cent credible interval depicted in the diagram and the 95 per cent confidence interval that is routinely favoured (the shortest one) coincide (see Chapter 10, section **c.6**). But despite yielding the same intervals in this particular case, the two methods of estimation are crucially different, for while Bayesian credible intervals describe a probability relative to the evidence that $\theta$ lies within the interval, confidence-interval estimates purport to express uncertainty about $\theta$ in non-probabilistic terms acceptable to classical statistics. But as we have argued at length, neither of the two main attempts to do this succeeds; and in the absence of any other suitable interpretation, we must conclude that classical confidence intervals are not estimates at all.

Bayesian credible intervals, we suggest, provide an intelligible interpretation and a rational explanation for the intuitions that appear to underlie classical confidence intervals.

## ■ c THE PRINCIPLE OF STABLE ESTIMATION

The inferences in the two examples we have given are very insensitive to variations in the prior distributions, as we said. However, we restricted the argument to cases of, in the first example, normal, and in the second, beta distributions, and the question arises whether this insensitivity would persist if these restrictions were relaxed. A theorem known as the Principle of Stable Estimation, due to Edwards, Lindman, and Savage (1963), shows that even a very considerable relaxation would preserve that insensitivity (the theorem is in fact a practically useful aspect of a very much more general result proved by Blackwell and Dubins, 1962). We shall indicate the scope of the Stable Estimation principle. It states that if the prior distribution satisfies certain conditions, which it specifies, the posterior distribution is approximately the same, to an extent that it specifies, as it would have been if the prior were uniformly distributed. Hence, provided the prior beliefs of different people meet the conditions of the theorem, the corresponding posterior beliefs at which each would arrive would be roughly the same.

In order to ascertain whether the conditions of the theorem apply, one must go through three steps. First, a 99 per cent or higher (the higher, the better) credible interval, covering some range *B,* should be calculated on the hypothesis that the prior distribution is uniform. Since a uniform distribution is only defined over an interval that is bounded at both ends, this will involve setting limits to the values that the parameter could have. Secondly, one must ascertain that within $B$, the actual prior is almost constant; more specifically, the range of probability values within $B$ should not be more than about 5 per cent (but the lower, the better) of the minimum probability value in $B$. Thirdly, one must verify that most (say 95 per cent, but the more, the better) of the posterior distribution corresponding to the actual prior is covered by $B$. As the authors pointed out, the third condition may be difficult to check in practice. For practical purposes, they proposed an alternative, which is actually stronger than the third condition, but which would normally be easier to verify. This is that nowhere outside $B$ should the actual prior be astronomically big compared with its value within $B$ (the authors note that a factor of up to 1000 would be tolerable). The Principle of Stable Estimation asserts that if the three conditions are met, the actual posterior distribution and that derived on the basis of a uniform prior are approximately the same. We shall omit the details of the approximation, save to point out that the better the conditions are met, the more satisfactory the approximation, and that in general a larger sample will satisfy the

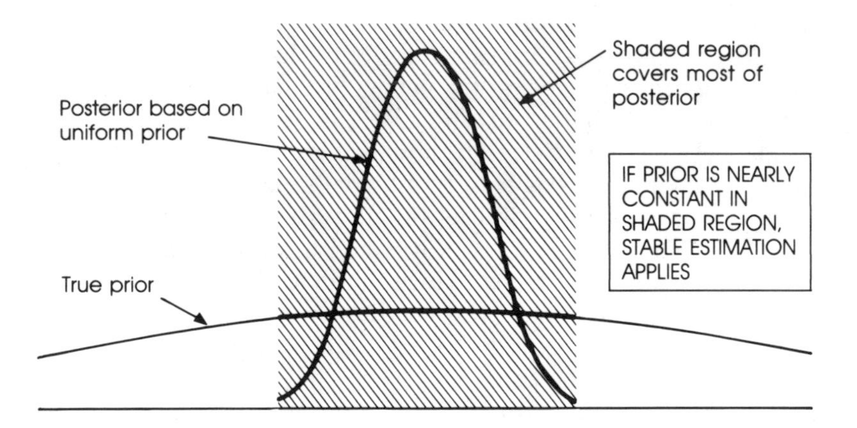


*Practical application of the Principle of Stable Estimation*

conditions better than a smaller one. The conditions under which the Principle of Stable Estimation applies are neatly illustrated by the preceding diagram, taken from Phillips (1973).

The Principle of Stable Estimation ensures that the relative insensitivity to variations in the prior distribution that we noted in the examples above is not confined to distributions belonging to any particular family. It thus facilitates a more general response to the fear that Bayesian inference is unacceptably subjective. The principle also removes the difficulty that would arise in applying Bayes's Theorem to parameter estimation if, as might at first appear, one had to start with an accurate and precise description of someone's prior distribution, for although people can often compare the probabilities of their beliefs very roughly, they find it difficult to assign precise values to those probabilities. The principle shows that provided the sample is sufficiently large, no great precision is required when describing the prior distribution in order to arrive at a correct and relatively precise posterior distribution.

## d DESCRIBING THE EVIDENCE

The result of an experiment is a physical phenomenon or state; on the other hand, evidence that is used in the appraisal of a hypothesis must be expressed as a statement. A question that naturally arises is, which aspects of the physical situation should the evidence statement incorporate? Clearly, not every aspect could be described, nor would anyone think a complete description desirable as an ideal, for some parts of an experimental result are plainly irrelevant. For instance, the colour of the experimenter's socks when sowing plant seeds would, presumably, not be worth recording if one were interested in the genetic structure of the plant. Clearly, evidence need not describe irrelevant facts such as these. On the other hand, it should omit no relevant information.

What constitutes relevance in this context, and how can one tell what is relevant and what is not? The answer to the first question is that a fact is relevant to a set of hypotheses when a knowledge of it makes a difference to one's appraisal of those hypotheses; in relation to Bayesian inductive theory, this means that a fact is relevant when it influences the probabilities of the hypotheses. The factors determining whether a fact is relevant in a particular case may be illustrated by reference to the type of hypothesis we have been considering in this chapter, namely, those that comprise an exhaustive and mutually exclusive set of hypotheses, each characterised by a particular value of some parameter $\theta$.

For simplicity, we shall relate the discussion to the example of an experiment that is designed to elicit a coin's physical probabilities of landing heads and tails. We regard this as an elementary representative of a type of experiment that crops up frequently in science. Suppose the coin has been tossed several times, and let us consider a number of descriptions of the result, each description being accurate but varying from others in its detail.

The first description, $e_1$, contains the least information. Intuitively, its evidential value is rather small, since it says nothing about $n$, the number of times the coin was tossed. A Bayesian inference from the result would require not only a prior distribution over $\theta$ (the physical probability of the coin to land heads), but one over $n$ as well. It is not necessary for us to enter further into this rather complicated case, save to note that unlike the classical approach, and more plausibly, we think, the Bayesian does not necessarily find $e_1$ entirely uninformative.

**TABLE XIII** Possible descriptions of the result 4 heads and 6 tails, obtained after tossing a coin 10 times

$e_1$: 4 heads
$e_2$: 4 heads, 6 tails, in a trial designed to have 10 throws
$e_3$: 4 heads, 6 tails, in a trial designed to cease after 6 tails
$e_4$: the sequence HHTHTTTHTT
$e_5$: The sequence HHTHTTTHTT obtained in a trial designed to terminate when the experimenter is called for lunch
$e_6$: 4 heads, 6 tails
$e_7$: the sequence HHTHTTTHTT, with the first throw being performed with the left hand, the second . . . and the sixth with the right hand
$e_8$: The experimenter intended to report the outcome either as the occurrence of 4 heads and 6 tails, or as its non-occurrence. It did occur.

Let us proceed to $e_2$, which states $n$ exactly and, moreover, gives the experimenter's stopping rule. We thus have sufficient information to calculate the posterior probability distribution in the accordance with Bayes's Theorem,

$$P(\theta \mid e) = \frac{P(e \mid \theta)P(\theta)}{P(e)}.$$

In the present instance, $e = e_2$, $P(e \mid \theta) = {}^nC_r\theta^r(1 - \theta)^{n-r}$, and $P(e) = \int {}^nC_r\theta^r(1 - \theta)^{n-r}P(\theta)d(\theta)$, where $n = 10$ and $r$ (the number of heads) $= 4$. The binomial factor, ${}^nC_r$, being a function of $n$ and $r$ only, is independent of $\theta$, and so cancels from top and bottom in Bayes's Theorem. Clearly, any other description of the trial for which $P(e \mid \theta) = K\theta^r(1 - \theta)^{n-r}$, where $K$ is independent of $\theta$, yields the same posterior distribution. This would be the case with $e_3$, which states a different stopping rule, and $e_4$, which describes the precise sequence of heads and tails in the outcome. In the former case, $K = {}^{n-1}C_{r-1}$ (*see* Chapter 9, section **c.5**), and in the latter, $K = 1$. Hence, in the Bayesian approach, it does not matter whether the experimenter intended to stop after $n$ tosses of the coin or after $r$ heads appeared in a sample; the inference about $\theta$ is exactly the same in both cases. As we have already noted, this is not so in the classical scheme.

$e_5$ is a case where the rule to stop the trial depends on some external event rather than on any feature of the sample. $e_5$ is equivalent to the conjunction $e_4$ & $l$, where $l$ states that lunch was ready just after the tenth throw of the coin, and, as before, $e_4$ describes the sequence of heads and tails in the outcome. In applying Bayes's Theorem, we need to consider the quantities $P(e_4 \& l \mid \theta)$, which can also be expressed as $P(e_4 \mid \theta)P(l \mid e_4 \& \theta)$. If the point at which lunch is ready is independent of $\theta$, as presumably it is, then $P(e_5 \mid \theta)$ equals $P(e_4 \mid \theta)P(l \mid e_4)$, hence it has the form $KP(e_4 \mid \theta)$, where $K$ is the same for all $\theta$. Then, as before, $K$ cancels from Bayes's Theorem, and so $e_5$ results in the same posterior probabilities as $e_4$. Therefore, as one would expect, this stopping rule has no inductive relevance.

Data are normally reported without mention of the stopping rule; for instance, $e_6$ merely states that the outcome contained 4 heads and 6 tails. But this poses a problem, for the probability of such a result depends not only on the parameter but also on the manner in which the trial was conducted, in particular, in regard to the stopping rule. Hence, without more information, one could not compute $P(e_6 \mid \theta)$, so that neither Bayes's Theorem nor even a classical test of significance could be applied. The difficulty is usually met by simply assuming that the sample size was fixed in advance, whether or not there is evidence that this was so, and then evaluating the probability of the evidence relative to $\theta$ accordingly. This arbitrary-seeming practice can be justified. The argument is this. Continuing with the coin example, we know that the 4 heads and 6 tails in the sample must have occurred in some order. Whatever that order was, its probability conditional on $\theta$ is $\theta^4(1 - \theta)^6$, so we may use this as the appropriate input into Bayes's Theorem. But we could also have used the term $K\theta^4(1 - \theta)^6$, without affecting the resulting posterior distribution, provided $K$ is the same for all $\theta$. Hence we are permitted to assume that the sample was drawn with a stopping rule that fixed its size in advance, for this is merely equivalent to assuming that $K$ is a constant equal to ${}^{10}C_4$.

The stopping rule may, in certain circumstances, contain relevant information, however. This would be so if one were relying upon a random sample to measure the mean height of a group of cooks who happened to be preparing lunch at the same time as the experiment was in progress. Suppose we know that tall chefs cook faster than short ones and that the trial was concluded as soon as lunch was ready. In this instance, exactly when the trial was concluded would depend on the unknown parameter and so would be some guide to the stature of the cooks. Ignoring the stopping rule in such a case would be overlooking relevant information. Hence, the stopping rule is not necessarily uninformative, but as this example suggests (we discussed another example in Chapter 9, section **c.6**), normally it would be. This concession should not be misunderstood. It does not mean that the scientist's *intention* to stop the trial at a particular point is of any inductive significance; hence, our position is quite different from that of the classical statistician. We are simply claiming that in estimating a parameter, one normally would derive all one's information from the composition of a suitable sample, but that sometimes events attending the sampling process also have significance as evidence.

The description $e_7$ in the list of possible descriptions of the coin-tossing trial seems to contain irrelevant information, an intuition endorsed by Bayes's Theorem, for $e_7$ is equivalent to the conjunction $e_4$ & $h$ where $e_4$ describes the sequence of heads and tails and $h$ states which hand was involved in each throw of the coin. As with an earlier example, if $h$ is independent of $\theta$ (which it usually would be), it is irrelevant to $\theta$. Similar considerations apply to the last example, $e_8$. Presumably, an experimenter's intention to ignore certain aspects of results which did not in fact eventuate would have no psychokinetic effect on the experiment and so would not affect the probability of the actual result. If so, information about that intention is not relevant and has no evidential significance in this context.

These conclusions, it seems to us, are unequivocally endorsed by most people's intuitions. It will be recalled, however, that classical tests and estimation procedures contradict them. Despite the classical approach having dominated the scene for above sixty years, these consequences of it still seem surprising and unacceptable.

## e SUFFICIENT STATISTICS

In an experiment to determine the parameters of what is believed to be a Bernoulli process, one would normally just count up the successes and failures in a random sample and ignore the order in which they occurred. Similarly, the mean of some population would normally be estimated by the mean of an appropriate sample, the specific details of each measurement being ignored. Such details are not invariably superfluous, however. If, for instance, a thousand throws of a coin produced an equal number of heads and tails, it would surely be useful to know that all the heads preceded all the tails, for the extra information might point to some change in the physical character of the coin which would not have been guessed from the summary data. The question whether the evidence should include a description of each measurement and its order of presentation in the sample, or whether some of the information may be ignored, is often dealt with through the theory of sufficiency.

It will be recalled that the Classical Theory of estimation and Fisher's version of significance tests assert that the statistics employed for an inference ought to be sufficient, on the grounds that only sufficient statistics contain all the relevant information. We have already pointed out that the notion of relevance to which this reason appeals is an intuitive one that has to be grafted onto classical views of statistical inference. The position is different for the Bayesian theory, where the natural notion of relevance (a datum is relevant to a hypothesis just in case it makes a difference to the final assessment of the hypothesis) automatically implies the inductive sufficiency of statistics that are sufficient in the technical sense. We showed this earlier (in Chapter 10, section **b.1**) with the demonstration that $p(\theta \mid t) = p(\theta \mid \mathbf{x})$ if and only if $t$ is sufficient relative to some datum $\mathbf{x} = x_1, \ldots, x_n$ and in the context of the set of hypotheses represented by possible values of $\theta$.

This conclusion reminds us that sufficiency is defined with respect, first, to a set of data and, secondly, to a set of hypotheses. In the Bayesian case, these hypotheses must exhaust all the possibilities, in the sense that $\Sigma P(\theta_i) = 1$. The sufficiency idea can be extended by regarding a statistic as 'almost sufficient' and so practically adequate, when $\Sigma P(\theta_i)$ is somewhat less than 1. It would be practically adequate in the sense that the posterior distributions relative to $\mathbf{x}$ and to $t$ would be similar enough for a given purpose (Pratt, 1965, pp. 169–71). The statistic would be inadequate, however, if some moderately plausible hypothesis, call it $h$, that is not represented by a numerical value of $\theta$, implied a moderately high value for $P(\mathbf{x} \mid t \,\&\, h)$ which, moreover, differed substantially from $P(\mathbf{x} \mid t)$. For in that case, $P(h \mid \mathbf{x})$ would diverge substantially from $P(h \mid t)$. For instance, if $t$ described the proportion of successes in a Bernoulli trial, and $\mathbf{x}$ the sequence of successes and failures, $h$ might attribute to the trial some variable probability of producing successes, or it might accuse the experimenter of cheating in some way. The existence of such hypotheses explains the fact, which we noted earlier, that the order of a sequence of observations may sometimes convey important information.

The classical account of sufficiency resembles the Bayesian one. It differs in two notable ways, however, both deriving from its failure, except in rare cases, to treat hypotheses as entities that may be qualified by a probability. First, according to the classical definition, $t$ is sufficient relative to $\mathbf{x} = x_1, \ldots, x_n$ if $P(t \mid \mathbf{x})$ is independent of $\theta$, where different values of $\theta$

represent different hypotheses. Unlike the corresponding Bayesian definition, $\theta$ does not appear in the conditional probability term, since it is not regarded as a random variable by classical statisticians. Secondly, and more importantly, the set of hypotheses relative to which sufficiency is defined in the classical scheme has to be circumscribed in an arbitrary way. Clearly, it ought to be limited, because if it contained every logical possibility, then it would, for instance, include a hypothesis (indeed, infinitely many hypotheses) that predicted with certainty the particular sequence $x_1, \ldots, x_n = \mathbf{x}$; hence, no statistic based on **x** (except, trivially, one that retained all the information in **x**) would ever be sufficient. On the other hand, the hypotheses relative to which sufficiency is defined cannot be selected on a probability criterion, for this would concede the case to the Bayesians. There appears to be only two other possibilities: classical statisticians may envisage some yet-to-be-described, non-probabilistic means by which hypotheses could be judged worthy of consideration, or the hypotheses could be picked in an arbitrary way, say, by tossing a coin or by a blind selection from a hatful of hypotheses. Neither seems equal to the situation. The second is hardly in keeping with what we understand by the scientific attitude; the first is difficult to make specific, particularly when one bears in mind the guiding principle of classical reasoning, namely, that inductive inferences should be purely objective.

## f METHODS OF SAMPLING

If you are interested in a population parameter $\theta$, and are given some sample data **x**, you need to compute the likelihood terms $P(\mathbf{x} \mid \theta)$, in order to derive, via Bayes's Theorem, a posterior probability distribution. Likelihoods also figure in classical inferences, though in quite different ways. Since classical inferences are supposed to be objective, the likelihoods they employ must also be, to which end classical statisticians generally require that experimental samples should be objectively random. As we explained (in Chapter 10, section **d**), this entails that elements from the population should be chosen for the sample using a physical randomizing device (e.g., the blindfold selection of names from a hat), in such a way that every element is endowed with precisely the same objective probability of being selected.

We have argued that the insistence on objective randomness in the sample-selection mechanism is indefensible as a general epistemic requirement. There was also the problem of what Stuart (1962, p. 12) called the Paradox of (Random) Sampling; it will be recalled that according to the Principle of Random Sampling, it is a general rule that if a sample was generated by experimenter I through a random mechanism, it may properly be employed for the purpose of estimation, while if the very same sample had been deliberately chosen in some non-random way by experimenter II, it would not be acceptable and would be quite uninformative. According to Stuart, we have to accept this counter-intuitive implication in order to avail ourselves of the objective methods of classical estimation.

The implication is counter-intuitive, but that is not to say that the sampling method is always irrelevant. Indeed, Bayes's Theorem indicates circumstances under which it should be taken into account, with results that seem intuitively plausible. Suppose that you wished to estimate the proportion, $\theta$, of *A*s in a population, that you have drawn a random sample of one (to cut the example to the bare bones), and that the element drawn was an *A;* and suppose you note that that element also possesses some other characteristic, say, *B*. So, for example, if you were interested in the proportion of the population unable to distinguish butter from margarine, your random sample might contain an individual who does lack this power of discrimination (*A*), and you might note that he is also above ninety years of age (*B*). The posterior probability of $\theta$ in the light of this sample information equals $\frac{P(AB|\theta)\ P(\theta)}{P(AB)}$.

Suppose now that the sample were to have been generated in a different way, namely, by first isolating all the *B*s and then making a random draw from this sub-group of the population. And suppose further that the element so selected was an *A*. Although the sample in this case is also an *AB,* because of the particular way in which it was sampled, the posterior probability of $\theta$ is $\frac{P(A \mid \theta.B)P(\theta)}{P(A \mid B)}$, which differs from the previous formula. So the inductive force of a given sample is not in general independent of the selection procedure. However, in the special case where $P(B \mid \theta) = (B)$, or, equivalently, $P(\theta \mid B) = (\theta)$—in other words, where $\theta$ and $B$ are probabilistically independent—the two formulas coincide.

This seems to us intuitively right, for the following reason. The number of $B$s drawn in a random selection from a population gives an indication of the relative size of the $B$ subclass, and if $B$ depends probabilistically on $\theta$, information about the proportion of $B$s would convey some information about $\theta$. On the other hand, if you knew that the sample was deliberately restricted to the $B$s alone, the mere fact that you had only drawn $B$s would carry practically no information about the proportion of $B$s in the population; hence that potential source of information about $\theta$ would be lost.

Although the Bayesian and classical positions agree that the method of sampling may be a relevant factor in any statistical inference, the former does not imply that non-random samples are uninformative, nor does it make the inconvenient demand that all samples must be random. But if samples are not random, how can the likelihood terms that are needed in Bayes's Theorem be evaluated? This is a question that has unfortunately not received its due attention from Bayesian statisticians, who, in theoretical expositions, generally skate round the issue by restricting their considerations to standard random-sampling methods. In addressing the question, we should first note that for a Bayesian, likelihoods are degrees of belief, like prior probabilities, and as such they are regulated by relatively weak rationality constraints. Let us then consider the simplest case where you have no knowledge of any factors correlated with the population characteristic whose measurement is sought. Suppose, for simplicity, that you are trying to discover a population's average height, and consider a person who is selected from that population, not necessarily by means of a physical randomizing device but using a selection method that in your view is 'blind', as far as height is concerned. For instance, suppose the people chosen were deliberately restricted to those whose names are entered in a certain register ($R$), but that no further selection criterion was employed; thus the sample is selected from the subpopulation at 'random', in the subjective sense that everyone on the register had the same subjective probability of being included in the sample. The posterior probability $P(\theta_R \mid \bar{x}_R)$, in which $\theta_R$ is the mean value in the subpopulation, can then be computed in the way we outlined above. But since you regard appearance on the register as uncorrelated with the population mean, your prior distributions of $\theta$ and $\theta_R$ would be substantially the same, and the conditional distribution $P(\theta \mid \theta_R)$ would be sharply peaked; hence the posterior distributions of $\theta$ and $\theta_R$ must be similar.

Often, however, there are factors that we believe to be correlated with the population characteristic of interest. The way to deal with such information, in our opinion, is to adopt a kind of stratified sampling. The method involves partitioning the population into what, according to our beliefs and information, are separate, homogeneous subgroups, and then sampling from each. If the sampling is random, the method coincides with a method approved of by classical statisticians ('stratified random sampling'), while if it is ad lib, it is essentially quota sampling, which market research and opinion polling companies adopt in practice. As has just been argued, if you have no knowledge of factors within each group that correlate with the population characteristic, then whatever the sampling method, the appropriate probabilities to apply to the sample are the same as would be dictated by strictly random sampling. We would then be in a position to make a Bayesian estimate of the characteristic of interest within each subgroup, from which a value for the whole population could be obtained by taking a weighted sum of these estimates (the relative sizes of the different strata in the population provide the weightings and, of course, must be known).

## g TESTING CAUSAL HYPOTHESES

The standard statistical methodology for testing and evaluating causal hypotheses was created by Fisher and is rooted in classical procedures of inference. Fisher's methodological principles are widely respected and applied, particularly in medical and agricultural trials. It will be recalled from our discussion in Chapter 11 that the novelty of Fisher's approach was to require a process known as randomization. We have argued that despite the weight of opinion that regards it as a sine qua non, the randomizing of treatments in a trial does not do the job expected of it and, moreover, that in the medical context, it can be both inconvenient and morally questionable. We shall present here (following Urbach, 1993) an outline of the Bayesian way of testing causal hypotheses, which we believe is soundly based and intuitively more satisfactory than the Fisherian method.

Consider the matter through a medical example. Suppose a new drug were discovered which, because of its structural similarity to a chemical produced in humans who suffer from and subsequently overcome depression, seems likely to be an effective treatment for that condition. Or suppose that the drug had given encouraging results in a pilot study among a small number of patients. (As we said earlier, without some indication that the drug is likely to be effective, a large-scale trial would be hard to defend, either economically or ethically.) A clinical trial to test the drug would normally take the following standard form. Two groups of sufferers would be constituted. One of them, the test group, would receive the drug, while the other, the control group, would not. In practice, the experiment would be more elaborate than this; for example, the control group would receive a placebo, that is, a substance which appears indistinguishable from the test-drug but which lacks any relevant pharmacological activity, and patients would not know whether they were receiving the drug or the placebo. Moreover, a fastidiously conducted trial would ensure that the doctor, too, is unaware of whether he or she is administering the drug or the placebo (trials performed with this restriction are said to be double blind). And further precautions might also be taken to ensure that any other factors thought likely by medical experts to influence recovery were equally represented in each of the groups. These factors are then said to have been controlled.

Why such a complicated experiment? Well, the reason for a comparison, or control, group is obvious. We are interested in the causal effect of the drug on the chance of recovery; so we need to know not only how patients react when they are given the drug but also how they would respond in its absence. The conditions in the comparison group are intended to simulate the latter circumstance. The requirement to match or control the groups, in certain respects, is also intuitive; but although always insisted upon, it is never derived from epistemological principles in the standard, classical expositions. However, selective controls in clinical trials do have an epistemological basis. It is provided by Bayes's Theorem, as we shall proceed to explain.

### g.1 A Bayesian Analysis of Clinical Trials

To simplify our exposition, we shall consider a particular clinical trial in which, say, 80 per cent of a specified number of test-group patients have recovered from the disease in question, while the recovery rate in the control group is 40 per cent: call this the evidence *e*. Ideally, we would like to be able to draw the conclusion that these percentages also approximate the probabilities of recovery for similar people outside the trial. This hypothesis, which is represented below as $H_\alpha$, says that the physical probability of recovery ($R$) for a person who has received the drug is around 0.80 and for one who has not received the drug is around 0.40, provided they also satisfy conditions *L, M,* and *N,* say. The latter conditions might, for example, specify that the patient's age lies in a certain range, that he or she has reached a certain stage of the illness, and so forth. (In the formulations below, ~Drug denotes the condition of the drug's absence.) For $H_\alpha$ to explain *e,* we also need to be able to assert that *L, M,* and *N* were satisfied by the subjects in both the test and the comparison groups, and that the former received the drug while the latter did not; this is the content of $H_\beta$. So the hypothesis claiming that the drug caused the observed difference in recovery rates is the combination $H_\alpha$ & $H_\beta$, which we have labelled *H,* and shall call the drug hypothesis.

$H_\alpha$: $P(R \mid L,M,N, \text{Drug}) \simeq 0.80$ *and* $P(R \mid L,M,N, \sim\text{Drug}) \simeq 0.40$

**(H)**

$H_\beta$: Patients in the test and comparison groups all satisfy conditions *L, M,* and *N.*

But there are hypotheses other than *H* that are also capable of explaining the evidence. One such alternative explanation is that the drug is ineffective and that condition *O,* say, was the true cause of the experimental findings. *O* might, for example, be the condition that the patient was confident of recovery; in other words, the alternative explanation attributes the difference in recovery rates to a psychosomatic effect. $H_\alpha'$ represents the hypothesis that under the conditions $L$, $M$, and $N$, the drug has no effect on the course of the disease, but that a confident attitude promotes recovery. By parallel reasoning to that above, the hypothesis that explains the evidence as a psychosomatic, confidence effect is the combination $H_\alpha'$ & $H_\beta'$, which we have labelled $H'$.

$$H_\alpha'\text{: } P(R|L,M,N, O) \simeq 0.80 \text{ and } P(R|L,M,N, {\sim}O) \simeq 0.40$$

**(H′)**

$H_\beta'$: $L,M,N$ and $O$ are in the test group; $L,M,N$ and ${\sim}O$ are in the comparison group.

The Bayesian is concerned to discover the probabilities of hypotheses and how they are updated in the light of new evidence via Bayes's Theorem. Let us examine how this updating works in the present case. To start with, and in the interest of simplicity, we shall suppose that $H$ and $H'$ are the only hypotheses with any chance of being true. In that event, Bayes's Theorem would take the following form:

$$P(H \mid e) = \frac{P(e \mid H)P(H)}{P(e \mid H)P(H) + P(e \mid H')P(H')}$$

$$= \frac{P(e \mid H)P(H)}{P(e \mid H)P(H) + P(e \mid H_\alpha' \;\&\; H_\beta')P(H_\alpha')P(H_\beta')}.$$

In the last equation, $H'$ has been spelled out explicity in terms of its constituents, and we have assumed that the two parts of the hypothesis are probabilistically independent, which seems reasonable and also simplifies the argument.

The probability of the drug hypothesis, for fixed $e$—that is, $P(H \mid e)$—could only be maximized if the second term in the denominator were minimized. And the only part of that term that can be altered by changing the conditions of the experiment is $P(H_\beta')$. $H_\beta'$ states, amongst other things, that patients in the test group were confident of recovery, while those in the control group were not. We can reduce the probability of this being the case—reduce $P(H_\beta')$, that is—by applying a placebo to the control group; for if the placebo has been well designed, patients would have no idea which experimental group they were in, so that a number of factors which would otherwise produce different expectations of recovery in the two groups would be absent. In many cases, the probability of $H_\beta'$ could be reduced further by ensuring that even the doctors involved in the trial are unable to distinguish the treatment from the placebo; such trials are called, as we have said, 'double blind'. By diminishing $P(H_\beta')$ in these various ways, the probability of $H$, for a given $e$, is increased.

The probability of the drug hypothesis would also be raised if the value of the numerator were increased. As we argued above, such an increase can only be brought about by adopting appropriate measures to raise $P(H_\beta)$. Such measures would involve matching conditions in the two groups so as to raise the probability that those conditions which the drug hypothesis says are relevant to recovery (namely, $L$, $M$, and $N$) are represented equally in the two groups. Now suppose one of these factors denotes the virulence of the disease, or the strength of the patient, or the like, which may well be hard to measure but which we have reason to think the doctor can, through long experience, intuit. We might also suspect that well-meaning doctors would be keen to secure the best available treatment for their patients. Hence, if doctors were given a hand in arranging the experimental groups, it seems likely that one of those groups would end up with a greater number of the sicker patients. This is where randomized allocation can play a role, for it eliminates those doctors' feelings and thoughts from the allocation process, and thereby makes it more probable than otherwise that the groups will be balanced on the factors mentioned.

We have greatly simplified this exposition by considering just one alternative to the drug hypothesis. Clearly, there are many such alternatives, each contributing an element to the denominator, which should therefore include as summands other terms of the sort $P(e|H'')P(H'') = P(e|H_\alpha'' \& H_\beta'')\, P(H_\alpha'')\, P(H_\beta'')$—again assuming independence. Each of these terms relates to some particular prognostic factor, and each needs to be minimized in order to make the posterior probability of $H$, the drug hypothesis, as large as possible, and this may be done by matching the experimental groups in ways suited to reducing $P(H_\beta'')$ as far as possible.

But not every conceivable prognostic factor can be matched across the groups; nor is there any need for this. Take some such factor whose causal influence is described in the hypothesis $H_\alpha''$ and suppose that hypothesis to be very improbable; in that case, the whole of the corresponding term in Bayes's Theorem would already be negligibly small and so the advantage of reducing it further through appropriate controls would be comparably small; and in most cases, that small advantage would be greatly outweighed by the extra cost and inconvenience of the more elaborate trial. For example, although one could introduce a control which made it less probable that the clinicians treating the two groups had different average shoe-sizes, such a precaution would have negligible impact on the posterior probability of the drug hypothesis, because the corresponding hypothesis attributing a causal influence from shoe-size to recovery is itself immensely improbable.

In summary, the Bayesian approach explains, first, why the experimental groups in clinical trials need to be matched in certain respects and, secondly, why they need not be matched in every conceivable respect. The classical theory of clinical trials, on the other hand, has no adequate account of how experimental groups should be matched, and standard expositions simply lay down the requirement of selective controls without argument, as something that is intuitively obvious.

The Bayesian analysis agrees with common sense in indicating that *the chief concern when designing a clinical trial should be to make it unlikely that the experimental groups differ on factors that are likely to affect the outcome.* Designs that achieve such a balance between the groups are termed 'haphazard' by Lindley (1982a, p. 439), though we prefer to think of them as adequately matched or controlled.

With this rule in mind, it is evident that randomizing the allocation of subjects to treatments might sometimes serve a useful purpose in clinical trials as a technique to improve the matching of the experimental groups, insofar as it prevents doctors from unconsciously or even deliberately placing disproportionately many ill patients in one of the groups. *But randomized allocation is not absolutely necessary; it is no sine qua non; it is not the only nor even always the best method of constructing treatment groups in a clinical trial.*

### g.2 Clinical Trials without Randomization

Bayesian and classical prescriptions for clinical trials differ in two respects of substantial practical importance. First, a Bayesian analysis permits the continuous evaluation of results as they accumulate, so that, for example, a trial can be immediately halted as soon as it becomes evident (through a Bayesian evaluation) that the drug has been ineffective. It will be recalled that, by contrast, when classical principles are strictly followed, a clinical trial that has been brought to an unscheduled stop cannot even be interpreted, whatever the results recorded at that stage. And sequential clinical trials, which were designed specially to allow interim analyses of trial results within the classical confines, are ineffective, we argued (Chapter 11, section **e**).

The second difference between Bayesian and classical principles relates to the formation of the comparison groups for a clinical trial, which classical statisticians insist must involve randomization. On the Bayesian view, randomization is optional, and the essential condition is for the comparison groups in a clinical trial to be adequately matched on factors believed to have prognostic significance. And for trial results to be useful, those beliefs should belong to acknowledged experts in the field. Techniques for eliciting expert medical opinion have been developed, for instance, by Kadane and his colleagues (1980, 1986), and promising clinical trials are being conducted using their results.

Freed from the absolute need to randomize, Bayesian principles, unlike classical ones, imply that decisive information about a drug can be secured from properly controlled retrospective trials. As we explained in Chapter 11, these are trials in which a group of patients is treated with the test-treatment and compared with a group of past patients who have already received an alternative therapy or none at all. Such comparison groups, which are constructed from medical records, are known as historical controls. Of course, and as Bayes's Theorem implies, retrospective trials employing historical controls, in order to be informative, must be adequately matched in respect of relevant prognostic factors, on the basis of expert opinion.

Retrospective trials are generally dismissed by classically minded medical statisticians as "inherently fallacious" (British Medical Association, 1957), and concurrent randomized trials are generally regarded as the "only way to assess new drugs" (McIntyre, 1991). But these views are not only indefensible on epistemic grounds, they are daily refuted in medical practice, where doctors' knowledge and expertise is accumulated in large part by continual, informal comparisons of current patients and similar patients of their experience.

Indeed, many classical statisticians concede that retrospective trials with historical comparison groups can in fact be useful and informative. This concession typically emerges in the face of the agonizing moral problems often presented in clinical studies. For example, a distinguished group of twenty-two statisticians (Byar et al., 1990) involved in investigations of possible AIDS therapies has argued that "some traditional approaches to the clinical-trials process may be unnecessarily rigid and unsuitable for this disease", and that under certain circumstances the traditional requirements of concurrent comparison groups and randomized allocation may be suspended. But although this attractive, moderate stance is sometimes advocated by classical statisticians, it has no classical basis, and so is not a classical position.

For consider the main epistemically relevant circumstance that is claimed by Byar et al. to justify a suspension of the traditional clinical trial structure: it is that "there must be a justifiable expectation that the potential benefit to the patient will be sufficiently large to make interpretation of the results of a non-randomized trial unambiguous". The authors do not say by what process such an unequivocal interpretation of results may be made; but no classical significance test can be intended, since as Byar has already been quoted as saying: "[i]t is the process of randomization that generates the significance test" (Chapter 11, section **c**), and there is no randomization here. Nor does any other classical method for interpreting trial results seem to be available. On the other hand, Byar and his co-authors may implicity be employing a Bayesian methodology, which *is* available: this is suggested by some of the other conditions which they say should be met before a retrospective trial can be justified, which presuppose an informal prior appraisal of the effects of the test treatment (e.g., "the scientific rationale for the treatment must be sufficiently strong that a positive result would be widely expected"). A Bayesian approach would also underwrite the supposition, which is provided with no other support, that other things being equal, the larger the apparent effect, the more decisive the inference.

The admissibility in principle of retrospective trials is not just of theoretical interest but of great practical significance too. First, such trials call for fewer new patients to be included in a trial, which is of particular importance when very rare diseases are the subject of study. And, of course, smaller trials are generally cheaper. Retrospective trials also avoid the ethical difficulties that are often seen in concurrent ones, because they do not expose subjects to ineffective placebos or what clinicians might judge to be inferior comparison treatments. These ethical benefits bring with them the further practical one of overcoming patients' natural and increasing reluctance to participate in such trials.

Retrospective trials, however, are not easy to set up. Historical controls that are adequately matched with current experimental groups can be compiled only with the aid of very thorough records of past patients, much more detailed than the records that are routinely kept, and more easily accessible. To this end, Berry, 1989, has proposed the establishment of a national data base containing patients' characteristics, diagnoses, treatments, and outcomes, which would be open to the public and contributed to by every doctor. Some modest work along these lines has been done, but the widespread yet erroneous opinion that historical controls are intrinsically unacceptable or impossible in principle has unfortunately discouraged efforts to overcome the purely practical obstacles that stand in the way of effective retrospective trials. Here is a case where the mistaken principles of classical statistical inference are adversely affecting human welfare.

**Summary.** Bayes's Theorem supplies coherent and intuitively satisfactory guidelines for the design of clinical and similar trials. This contrasts with the classical principles governing such trials. One striking difference between the two approaches is that the second simply takes the need for controls for granted, while the first explains that need and, moreover, distinguishes in a plausible way between factors that have to be controlled and those that do not. Another difference is that the Bayesian approach does not make the random allocation of subjects to treatments a universal, absolute requirement. We regard this as a considerable merit, since we have discovered no good reason for regarding a random allocation as an indispensable precondition and several good reasons for not so regarding it (see Chapter 11).

## ■ h CONCLUSION

The Bayesian way of estimating parameters permits different degrees of confidence to be associated with different values and ranges of values, as classical statisticians sought, unsuccessfully, to do. It does this through a single principle, Bayes's Theorem, which applies to all inferential problems. Hence, the Bayesian treatment of statistical and deterministic theories is the same and is backed by the same philosophical idea.

Bayesian estimation is also intuitively correct. Thus it accounts for the intuitive sufficiency-condition and for the natural preference for maximum precision in estimators, while finding no place for the criteria of 'unbiasedness' and 'consistency', which, we have argued, have nothing to recommend them. The Bayesian method also avoids one of the most perverse features of classical methods, namely, a dependence on the outcome space and hence on the stopping rule.

There is a subjective element in the Bayesian approach which offends many, but this element, we submit, is wholly realistic. Perfectly sane scientists with access to the same information often do evaluate theories differently, although, as the Bayesian predicts, they normally approach a common view as the evidence accumulates (for further discussion of this point, see Chapter 15, section **i**). The Bayesian also anticipates the possibility of people whose predispositions either for or against certain theories are so pronounced and different from the norm that their opinions remain eccentric, even with a large volume of relevant data. You might take the view that such eccentrics are pathological, that each theory has a single value relative to a given body of knowledge, and that responsible scientists ought not to allow personal, subjective factors to influence their judgment. But then you would have to show why this should be the case and how it is possible. Many have attempted this, but none has succeeded.

## EXERCISES

1. Suppose you are examining a theory, *a,* which you know follows from another theory, *b*. Show that if *b* is now refuted, your confidence in *a* should, on the Bayesian view, be reduced. Show also that this reduction is greater, the larger the initial probability of *b*. (Hint: you should be interested in $P(a \mid \sim b)$.) Can you think of an example of such a situation? (Polya, 1954, vol. 2, p. 122)

2. Suppose that 1 per cent of the population suffers from a certain disease. A randomly selected person comes out positive on a diagnostic test for the disease. The test has a sensitivity of 95 per cent, that is, 95 out of 100 people with the disease respond positively. And it has a 95 per cent specificity, that is, 95 out of 100 without the disease give a negative response to the test. Is it likely, in the light of the information provided, that the person has the disease? Consider first your intuitive response, then do the calculation (use Bayes's Theorem).

3. "While recycling of waste products is a laudable ecological activity, medical investigators need to be warned against re-use of observations. Whenever data through inductive reasoning have been used to propose a hypothesis, new and independent observations are necessary to test it. If data—through statistical analysis—are re-used to test the very hypothesis which they served to generate, circularity and erroneous conclusions may be the result" (Andersen, 1980, p. 39). Is this conclusion correct? Is it compatible with a Bayesian view?

4. Show that if the prior probability over the mean of a normal population is uniform, on the basis of a random sample from the population, the shortest, posterior credible interval for that mean and the shortest confidence interval coincide,

whatever the size of the random sample. Explain how the Bayesian and the classical conclusions, nevertheless, differ substantially.

**5.** A beta distribution for $x$ is given by

$$P(x) = \frac{1}{B(u,v)} x^{u-1}(1-x)^{v-2},$$

where

$$B(u,v) = \frac{\Gamma(u)\Gamma(v)}{\Gamma(u+v)},$$

and, when $w$ is an integer, $\Gamma(w) = (w - 1)!$ (When $w$ is not an integer, the value of the gamma function can be obtained from tables in mathematical handbooks.)

Graph the beta distributions for which $u = 2, v = 1$; $u = 3, v = 5$; and $u = 7, v = 4$. Suppose a coin were tossed 20 times and yielded 7 heads and 13 tails. What would be the posterior distributions over $p$, the coin's physical probability for landing heads, corresponding to these three beta prior distributions? Compare the results that would be obtained if 40 coin tosses had led to 14 heads.

**6.** The Hippocratic Oath, which is supposed to summarise a physician's moral responsibilities towards his patients, includes the following: "The regimen I adopt shall be for the benefit of my patients according to my ability and judgment, and not for their hurt or for any wrong". Is this commitment consistent with a doctor recommending to a patient that he or she participate in a randomized clinical trial?

**7.** A doctor was formally charged by the General Medical Council of the United Kingdom with professional misconduct. Mrs *A* complained about the doctor's lewd language and behaviour towards her during a medical examination connected with the fitting of a contraceptive coil. Mrs *B* complained in strikingly similar terms that the doctor had used obscene language and acted in a sexually suggestive manner while examining her foot. The Council ruled that the evidence of Mrs *A* and of Mrs *B* corroborated each other;

since the two women gave independent accounts of separate incidents, the circumstances were such as to exclude any danger of a jointly fabricated account, and there was sufficient similarity between them. This ruling was upheld by three Appeal Court judges after the defence took the case to the Privy Council.

What basis, if any, is there in Bayesian reasoning for this ruling?

One of the Appeal Court judges was reported in a British newspaper to have remarked during the summing up, that "corroboration by similar fact evidence appeared in a variety of forms. In order to be admissible the similarity of the evidence had to be either unique, in which case its probative value would approach that of a fingerprint, or it had to be striking, when its probative value would depend on how striking the similarity was. Similar fact evidence could not be defined by a degree of similarity measurable on a scale or by reference to the place or time at which it appeared" (*The Guardian,* 5 July 1989, p. 47). Comment.

**8.** The chief prosecution witness in a trial was under contract with a newspaper to write a story about the case. The fee was to be $40,000 if the accused was found guilty and $5,000 if innocent. This arrangement was exposed during cross-examination by counsel for the defence, and as a result, the jury brought in an acquittal. Could the jury members have been reasoning in a Bayesian way?

**9.** **The Monty Hall Problem.** You are on a television game show and are offered the opportunity of winning a car. You are informed that the car is behind one of three doors and that there is nothing behind the other two. You are asked to point to a door; the game show host then opens one of the other doors to reveal no car. You are now invited to open one of the unopened doors yourself and to keep whatever you find there—either a car or nothing at all. (Monty Hall used a version of this game in his television show.)

Assuming that you wish to win the car, should you open the door you first pointed to or switch to the other unopened door, or does it make no difference?

The answer is extremely surprising, namely, you would increase your chance of winning the car from $\frac{1}{3}$ to $\frac{2}{3}$ by switching to the other unopened door.

Let *a*, *b*, and *c* stand for the propositions that the car is behind doors *A*, *B*, and *C*, respectively. Show that the lower *P*(*a*), the more sensible would it be first to pick door *A* and then to switch to the remaining unopened door.

# PART VI

# *Finale*

Our next, and last, chapter is also one of the longest in this book, fittingly enough, since it sets out the objections which have been raised against the theory of method we have advanced in the previous chapters, and those objections are many and various. Without more ado, we shall pass to their consideration and rebuttal.

CHAPTER 15

# The Objections to the Subjective Bayesian Theory

## a INTRODUCTION

In the preceding chapters we have developed the theory of subjective or personalistic Bayesianism as a theory of inductive inference. We have shown that it offers a highly satisfactory explanation of standard methodological lore in the domains of both statistical and deterministic science; and we have also argued at length that all the alternative accounts of inductive inference—like Popper's or Fisher's—achieve their explanatory goals, where they achieve them at all, only at the cost of quite arbitrary stipulations. However, the subjective Bayesian theory itself has been the object of much critical attention, to such an extent that it is still regarded in some influential quarters as vitiated by hopeless difficulties. These difficulties, in our view, stem from misunderstanding, and in this final chapter we shall do our best to dispel them.

Of the standard criticisms some—due largely to Popper and his followers—are answered relatively simply and quickly, and we shall deal with these first.

## b THE BAYESIAN THEORY IS PREJUDICED IN FAVOUR OF WEAK HYPOTHESES

Discussing theories of inductive inference which assess the empirical support of hypotheses by changes in their probabilities on receipt of the relevant new data, Watkins (1987, p. 71) asserts that such theories are "prejudiced" in favour of logically weaker hypotheses. This is a favourite charge of the Popperian school and is frequently made by its eponymous founder; for example, Popper (1959, p. 363; his italics) writes that "[scientists] have to choose between high probability and high informative content, since *for logical reasons they cannot have both*".

Such a charge is quite baseless. There is *nothing* in logic or the probability calculus which precludes the assignment of even probability 1 to any statement, however strong, as long as it is not a contradiction, of course. The only other way in which probabilities depend on logic is in their decreasing monotonically from entailed to entailing statements. But this again does not preclude anybody from assigning any consistent statement as large a probability as they wish. Popper's thesis that a necessary concomitant of logical strength is low probability is simply incorrect.

Glymour attempts to argue a variant of Popper's objection, but this, too, is easily rebutted. Glymour claims that since the observable consequences of scientific theories are at least as probable as the theories themselves, then in a Bayesian account one is unable to account for our entertaining theories at all:

> On the probabilist view, it seems, they are a gratuitous risk. The natural answer is that theories have some special function that their collection of observable consequences cannot serve; the function most frequently suggested is explanation. . . . [But] whatever explanatory power may be, we should certainly expect that goodness of explanation will go hand in hand with warrant for belief, yet if theories explain and their observational consequences do not, the Bayesian must deny the linkage. (Glymour, 1980, pp. 84–85)

The Bayesian certainly does want to justify the quest for theories in terms of a desire for explanation that a congeries of observational laws cannot by itself provide; but he would also, for very good reason, deny the linkage Glymour alleges between explanatory power and warrant for belief. Indeed, counter-examples to the claim that any such linkage exists are only too easy to find: a tautology, to take an obvious one, has maximal warrant for belief and minimal explanatory power. This does not, of course, imply that what we take to be good explanations do not tend to have correspondingly high probabilities on the available evidence. They do. But Glymour's premiss makes the additional claim that an increase in "warrant for belief" should imply an increase in explanatory power. That premiss is clearly false, and Glymour's objection collapses.

It is strange that Glymour and the Popperians should converge, from quite different starting points, in charging Bayesians with an implicit denial of the value of deep explanatory theories. Glymour thinks that good explanatory theories by that token justify a correspondingly large claim to belief, whereas the Popperians assert that such theories merit the lowest possible degree of belief. Whatever their starting points, however, the charge of Glymour, Popper, et al., that Bayesians must in principle undervalue theories is patently false. Perhaps a simple analogy will dispel any lingering doubts that may remain. A jury has always at least two mutually inconsistent hypotheses to consider: that the accused is guilty is one, and that the accused is not guilty and there is some alternative explanation of the known facts is the other. The jury has to determine which of these is the more probable hypothesis, given the evidence. Imagine their surprise at being informed that they are thereby committed, on their return to the court, to announcing that their favoured conclusion is the restatement of the evidence! (*see also* Horwich, 1982, p. 132). Scientists, like the court, want information of a specific sort combined with the assurance that it is credible information; and these demands *can,* despite Popper's claim to the contrary, simultaneously be met. The Bayesian theory tells us how.

## c THE PRIOR PROBABILITY OF UNIVERSAL HYPOTHESES MUST BE ZERO

Popper, we noted in the previous section, asserts that it is impossible for a hypothesis to possess both high informative content and high probability. In particular, he asserts that the probability of a universal hypothesis must, for logical reasons, be zero (1959a, appendices *vii and *viii). He occasionally remarks (for example, 1959a, p. 381) that the constraints imposed by the probability calculus alone require that the only consistent assignment of a probability to such a hypothesis is zero.

Were Popper correct, then that would be the end of our enterprise in this book, for the truth of Popper's thesis would imply that we could never regard unrestricted universal laws as confirmed by observational or experimental data, since if $P(h) = 0$, then $P(h \mid e) = 0$ also, whatever finite sample data $e$ may consist of.

Popper's thesis is quite untrue, however. Even in Carnap's so-called continuum of inductive methods (Carnap, 1952; *see* our discussion in Chapter 4), one of those methods (corresponding to $\lambda = 0$), assigns, in an obviously consistent way, positive probabilities to a class of strictly universal hypotheses over an infinite domain. And Hintikka's systems of inductive logic almost invariably assign positive prior probabilities to consistent universal sentences, as we also noted in Chapter 4, whether the domain of individuals is finite or infinite. It is even possible to assign positive probabilities to *all* the non-contradictory sentences in a language powerful enough to include all of science and mathematics (Horn and Tarski, 1948, theorem 2.5).

Popper's arguments for his zero-probability claim are really designed to show something considerably less ambitious than the false thesis that there can be no consistent assignment of a non-zero probability to a universal hypothesis. What they aim at showing is that the assignment of positive probabilities to universal hypotheses involves a quite unacceptable degree of arbitrariness. He has three main arguments. We shall review them briefly and conclude that none succeeds (the discussion follows Howson, 1973).

**1.** Popper points out that only one among the $2^n$ state descriptions in $L(A,n)$ (*see* Chapter 4, section **c.1**) satisfies the universal sentence $\forall x\, A(x)$. Hence, the proportion of 'possible worlds' satisfying that sentence is zero in the limit as $n$ tends to infinity. This well-known property of $m^\dagger$ is maintained, moreover, even if we expand the language $L$ to incorporate any number of predicates. But Popper fails to provide any reason why we should adopt that particular a priori distribution. As we argued in Chapter 4, no equiprobability distribution over any partition of logical space qualifies as representing genuine epistemic neutrality. $m^\dagger$, for example, assigns the statement 'There are universal laws' the a priori probability 0, and the statement 'There are no universal laws' probability 1, *an assignment which is irrevocable however strong the evidence might be that phenomena exemplify lawlike behaviour.*

To be fair to Popper, he is not entirely wholehearted in his endorsement of $m^\dagger$, but he claims that $m^\dagger$ *is* correct in making all the sentences $A(a_i)$ probabilistically independent with constant probability; and this is clearly enough to imply that in the limit as $n$ tends to infinity $P[\forall x\, A(x)]$ is equal to 0. The $A(a_i)$ should be probabilistically independent, according to Popper, because "every other assumption would amount to postulating *ad hoc* a kind of after-effect" (1959, p. 367). Popper nowhere argues for the constant probability assumption, and indeed, such an assumption appears to be quite arbitrary. Nor does he notice that independence, by an argument exactly parallel to his own, should amount to postulating *ad hoc* a *lack* of after-effect. If one postulate is unaceptable, so ought to be the other. But every distribution $P$ will either make the $A(a_i)$ independent or it will not, so either way $P$ will be unacceptable.

**2.** Let $e^n = e_1 \& \ldots \& e_n$. If $h \vdash e_i$ for each $i$, $1 \leq i \leq n$, then $P(h \mid e^n) = \frac{P(h)}{P(e^n)}$. $P(e^n) \leq P(e^{n-1})$, so if $P(h) > 0$, then $P(e^n)$ must tend to a non-zero limit as $n$ tends to infinity. Since $P(e^n) = P(e_n \mid e^{n-1}) \ldots P(e_2 \mid e_1)P(e_1)$, it follows that $P(e_n \mid e^{n-1})$ tends to 1. Popper now invites us to consider all the 'grue' variants $h_k$ *of* $h$, where $h_k$ entails $e_1, e_2, \ldots, e_k$, and also $\sim e_{k+1}, \sim e_{k+2}, \ldots$ If each $P(h_k) > 0$ also, and it seems quite unjustified to deny this and concede it for $h$, then $P(c_{kn} \mid c_k^{n-1})$ also tends to 1, where $c_{ki} = e_i$ if $i \leq k$, and $c_{k_i} = \sim e_i$ if $i > k$. Hence $P(c_{kn} \mid c_k^{n-1})$ tends to 1. In that case, claims Popper, there must be an $m$ such that both $P(e_m \mid e^{m-1}) > \frac{1}{2}$ and $P(\sim e_m \mid e^{m-1}) > \frac{1}{2}$ (1959a, p. 371), which is a contradiction. But this does not follow at all; it presupposes (*see* exercise 3 at the end of this chapter) that the sequence of functions $f_n(k) = P(c_{k\,n} \mid c_k^{n-1})$ converges *uniformly* to 1 over the set $k \geq 1$, which there is no reason whatever to believe true.

**3.** Consider the sequence of polynomial theories

$$h_n\text{: } y = a_n x^n + a_{n-1} x^{n-1} + \ldots + a_1 x + a_0,$$

where $x$ can take any value and the $a_i$ are undetermined or adjustable *non-zero* real parameters, to be evaluated from the data. Popper claims (1959a, pp. 381–82) that we must accept that $P(h_{i+1}) \geq P(h_i)$ for all $i$, since the smaller the index $i$, the more opportunities there are for falsifying $h_i$ (in general, it will require $n$ observations to determine the $n$ parameters in $h_n$). Given that the $a_i$ are non-zero, the $h_i$ are mutually inconsistent, and so $\Sigma P(h_i) \leq 1$, from which it follows that $P(h_i) = 0$ for all $i$. This argument too fails, because Popper's premiss equating testability with improbability is false. That fewer independent observations are required to test $h_i$ than $h_{i+1}$ does not imply that $h_{i+1}$ should be regarded as less likely to be false than $h_i$. As we have already argued in Chapter 7, section **j.3**, in supposing otherwise Popper is confusing pragmatic with epistemological considerations.

So none of Popper's arguments for $P(h) = 0$, where $h$ is a universal generalisation, succeeds. Moreover, Popper's recommendation actually turns out to be impossible to implement consistently. For as we saw above, the assignment of probability 0 to universal sentences in $L(A,n)$ entails that the meta-level *universal* sentence 'There are no laws' must be assigned probability 1. So Popper's position is worse than being arbitrary; it is incoherent.

Nevertheless, Popper has called our attention to a fact which deserves some comment, namely, that the history of science is the history of great explanatory theories eventually being refuted. In view of this, ought we not rationally to expect all theories to be eventually refuted? It is far from clear that such bleak pessimism really is the lesson taught by the history of science. The mere fact that succeeding extensions of the observational base of science have caused the demise of many an explanatory theory does not demonstrate the appropriateness of total scepticism, nor does it even make it plausible. If up till now I have failed to find the thimble, I do not conclude, and certainly ought not to conclude, that further quest is hopeless. Of course, science is not hunt the thimble, but this does not destroy the point of the analogy, which is that a number of past failures to discover the truth does not by itself imply that one will not one day be successful.

Pessimism on that particular score is certainly not something to which many practitioners of science subscribe. There is a great deal of biographical and anecdotal evidence which suggests that, on the contrary, some very illustrious scientists are, if anything, overoptimistic. Einstein's confidence sometimes bordered on the hubristic: when, after the reports of the

1919 eclipse expedition, someone asked him what he would have felt had the result not confirmed General Relativity, he is said to have replied, "Then I would have to pity the dear Lord. The theory is still correct". Even where there is doubt about a theory, that doubt is often accompanied by a belief that some substantial part of the theory is true or approximately true. In such cases, it *is believed* that a suitably modified form of that theory will turn out to be true. To sum up, there is no evidence that scientists regard general theories as invariably false and no evidence that they ought to.

To sum up: (a) Popper's logical arguments for the probability of laws being zero are all invalid; (b) it is both irrationally dogmatic and incoherent to make these probabilities uniformly zero; and (c) the history of science offers no evidence or encouragement for the view that they should be zero.

## d PROBABILISTIC INDUCTION IS IMPOSSIBLE

This dramatic claim is made by Popper and Miller (1983), who also supply what purports to be a rigorous proof of it. This proceeds as follows. According to Bayesian theories of support or confirmation, whether they are subjectively based or not, evidence $e$ supports hypothesis $h$ if and only if $P(h \mid e) > P(h)$. Suppose that $h$ entails $e$, modulo background information including initial conditions and so forth. It follows from theorem 17, Chapter 2, that $e$ supports $h$ if and only if $P(h) > 0$ and $P(e) < 1$. Suppose that these latter conditions are satisfied also, so that $h$ is (it seems) supported by $e$. Popper and Miller demonstrate that if, in addition, $P(h \mid e) < 1$, then $\sim e \vee h$ is *counter-supported* by $e$, in the sense that its posterior probability relative to $e$ is *less* than its prior probability (the proof is very straightforward, and we leave it as exercise 4 at the end of this chapter).

This simple theorem of the probability calculus is given a dramatic significance by Popper and Miller, for they claim that $\sim e \vee h$, for reasons which we shall give later, represents *the excess or inductive content of* h *over* e, or that part of $h$'s content going beyond $e$, and they interpret their result as stating that this excess content is always counter-supported by $e$. All support, conclude Popper and Miller, is really *deductive* support (since $e$ entails $e$); the support given by $e$ to the genuinely inductive content of $h$ is always negative.

While various elements in Popper's and Miller's argument have been challenged, nobody (apart from Rivadulla, 1991) seems to have noticed that, if correct, their argument proves that *counter-induction is valid:* it is surely just as bad, from their anti-inductivist point of view, to show that $\sim e \vee h$ is *counter-supported* by $e$ as if they had proved that it was supported by $e$. *Both* are species of inductive inference, since in both cases $e$ gives information about what is claimed to transcend it. In one, $e$ tells us that the allegedly inductive content of $h$ is more likely to be true than it was before, and in the other, $e$ tells us that that content is less likely to be true.

Popper and Miller obtain their odd result only because they adopt an eccentric interpretation of the idea of the excess content of one set of sentences with respect to another. Suppose, with Popper and Miller, we define the content of $h$ to be its class *Cn(h)* of (non-tautologous) consequences. We want to define the excess content of $h$ over $e$. How do we do it? There is a fairly standard answer for defining the excess of $A$ over $B$ where $A$ and $B$ are any two sets: the excess of $A$ over $B$ is the largest subset of $A$ which includes nothing in $B$, and this is, of course, just the set-theoretic difference $A - B$. But Popper and Miller define the content of $h$ excess to that of $e$ to be the largest subset of *Cn(h)* which contains nothing in *Cn(e) and which is also deductively closed*. And this is equal to $Cn(\sim e \vee h)$.

But *why* stipulate that the difference between *Cn(h)* and *Cn(e)* is deductively closed? Popper and Miller have, as they admit, chosen to represent the difference of *Cn(h)* with respect to *Cn(e)* in the algebraic context of Tarski's well-known 'calculus of deductive systems', which requires that the results of performing operations of sum, difference, and product on deductively closed sets, like *Cn(h)* and *Cn(e), must themselves be deductively closed sets*. Indeed, the Tarski difference between *Cn(h)* and *Cn(e)* is $Cn(\sim e \vee h)$. But we repeat, why choose this method of defining the difference between *Cn(h)* and *Cn(e)*, rather than that represented by the set-theoretical difference, of which $Cn(\sim e \vee h)$ is, it is not difficult to see, a (small) proper subset? (In fact, $Cn(\sim e \vee h)$ contains only those consequences of $h$ which are also consequences of $\sim e$.) It is like looking through a distorting lens which makes a good deal of

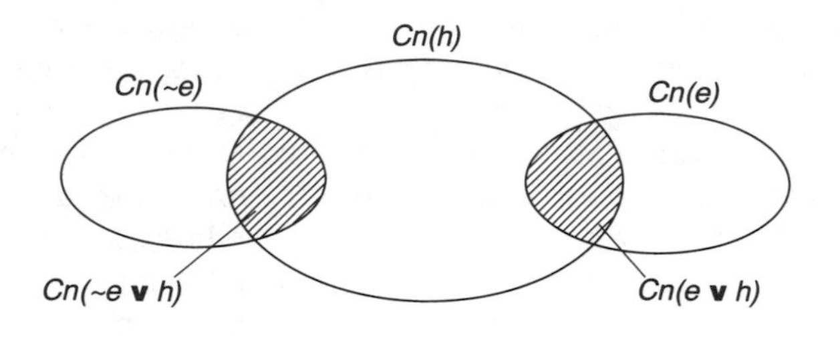


**FIGURE 1**

what is actually present—the consequences of $h$ which are consequences neither of $e$ nor of $\sim e$—disappear from view (figure 1; $e$ and $h$ are represented as logically independent).

The same distorting effect is achieved with Popper's well-known *numerical measure* of content $ct(a) = P(\sim a)$. It is easy to verify, and we shall leave it as an exercise, that $ct(e) + ct(\sim e \vee h) = ct(h)$ when $h$ entails $e$. The reason for this is that $P(\sim a)$ determines a measure $P[M(\sim a)]$ on the algebra of extensions $M(a)$ of sentences of $a$ (in the terminology of Chapter 2, section **f**). It is straightforward to show that this algebra, under the mapping which associates the consequence-class $Cn(a)$ of $a$ with the set of models of $\sim a$, is isomorphic to the Tarski algebra of consequence classes. So we can regard Popper's content measure as a measure on the Tarski algebra itself, inheriting the distorting features of that algebra which we have noted (*see also* Howson, 1993).

Howson and Franklin (1986) investigate numerical content-measures which reflect more faithfully the structure of the underlying consequence classes than does $ct$ and show that in this respect more adequate measures are afforded by $\mathrm{Inf}(a) = -\log P(a)$, the so-called information measure (its expected value is Shannon's entropy), and $[1 - P(a)]:P(a)$, or the odds against $a$. It turns out that any adequate measure based on a probability function $P$ must be a decreasing function $f(P)$ which, unlike $1 - P$, is strictly convex in some subinterval of the unit interval; the reader is referred to Howson and Franklin's paper for further details.

Popper and Miller, in one of their publications (1987), defend their adoption of $Cn(\sim e \vee h)$ as the excess content of $h$ over $e$ by pointing out, correctly, that in any larger class there will be statements which will share non-tautologous consequences with $e$, and which consequently, in their terminology, will be to some extent 'deductively dependent' on $e$. For example, if $a$ and $b$ are logically independent, then setting $h = a \,\&\, b$ and $e = a$, $C_n(\sim e \vee h)$ will not even contain $b$; however, this is just as it should be, if Popper and Miller are correct, because $b$ will share non-tautologous consequences with $a$.

But this is to impose the condition that the difference between $Cn(h)$ and $Cn(e)$ should be a *hereditary* one, in the sense that not only should the difference itself be a set disjoint from $Cn(e)$, but so should the set of consequences of every member of that set. Such a stipulation is quite arbitrary, and we recommend that it be ignored. If it is, Popper and Miller's strange version of induction ceases to be provable: which is as it should be.

## ■ e THE PRINCIPAL PRINCIPLE IS INCONSISTENT (MILLER'S PARADOX)

Miller (1966) produces an interesting argument which purports to demonstrate that the principle which, following Lewis, we have called the Principal Principle, is inconsistent. According to that principle, on which the Bayesian analysis of statistical inference rests, the (subjective) probability that an event described by the sentence $a$ will occur at a trial of type $T$ is equal to $r$, if our data are confined to the information that the physical probability of $a$, relative to the conditions $T$, is $r$. We can write this concisely as the equation

**(1)** $P[a \mid P^*(a) = r] = r,$

where $P$ is a degree-of-belief probability function and $P^*$ the physical probability function.

Miller's argument is as follows. Let $r$ be $\frac{1}{2}$. Then by (1)

$$P[a \mid P^*(a) = \tfrac{1}{2}] = \tfrac{1}{2}.$$

But clearly, $P^*(a) = \frac{1}{2}$ if and only if $P^*(a) = P^*(\sim a)$; and the probability calculus tells us that we can substitute equivalent statements, whence we obtain

$$\textbf{(2)}\ P[a \mid P^*(a) = P^*(\sim a)] = \tfrac{1}{2}.$$

However, we can also instantiate (1) thus:

$$\textbf{(3)}\ P[a \mid P^*(a) = P^*(\sim a)] = P^*(\sim a),$$

and combining (2) and (3) we infer that $P^*(\sim a) = \frac{1}{2}$, which is odd, since no factual premiss of any kind has been employed in the derivation. While this result may not have the form of an outright contradiction, it very quickly leads to one. For we can repeat the reasoning above with the two substitution instances $P^*(a) = \frac{2}{3}$, so that $P^*(a) = 2P^*(\sim a)$, whence we would infer that $P^*(a) = \frac{2}{3}$, in explicit contradiction to $P^*(a) = \frac{1}{2}$.

Were Miller's derivation formally sound, the consequences for the Bayesian theory of statistical inference would be little short of disastrous, for the characteristic and often striking properties of the posterior probabilities of statistical hypotheses are due to the behaviour of the *likelihood function* $g(i) = P(e \mid h_i)$, where $\{h_i\}$ is a family of alternative hypotheses about the value of some physical probability distribution. But, as we also noted in Chapter 13, odds different from those based on the Principal Principle are demonstrably unfair, and this tells us that *something* must be wrong with Miller's argument. The question is, what? Miller's error is difficult to spot because it is concealed by the notation which, precisely in consequence of its being well adapted to smooth exposition and development of the theory, does not make explicit all the distinctions which are nevertheless implicit.

The erroneous step in Miller's derivation is to take (3) to be a substitution instance of (1). (3) is *not* a substitution instance of (1); it makes a quite different *type* of assertion from (1), and it will help the reader see why to turn back to and re-read

Chapter 2, section **d,** where random-variable statements are introduced and their meaning discussed, for (1), though it may not look like it, is an equation involving random variables. This fact is obscured by our tendency to regard $P^*(a)$ as a number, or scalar. But in the context of a discussion in which '$P^*(a) = r$' is itself a statement assigned a probability value (by the function $P$), $P^*(a)$ is *not* a number: it is something which takes a range of possible values—those possible values being, of course, all the real numbers in the closed unit interval. And a quantity which takes different values in different possible states of the world is a function, and when it is a function over whose values there is a probability distribution, it is a *random variable.*

Let us accordingly write $P^*(a)$ in (1) explicitly as a random variable $X$. So (1) says

**(4)** $P(a \mid X = r) = r,$

for all $r$, $0 \leq r \leq 1$. Now recall, from Chapter 2, section **f,** that we can replace the sentences in (4) by the sets of possible worlds making them true. So we can rewrite (4) as

**(5)** $P[M(a) \mid M(X = r)] = r.$

Looking back at Chapter 2 again, we see that $M(X = r) = \{w{:}X(w) = r\}$, where the $w$'s are the members of the outcome space of the stochastic experiment relative to which the distribution $P^*$ is defined. So (5) can be written

**(6)** $P[M(a) \mid \{w{:}X(w) = r\}] = r.$

But (3) has the form

**(7)** $P[M(a) \mid \{w{:}X(w) = Y(w)\}] = Y(w),$

where $Y$ is another random variable equal to $1 - X$; $Y$ is of course $P^*(\sim a)$ and $P^*(\sim a) = 1 - P^*(a)$. But it is now obvious that (7) is not a legitimate substitution instance of (5); it contravenes the logical rule that terms involving so-called free variables, like $w$, must not be substituted into contexts in which those free variables become bound, as the operator $\{w{:} \ldots\}$ binds $Y(w)$. (*See* Mendelson, 1964, p. 48, for the statement of this rule for logical quantifier operators.) It follows that (3) is not a legitimate substitution instance of (1) and the derivation of Miller's paradoxical conclusion cannot proceed. The Principal Principle is consistent.

## f THE PARADOX OF IDEAL EVIDENCE

Suppose you are contemplating a long sequence of tosses—say, 1000—of a coin, about which you initially know nothing except that it looks like an ordinary coin. You are asked on two separate occasions what your degree of belief in the outcome of the 1000th toss is (i) now, before the tosses commence, and (ii) after the outcomes of 999 have been recorded and the 1000th toss is about to take place. In view of the sparseness of your initial data, your degree of belief *now* that it will land heads at the 1000th toss is, let us suppose, one half, more or less. Now suppose that the 999 tosses record an observed number of heads around 500—say, 503. You compute, using standard Bayesian calculations, the conditional probability on this data of a head at the 1000th toss and find that it is still the value, approximately one half, specified in your prior distribution over $\{H,T\}$. Your degree of belief in the coin's landing heads at the 1000th toss, obtained by conditionalisation from this conditional probability after observing the outcomes of the 999 tosses, *is therefore unchanged by the very extensive statistical data now available;* indeed, it is probabilistically independent of it.

Popper, and many others, regard this as highly damaging to a theory like the Bayesian theory which, since it registers states of belief by a *single* number, apparently cannot distinguish your posterior degree of belief, based as it is on very weighty evidence, from your prior belief, based on virtually none at all (we shall hear this same criticism echoed by Shafer later in this chapter, section **k**). According to Popper, the example above, which he terms 'the paradox of ideal evidence' (1959a, p. 407), points to the need for at least a *two-dimensional* representation of your belief state, where one of the dimensions is the ordinary probability of $h$ in the light of the evidence, and the other represents in some way the *weight* of the evidence on which that probability is based. Your initial degree of belief in a head was based on sparse and imprecise data, and therefore its weight index would be very small. Your later degree of belief, though numerically the same as your first, was based on very weighty data, by contrast. The latter subjective probability, therefore, according to the proponents of this weight of evidence theory, is more *epistemically reliable* (Gärdenfors and Sahlin, 1989, p. 321).

But it has been pointed out many times that the standard Bayesian account makes just these sorts of discriminations anyway, without needing to introduce a two-dimensional representation of belief states. The objective probability of heads in Popper's coin-tossing experiment is a random variable $X$ in the Bayesian theory, taking as values all the real numbers $x$, $0 \leq x \leq 1$. Let $P_0(x)$ and $P_{999}(x)$ be the prior and posterior subjective-probability density distributions, respectively, over these values, i.e., before any tosses have taken place and after 999 have taken place. That we tend not to regard the initial data as providing grounds for discriminating very much between the values of $X$ in a fairly broad band around the value $x=\frac{1}{2}$ will be reflected in the prior density $P_0(x)$ being rather diffuse. The data specifying the outcomes of the 999 tosses will, on the other hand (as we saw in Chapter 13, section **e**), cause the likelihood function $x^{503}(1-x)^{496}$ to be sharply peaked in the region $x=\frac{1}{2}$. Hence the posterior density $P_{999}(x)$ will concentrate virtually all the posterior probability in that region. So we see that the difference between the two 'weights of evidence' is reflected in the large variance of the prior density of $X$ and the very small variance of the posterior density, and we could, if we so wished, define the respective weights as inversely proportional to these variances.

This is not all. Let $h$ be the sentence 'A head will occur on the 1000th toss'. We shall leave it as an exercise for the reader to show that the prior probability $P_0(h)$ and the posterior probability $P_{999}(h)$ of $h$ are respectively equal to the expectations $P_0(h) = E_0(X)$, $P_{999}(h) = E_{999}(X)$, where $E_0(X)$ is the expected value of $X$ relative to the prior density $P_0(x)$ and $E_{999}(X)$ is the expected value of $X$ relative to the posterior density $P_{999}(x)$. There is a clear sense in which the posterior probability $P_{999}(h)$ is epistemically more reliable than $P_0(h)$, even though they are (approximately) numerically equal. For the mean of $X$ relative to increasing evidence will be increasingly insensitive to change: the relative frequency of heads will eventually converge to some value, to which your successive subjective probabilities of heads conditional on the observed outcomes will also converge. Therefore those subjective probabilities, *even if they remain numerically much the same,* are increasingly unlikely to alter as a result of the acquisition of more data. They are, in short, more reliable. Even where it is not assumed that the sequence of relative frequencies converges (if it is not generated by a collective), it follows from Dawid's well-known theorem on calibration (Dawid, 1982) that still you must, on pain of inconsistency, believe (in fact, *be certain*) that your conditional probabilities on accumulating data will converge on the average to the observed relative frequency, and hence be increasingly reliable in the sense of being unlikely to be disturbed significantly (we shall discuss Dawid's theorem at greater length in section **m** below).

The 'paradox' of ideal evidence is like many of the other objections to the Bayesian approach, in that on closer examination it reveals the strength, rather than the weakness, of that approach. In this particular case, we can see that the Bayesian theory, without any further assumptions or modifications being necessary, makes just those discriminations which its critics charge it with lacking.

## ■ g HYPOTHESES CANNOT BE SUPPORTED BY EVIDENCE ALREADY KNOWN

### g.1 $P(h \mid e) = P(h)$ if e Is Known When *h* Was Proposed

In a much-discussed essay, severely critical of the Bayesian approach, Clark Glymour observed that

> Newton argued for universal gravitation using Kepler's second and third laws, established before the *Principia* was published. The argument that Einstein gave in 1915 for his gravitational field equations was that they explained the anomalous advance of the perihelion of Mercury, established more than half a century earlier. . . . Old evidence can in fact confirm new theory, but according to Bayesian kinematics it cannot. (1980, p. 86)

By "Bayesian kinematics" Glymour here means simply the principle of Bayesian conditionalisation. Glymour's objection is grounded on the fact that relative to a stock of background information including $e$, $P(e)$ is 1, whence $P(e \mid h)$ is 1 also, so that it follows immediately from Bayes's Theorem that $P(h \mid e) = P(h)$. Thus $e$ does not raise the prior probability of $h$ and hence, according to the Bayesian, does not confirm it.

Glymour's argument has the rather strange consequence that *no* data, whether obtained before or after the hypothesis is proposed, can, within a Bayesian theory of confirmation, confirm *any* hypothesis, for even if the hypothesis $h$ is proposed before evidence $e$ is collected, then by the time someone comes to do the Bayes's Theorem calculation, the terms $P(e)$, $P(e \mid h)$ must again be set equal to 1, since by that time $e$ will of course be known and hence in background information. But few would infer in such a case that according to the Bayesian theory $e$ did not support $h$. And not even the most committed opponent of that theory claims it to be damaged by this demonstration, for it is clear that the theory has been incorrectly used. It is equally clear where the mistake lies, namely, in relativising all the probabilities to the *totality* of current knowledge. They should, of course, have been relativised to current knowledge *minus e*. The reason for the restriction is, of course, that *your current assessment of the support of* h *by* e *measures the extent to which, in your opinion, the addition of* e *to your current stock of knowledge would cause a change in your degree of belief in* h.

It might be objected that the relativisation of the probabilities in Bayes's Theorem calculations to what is strictly a fictitious state of background information is simply ad hoc: it is a device which avoids the otherwise embarrassing necessity of setting $P(e)$ and $P(e \mid h)$ equal to 1, but it does so at the cost of being in conflict with core Bayesian principles. This charge is untrue. Core Bayesian principles simply state the conditions—obedience to the probability calculus—for a set of degrees of belief, relative to a stock of background information, to determine a corresponding set of odds which are not demonstrably unfair. There is absolutely nothing in this which asserts that in computing levels of support, one's subjective probabilities must define degrees of belief relative to the totality of one's current knowledge. On the contrary; the support of $h$ by $e$ is gauged according to the effect which one believes a knowledge of $e$ *would now have* on one's degree of belief in $h$, on the (counter-factual) supposition that one does not yet know $e$.

A more serious objection to this sort of relativisation of the probabilities in calculations of support is that in general there simply is no uniquely determined set $K - e$ which is the result of deleting the information represented by a sentence $e$ from a database $K$—except, of course, in the case where $e$ is not in $K$, so that then $K - e$ is just $K$. Even if $e$ is separately represented in $K$, matters are not straightforward, for the result of simply removing $e$ from $K$ will depend on how $K$ is axiomatised. For example, $K_1 = \{a,e\}$ and $K_2 = \{e \rightarrow a,e\}$ represent exactly the same stock of information, since they have the same consequences, but subtracting $\{e\}$ set-theoretically from each will leave $\{a\}$ in the first case and $\{e \rightarrow a\}$ in the second.

Much has been written on this problem (the contemporary state of the discussion is nicely summarised in Gärdenfors, 1989), and it seems that there is no *uniformly* satisfactory way to characterise the operation of deleting $e$ from $K$. One briefly canvassed idea was to define $K - e$ as the so-called *full meet contraction function* on $K,e$, namely, the intersection of all those subsets of $K$ which do not entail $e$. In the case where $K$ is the deductive closure $Cn(k)$ of some sentence $k$ (which might, of course, be some arbitrarily large conjunction), this results in the familiar set $Cn(\sim e \vee k)$. $Cn(\sim e \vee k)$, recall, is Popper and Miller's characterisation of the excess content of $k$ over $e$. We found good reason to reject $Cn(\sim e \vee k)$ for that role, and $Cn(\sim e \vee k)$ has also been uniformly rejected as supplying a meaning for $K - e$, on the ground that it is too small, containing as it does consequences only of $\sim e$ and $k$ (Gärdenfors, 1989, p. 79; Makinson, 1985, p. 359).

It is not the case, however, that in the sorts of situations of interest to us $K - e$ cannot be given a definite and satisfactory meaning. Indeed it can, for we have the paradigm example of how it can be done in Bayes's Theorem calculations using newly recorded evidence. Nobody, and this includes Glymour himself, denies that a perfectly proper way of deleting $e$ from the current $K$ in this case is simply to subtract it set-theoretically from the set of sentences represented by $K$, in which $e$ will be an isolated point, or to put it more grandly, an independent 'axiom'. That $K$ could be axiomatised differently, as the set $\{a,e\}$ could be reaxiomatised $\{e \rightarrow a,e\}$, raises no difficulty of principle or practice, for it is simply a fact that the agent represents his or her own knowledge in a particular way. Given this, and given that *e* is usually (though not, of course, always) an independent item in this representation, the state of the individual's knowledge-base on the counter-factual assumption that *e* is not part of it is uniquely determined. Where *e* is not an independent item in *K,* the situation is less easy to deal with, and we may just have to conclude that the notion of what the probability of *e,* and of *e* given *h,* would be, were one counter-factually assumed not to know *e,* is not defined. But that such counter-factually based probabilities do not always exist should not blind us to the fact that in many cases they do, and consequently that in many cases the notion of support relative to known evidence is also perfectly meaningful.

Glymour considers this type of response quite sympathetically, but contends that nevertheless in

> actual historical cases . . . there is no single counterfactual degree of belief in the evidence ready to hand, for belief in the evidence sentence may have grown gradually—in some cases it may have even waxed, waned, and waxed again. (1980, p. 88)

He cites as an example the data on the perihelion of Mercury; there were different values obtained for this, over a period of several decades, by different methods, and employing mathematical techniques sometimes without rigorous justification. Glymour contrasts this situation with the results of tossing a coin a specified number of times, where he thinks it does make sense to talk of the probability of that outcome, as if it had not yet occurred. But in the case of Mercury's estimated perihelion advance, "there is no single event, like the coin-flipping, that makes the perihelion anomaly virtually certain" (Glymour, 1980, p 88).

But whether there is as much epistemic warrant for the data in 1915 about the magnitude of Mercury's perihelion advance as there is about the number of heads we have just observed in a sample of a hundred tosses of a coin is beside the point. We may be more tentative about some data, and about other data, less. The Bayesian theory we are proposing is a theory of inference from data; we say nothing about whether it is correct to accept the data or even whether your commitment to the data is absolute. It may not be, and you may be foolish to repose in it the confidence you actually do. The Bayesian theory of support is a theory of how the acceptance as true of some evidential statement affects your belief in some hypothesis. How you came to accept the truth of the evidence, and whether you are correct in accepting it as true, are matters which, from the point of view of the theory, are simply irrelevant. Glymour's disquisition on the frailty of much scientific data is therefore, however valuable in its own right, beside the point of evaluating the adequacy of the Bayesian theory of inference.

In the course of his discussion, Glymour proposes an altogether different explanation of why it is that old evidence appears to confirm a new theory: it may not be the "old result that confirms a new theory, but rather the new discovery that the new theory entails (and thus explains) the old [result]" (Glymour, 1980, p. 92). This observation prompts Glymour to propose a new criterion for old evidence $e$ to be taken as confirming a new theory $h$, namely that

**(8)** $P[h \mid e \,\&\, (h \vdash e)] > P(h)$,

where the probability calculus is weakened appropriately, by replacing the conditions on axioms 2 and 3 by 'if it is known that $t$ is a tautology, then . . .', and 'if it is known that $a \vdash \sim b$, then . . .' (we have modified Glymour's notation in (8) slightly and omitted explicit reference to background information). This emendation of the classical theory is sympathetically endorsed by Niiniluoto (1983), and is further examined by Garber (1983), though the same idea seems first to have been proposed and developed by I. J. Good (starting with Good, 1977), who calls the resulting notion of probability "dynamic" or "evolving" probability.

Despite the considerable interest provoked by Glymour's novel explanation of how old evidence confirms new theories, we believe that it is refuted by the observation that scientists often build their theories around some particular item of data, which is thereafter cited as being among the evidence in support of those theories. Einstein, and many others, believed the invariance of the speed of light provided powerful evidence for the Theory of Special Relativity, despite the fact that the theory was constructed to explain the phenomenon. Newton, and many others, believed that the explanation of Kepler's Laws by his theory of gravitation was one of the most important pieces of evidence for that theory, though the entailment of an approximate form of Kepler's Laws was the principal constraint on the form of the theory. In such cases, therefore, the ground for believing the theory supported by the evidence cannot be the discovery that the theory entails it, for that is exactly what the theory was designed to do.

### g.2 Evidence Doesn't Confirm Theories Constructed to Explain It

To fully sustain our objection to Glymour, however, we have to answer those who believe such evidence inadmissible, and there is a body of opinion which does. We have already (in Chapter 7, section **j.4**) encountered—and showed to be untenable—the doctrine that theories, to be acceptable, must be supported by evidence independent of that which they were constructed to explain. Here, however, we face a more radical doctrine: *such evidence fails to support at all.* This view, which has attracted considerable following among philosophers (though not among scientists), is invariably defended by the following argument, recently stated by Giere, but which goes back to Peirce if not earlier:

> if the known facts were used in constructing the model and were thus built into the resulting hypothesis . . . then the fit between these facts and the hypothesis provides no evidence that the hypothesis is true [since] these facts had no chance of refuting the hypothesis. (Giere, 1984, p. 161)

Despite beguiling many, the argument is radically unsound. Whatever plausibility it might seem to possess is quickly removed by noting that no *fact* has a "chance" of refuting anything. If *e* is a factual statement and *h* a hypothesis, then *e* either refutes *h* or it does not, and it does or does not do so whether *h* was designed to explain or embody *e* or not. Giere, and all the many other people (like Worrall, 1989) who have been swayed by this argument, are confusing an experimental set-up, *E,* with one of its possible outcomes, *e*. This is similar to identifying a random variable with one of its values.

Once the distinction is made, the argument collapses, for even if $h$ was constructed from one particular outcome of $E$, $E$ could, if it was a well-designed experiment, have produced another inconsistent with $h$.

Of course, had it done so, it would not have been this $h$ that you considered, but another, $h'$, designed to accommodate that other outcome. Can one not reformulate the objection, then, in something like the following way: there is no chance—if you are careful—of the $h$ you produce being inconsistent with the $e$ you want explained by it. This is undeniably, indeed trivially, true. *But it still does not follow that* h *is necessarily unsupported by* e. Recall that it is necessary for $h$ to be supported by $e$ that, in Bayesian terms, $P(e \mid h)$ is larger than $P(e \mid {\sim}h)$, and there seems to be no reason to suppose that, just because you contrived $h$ to explain $e$, $P(e \mid h)$ is not larger than $P(e \mid {\sim}h)$.

Redhead, however, has recently taken up the argument for this 'null support thesis' by claiming that in the circumstances envisaged $P(e \mid {\sim}h)$ will not only not be smaller than $P(e \mid h)$, *but it will necessarily be equal to 1* (1986). Redhead's argument is that by "filtering" out all the hypotheses which do not explain $e$, you are in effect assigning them all probability 0. We leave it as an exercise to show that such an assignment entails that $P(e \mid {\sim}h)$ is equal to 1. But there is a fatal flaw in this argument. For it follows from $P(e \mid {\sim}h) = 1$, together with $P(e \mid h) = 1$, that $P(e)$ itself must be equal to 1 (since $P(e) = P(e \mid h)P(h) + P(e \mid {\sim}h)P({\sim}h) = P(h) + P({\sim}h) = 1$). This, as we know, would imply that no hypothesis at all, built to explain $e$ or not, could ever be supported by it, and that, we know, is false. And accepting that, as we must, means rejecting the assignment $P(e) = 1$, and, by implication, $P(e \mid {\sim}h) = 1$.

There are also simple counter-examples to Redhead's thesis. Suppose we are removing tickets from an urn. We remove all $n$ and discover that $r$ of the tickets are red. Let $e$ record this observation. We now formulate the hypothesis $h$: 'There were $r$ red tickets in the urn'. Clearly, $P(e \mid {\sim}h)$ is so small as to be practically zero, yet $h$ was constructed to be the most plausible hypothesis which explains $e$.

### g.3 The Principle of Explanatory Surplus

We shall now consider a slightly more-sophisticated version of the thesis that evidence on which a theory is constructed cannot be used as evidence for it. Suppose that $h$ is constructed with the help of a body $e$ of data. Suppose also that we could decompose $e$ into two parts, that which was actually used in the construction of the theory ($e'$) and the remainder, $e''$. It remains a widely held view that while $e$ may well support $h$, it does so only because the 'explanatory surplus' $e''$ supports $h$: $e'$ never does (the term 'explanatory surplus' is due to Gillies, 1989; both he and Worrall, 1978, are philosophers who have recently held this position). It might be, for example, that $e'$ reports the minimum necessary number of observations to fix a set of adjustable parameters in $h$. There is no guarantee that the surplus data $e''$ will satisfy $h$; $e''$ might even falsify it, and that is why $e''$ but not $e'$ can be regarded as potentially supporting evidence for $h$.

It is clear that we are back with the argument that only data which has a 'chance' of conflicting with $h$ can support it. We declared the argument unsound, because the experiment $E$ may well have possible outcomes inconsistent with $h$. But if $h$ is guaranteed, *whatever* outcome $E$ delivers, to be consistent with that outcome—because $h$'s parameters are determined by that outcome, for example—does that not show that the argument is valid for such cases? No. The appearance to the contrary arises because we are confusing two things: (i) the hypothesis $h$ before its parameters are determined from the non-surplus data $e'$; and (ii) the hypothesis with its parameters evaluated, call it $h'$. It is true that $h$ is not in general supported by $e'$, because $e'$ will in general give no information about the truth-value of $h$. *But it is $h'$, not $h$, which was constructed with the help of $e'$.* And as regards $h'$ we repeat what we said earlier: the data source $E$ may well have possible outcomes inconsistent with $h'$, and therefore be in principle capable of falsifying $h'$. And $e'$ might well support $h'$.

Indeed, it is easy to think of cases where it will. Consider the following simple example, where $h$ is a pure linear hypothesis $y = ax+b$, with the two parameters $a$ and $b$ undetermined. Let $e'$ be two independent joint observations of $y$ for specified values of $x$, determining $a$ and $b$. $h$ is not supported at all by $e'$ if $e'$ is totally independent of $h$; $e'$ could be any pair of joint values. Suppose this is the case, so that $P(h \mid e') = P(h)$. The $x$-values are part of the experimental specification, so that relative to this we have $h' \Leftrightarrow h \,\&\, e'$. It is now easy to show that while $h$ is not supported by $e'$, $h'$ certainly is. First, we have that $P(h' \mid e') = P(h \,\&\, e' \mid e') = P(h \mid e) = P(h)$. We can assume that the specific values of $y$ revealed by $e'$ have probability less than 1, in which case it follows immediately that $P(h' \mid e') > P(h')$. Q.E.D.

We can also see from this example why the practice of merely saving the phenomena is so universally disparaged. If $h$ is constructed with enough free parameters to be able to accommodate whatever data are to be explained, but there is no independent reason to believe $h$ true, that is just another way of saying that $P(h)$ is very small. But as we see from the reasoning above, $P(h' \mid e') = P(h)$, i.e., $P(h' \mid e')$ is exactly as large as the prior probability of $h$, which is reckoned to be small. So although $P(h' \mid e')$ will in general exceed $P(h')$, the support $e'$ gives to $h'$ as measured by the difference between $P(h' \mid e')$ and $P(h')$ can never be considerable.

## ■ h PREDICTION SCORES HIGHER THAN ACCOMMODATION

It is not true that a hypothesis constructed to accommodate some evidence $e$ is never supported by $e$. Nor is it even true that a hypothesis is unsupported by just the part of the data actually used in its construction. But an influential tradition, including Leibniz and later Whewell, has claimed that independent prediction of data nevertheless confers *more* support, for given data $e$ and hypothesis $h$, than if $h$ had merely accommodated $e$. Furthermore, it is often claimed, the Bayesian theory lacks the means even to discriminate between evidence incorporated as a deliberate constraint in the construction of a hypothesis $h$ and evidence independently by $h$. Hence, since independent prediction is—it is alleged—such an important methodological criterion by which theories are evaluated as to their empirical adequacy, the Bayesian theory must be seriously inadequate.

But it is false that the Bayesian theory cannot make the discrimination between evidence accommodated and evidence independently predicted. It does so in a variety of ways, one of which turns on prior probabilities. Consider, for example, the case of a hypothesis $h$ with some or all of its parameters left undetermined. Suppose there is some data $e$ which a rival hypothesis $h''$ predicts directly, but which $h$ can only account for once its parameters have been fixed by $e$, and $e$ is just sufficient to do this. Let $h'$ be the result of fixing those parameters. If $h$ and $h''$ start off with equal prior probabilities, and $e$ is independent of $h$, it is easy to show that the adjusted form $h'$ of $h$ will in general obtain much less support from $e$ than will $h''$. For the difference between the posterior and prior probabilities of $h''$ is $P(h'' \mid e) - P(h'') = \frac{P(h'')[1 - P(e)]}{P(e)}$, since $h''$ entails $e$. Similarly, the difference between the posterior and prior probabilities of $h'$ is $\frac{P(h')[1 - P(e)]}{P(e)}$. But if the prior probability of $e$ is less than 1, as it standardly will be, we shall certainly have $P(h) < P(h')$, for by the reasoning of the previous section $P(h') = P(h \,\&\, e) = P(h)P(e)$. But by assumption, $P(h'') = P(h)$, so the incremental support of $h''$ exceeds that of $h'$, the adjusted hypothesis.

The disparity in supports in this example between the predicting and the accommodating hypotheses reflects the disparity in prior probabilities of $h'$ and $h''$. But this is not always the reason. In the following example, discussed by Maher (1988), the enhanced support for the independently predicting hypothesis depends crucially on incorporating into background information the fact that the prediction was independent of knowledge of the evidence. In this example, an individual predicts the outcomes of a coin tossed 100 times. Compare the following two scenarios: (i) The subject predicts all the outcomes in advance, and it is discovered that after 99 tosses none are wrong. Let $h$ be the prediction of the whole 100 tosses, and $e$ the description of the outcomes of the first 99. (ii) the subject again 'predicts' $h$, but only *after* learning $e$. We should be inclined to repose more trust in the prediction $h$ in (i) than in (ii), because we believe there is evidence that the subject in (i) has some privileged knowledge of the apparatus. By contrast, there is no evidence that the subject in (ii) has such knowledge.

Let us see how this reasoning can be expressed formally. Let $H$ be the hypothesis that the subject in each case has privileged knowledge of the apparatus. Consider (i), where the fact that the subject predicts the statement $h$ goes into background information (i.e., into what we knew before learning $e$). Assuming that $H$ has a non-zero prior probability, however small, it will have that probability raised to somewhere close to 1 by $e$. For given that background information includes the information that $e$ was predicted, $P(e \mid {\sim}H)$ can be interpreted as the probability that the subject got $e$ right by chance, and for the whole 99 outcomes this probability will be exceedingly small. By Bayes's Theorem $P(H \mid e)$ is therefore close to 1. Now $H$ entails $h$ which entails $e$, since background information includes the information that the subject, who is right according to $H$, predicted $h$ which implies $e$. It follows that

$$P(h \mid e) = P(H \mid e) + P(h \mid {\sim}H \,\&\, e)P({\sim}H \mid e),$$

and so $P(h \mid e)$ will also be close to 1.

In (ii), the background information relative to which $P$ is computed is now that the subject knows the outcomes of the first 99 tosses and predicts that the next toss will be a head (say). Relative to this information, $H$ no longer implies $h$ or $e$, though $H \,\&\, e$ entails $h$. The probability of $e$ conditional on $H$ and on ${\sim}H$ is the same, we can suppose, as its probability conditional on the hypothesis of chance. This being so, it is straightforward to show by Bayes's Theorem that $P(H \mid e) = P(H)$. Hence

$$P(h \mid e) = P(H) + P(h \mid {\sim}H \,\&\, e)P({\sim}H),$$

which is approximately equal to $P(h \mid \mathit{chance} \,\&\, e)$ if $P(H)$ is small (an extended discussion is in Howson and Franklin, 1991).

## i THE PROBLEM OF SUBJECTIVISM

### i.1 Entropy, Symmetry, and Objectivity

Possibly the most popular of all the objections to the subjective Bayesian theory is that it is too subjective. Fisher, in his remark which we quoted in Chapter 4, section **c.2**, that results concerning the measurement of belief "are useless for scientific purposes", summed up what many thought and still think to be a crucial objection. Science is objective to the extent that the procedures of inference in science are. But if those procedures reflect purely personal beliefs to a greater or lesser extent, as they appear to do if they are constrained only to follow Bayes's Theorem, with no condition other than mere consistency being imposed on the forms of the priors, then the inductive conclusions so generated will also reflect those purely personal opinions. Echoing Fisher, E. T. Jaynes claims that

> the most elementary requirement of consistency demands that two persons with the same relevant prior information should assign the same prior probabilities. Personalistic doctrine makes no attempt to meet this requirement . . . the notion of personalistic probability belongs to the field of psychology and has no place in applied statistics. Or, to state this more constructively, objectivity requires that a statistical analysis should make use, not of anybody's personal opinions, but rather the specific factual data on which those opinions are based. (Jaynes, 1968, p. 228)

Jaynes developed his ideas well beyond the programmatic stage. One of his most influential suggestions is embodied in the *Principle of Maximum Entropy:* the distribution allegedly containing no information beyond that contained in 'the specific factual data' is the distribution, where it exists, which has maximum *entropy* subject to the constraints imposed by those data. Formally, the entropy of a discrete distribution is $-\Sigma p_i \log p_i$ (0log0 is equal to 0). Intuitively, the entropy of a distribution is its degree of diffuseness, so that the more concentrated it is, the smaller is its entropy. The uniform distribution, $p_i = n^{-1}$ is the maximum entropy distribution for the constraint that $X$ takes $n$ values.

The principle of maximum entropy turns out to be far from unproblematic in practice (and also in principle, but we shall come to that shortly). For many sets of constraints, maximum entropy distributions do not exist, and where they do, they can lead to strongly counter-intuitive results (Shimony, 1985). Also, there is a technical problem in extending the principle to the continuous case: if $X$ is continuously distributed, then the limit of the sequence $-\Sigma p_i \log p_i$, obtained by chopping up the range of $X$ into smaller and smaller subintervals such that $p_i$ is the probability that $X$ lies in the $i$th subinterval, is infinity. Shannon, who introduced entropy as a measure of uncertainty, defined continuous entropy as the integral from minus infinity to plus infinity of $-P(x)\log P(x)$, where $P(x)$ is the probability density of $X$. This quantity is not the limiting case of the discrete entropy, however, nor (as is well-known) is it invariant under change of variable (in fact, it is the expected value of the logarithm of a probability *density*, or probability per unit of $X$).

Jaynes was well aware of the technical problems posed by extending the entropy functional to continuous distributions, and he proposed a version of the cross-entropy we encountered in Chapter 6 (Jaynes, 1983, p. 59), though, containing as it does a strange point-density function, it can scarcely be considered a more successful solution than Shannon's. Jaynes has, however, advanced a quite distinct prior-determining criterion for continuous distributions. This follows an earlier idea of Jeffreys's, which was to choose that prior for a given problem which satisfies suitable invariance conditions. The invariance conditions suggested by Jaynes are those allegedly implicit in the problem at hand. For example, if nothing is specified about the position of some parameter other than that it lies somewhere on the real line, then according to Jaynes this means that the prior distribution over its possible locations should be translation invariant. The resulting distribution turns out to be uniquely specified by this condition, and it is the uniform density distribution over the line.

That density does not, however, integrate to one; it diverges. So does the density $x^{-1}$ of the distribution, claimed to be obtained by assuming scale-invariance, for a variable $X$ which takes non-negative values only. Such distributions are called *improper,* and though they infringe the axiom of the probability calculus which demands that the probability of certainty is one, they are nevertheless considered legitimate by many, usually for the reason that they can be regarded as handy approximations of 'proper' distributions. It is also often cited in their defence that in certain cases improper priors can be combined in Bayes's Theorem with ordinary likelihoods to generate perfectly proper posterior distributions. This is true, but it doesn't justify their use. Improper priors are strictly inconsistent with the probability axioms, and there can be no guarantee, therefore, that they will not lead to contradictions elsewhere, as the so-called marginalisation paradoxes discovered by Dawid, Stone, and Zidek (1973) bear witness.

There are other difficulties with using the symmetries implicit in the problem at hand to determine the appropriate prior distribution. Determining exactly what these symmetries are is by no means as objective a matter as it is made out to be. Even having determined what you think they are, the resulting constraints may fail to determine a distribution uniquely (underdetermination) or at all (overdetermination: the constraints are inconsistent). Even the apparently straightforward requirement of scale-invariance does not unambiguously determine the density $x^{-1}$ (Milne, 1983).

Despite the technical and other problems to which they frequently lead, Jaynes's ideas have been very influential, and the ideal that he enunciates, of conjuring up a prior distribution which, in the context of a given problem, expresses only the 'specific factual data', continues to inspire. Thus Rosenkrantz, an enthusiastic supporter of Jaynes, defends the uniform prior distributions that often arise within Jaynes's theory by pointing out an analogy with current cosmological practice:

> Steady-date cosmologists, to take one of myriad instances, start off by assuming the laws of physics are the same in temporally and spatially remote regions of the universe. This, they urge, is surely the simplest assumption. But it is more than that. To assume that different laws obtained a billion years ago would be entirely arbitrary; it would be to import knowledge we do not in fact possess. (Rosenkrantz, 1977, p. 54)

*But we don't know that the laws were the same either*. Rosenkrantz has failed to see that *any* assumption 'imports knowledge', the assumption that things were essentially the same just as much as the assumption that they were not. So it is with a uniform or any other prior distribution, as we argued at length in Chapter 3. And this objection strikes at the heart of Jaynes's programme. *No prior probability or probability-density distribution expresses merely the available factual data; it inevitably expresses some sort of opinion about the possibilities consistent with the data*. Jaynes's and Rosenkrantz's 'objective' priors may well not embody the opinions of any one person; but they are as far removed from the data as if they did.

Nor, it seems to us, is it a tenable claim that the distribution which maximises entropy is "the one which is maximally noncommittal with regard to missing information" (Jaynes, 1957, p. 623). Any distribution, in our opinion, is as informative as any other insofar as it supplies a definite probability to every Borel set. In particular, the flat distributions determined by the Principle of Maximum Entropy, where there are no constraints other than those supplied by the probability axioms, are not less committal than any other; they merely have a different shape.

### i.2 Simplicity

Another frequently canvassed 'objective' criterion for determining priors, at any rate up to a rank ordering, is that which states that hypotheses should be ordered in prior probability according to how *simple* they are. There is no claim to informationlessness for the priors admitted by this criterion, however. Jeffreys, a working scientist when not writing works on general methodology, famously incorporated such a Simplicity Postulate into his Bayesian theory. But he did so, not because he felt that he was thereby not trespassing beyond the data, but because he believed that simplicity was, as a matter of empirical fact, the dominant factor in people's prior evaluation of theories.

That this is so is doubtful, however, if only because it is notoriously difficult to know exactly what people refer to under the name of simplicity, and there seems no good reason to suppose that they are all referring to the same thing. Some people, for example, maintain that simplicity resides in an organic unity exemplified by the fundamental principles of the simple theory. Others say that it resides in the fewness of the adjustable parameters which the theory introduces (this was Jeffreys's view). Yet others say it resides in the ease with which computations can be done within the theory. Some, like Einstein on occasion, when they cite simplicity as a principal virtue, seem to appeal to some inarticulable personal aesthetic. But all these notions, even where any clear sense can be made of them, are to a great extent independent of each other.

And it is certainly not always easy to make clear sense of them, even of the apparently perspicuous (and popular) idea that a linear hypothesis is simpler than a nonlinear one, for whether an equation is linear or not depends on the choice of coordinates: the equation of the unit circle is the quadratic

$x^2 + y^2 = 1$ in Cartesian coordinates, but it takes the linear form $r = 1$ in polar coordinates, to choose a well-known example. Even Jeffreys's own coordinate-independent characterisation of simplicity as paucity of independent parameters is, on inspection, far from ambiguous. Newton's theory, for example, might be thought to possess very few undetermined parameters—some people claim that it contains only one, the gravitational constant. But as applied, say, in the kinetic theory of gases, it contains of the order of $10^{23}$ undetermined parameters, and when further degrees of freedom are added, the number rises correspondingly. (On the other hand the charge, made by Popper [1959a] and recently repeated by Watkins, that Jeffreys's Simplicity Postulate is inconsistent is not true and arises from misunderstanding what the postulate says [Howson, 1988]).

Even if there were a univocal notion of simplicity, and even if people were invariably to favour simpler hypotheses, this would still not, we believe, warrant the adoption of a Simplicity Postulate as an axiomatic component of our Bayesian theory. For reasons we have already been at pains to emphasise, we believe that the addition of *any* criterion for determining prior distributions is unwarranted in a theory which purports to be a theory of consistent degrees of belief, and nothing more. How you should determine your prior distributions is simply not something that comes within the scope of such a theory; nor, as we shall argue shortly, is it even desirable that it should.

The alternative to including rules for determining priors is *not,* however, as Jaynes, Fisher, and others (including, regrettably, even some Bayesian personalists) believe, a Bayesian theory condemned to be no more than a record of the whims of individual psychology. That quite fallacious inference has been possibly more damaging to rational methodology than any other. The charge of excessive subjectivism rests largely, as we have seen, on the fact that certain quantities, specifically the prior distributions, are not determined within the theory. But this no more means that the theory is subjective than it means that deductive logic, which does not determine the truth-values of the premisses of a deduction, is unduly 'subjective'.

The analogy with deductive logic is entirely appropriate here. Deductive logic is both a theory of (truth-value) consistency, and thereby also a theory of deductively valid inferences from premisses whose truth-values are exogenously given. Inductive logic—which is how we regard the subjective Bayesian theory—is a theory of (degree of belief) consistency, and thereby also a theory of inference from some exogenously given data and prior distribution of belief to a posterior distribution. And as far as the canons of correct inference are concerned, neither logic allows freedom to individual discretion: both are quite *impersonal* and objective.

Nor is the subjective Bayesian theory empty of definite and valuable information. Quite the contrary: the emphasis of the preceding chapters has been very much on just how great is its power to explain and justify methodological practice. In addition it does, to as great an extent as is desirable, incorporate Jaynes's requirement that "two persons with the same relevant prior information" assign the same prior probabilities. It does so asymptotically, as their data garnered from experience grow without bound. Even then, as we point out in Chapter 14, it characteristically does not take all that much sample data to diminish the different distributions to the point where they are practically identical. Experience is allowed to dominate prior beliefs, in other words, though in a controlled way; disagreement is not eradicated at once, but its effect usually falls off quickly.

All this is as it should be. People do have diverse opinions, and this diversity is a source of strength, not weakness, just as—to employ a metaphor Popperians will appreciate—diversity in a gene pool is a source of strength, a token of its ability to adapt quickly to a changing environment. That such diversity is explicitly allowed for in the personalist Bayesian theory is therefore a point strongly in its favour, not against it. The prescription of the same 'objective' prior probability for everybody in the same knowledge state is a prescription for stagnation and eventual catastrophe, as is the suppression of dissent quite generally. A tolerance of diversity enables a society of theories as much as of people to anticipate unexpected developments: some 'crank' or other, like Einstein, will have foreseen them.

## ■ j PEOPLE ARE NOT BAYESIANS

In their summary of an influential piece of empirical work, Kahneman and Tversky deliver themselves of the following judgment:

> The view has been expressed . . . that man, by and large, follows the correct Bayesian rule, but fails to appreciate the full impact of evidence [they cite W. Edwards, 1968], and is therefore conservative.
>
> The usefulness of the normative Bayesian approach to the analysis and the modeling of subjective probability depends primarily not on the accuracy of the subjective estimates, but rather on whether the model captures the essential determinant of the judgment process. The research discussed in this paper suggests that it does not. . . . In his evaluation of evidence, man is apparently not a conservative Bayesian: he is not Bayesian at all. (Kahneman and Tversky, 1972, p. 46)

It has been the burden of the foregoing chapters that a Bayesian theory is capable of explaining standard modes of scientific inference where other theories are not. Yet the empirical studies Kahneman and Tversky refer to are taken by these authors to indicate very strongly that people do not use Bayesian reasoning where the Bayesian theory appears to say that they should.

Other studies seem to reinforce Kahneman and Tversky's conclusions. Let us briefly consider a recent one (Cosmides and Tooby, 1992). A questionnaire informs respondents that disease $D$ has a prevalence of 0.1% in the population. Test $T$ to detect $D$ has a false positive rate of 5%. What is the chance of someone having $D$, given that they have tested positive?

A majority of people gave the answer 95%. It turned out that among these were some who did not understand what 'false positive' meant. The false positive rate is the percentage of those who do not have the disease who test positive. Also omitted from the original questionnaire was the information about the true positive rate (100%). With this information supplied, the allegedly correct answer for the chance that someone does have $D$ who tests positive is slightly under 2%, and it is obtained by transforming the data into probabilities in the following way. Let $e$ be 'a person tests positive', and $d$ be 'they have $D$'; if $P(e \mid d) = 1$, $P(e \mid \sim d) = 5\%$, and $P(d) = .1\%$, then $P(d \mid e)$ is easily calculated by Bayes's Theorem to be approximately $\frac{1}{51}$.

Even when the additional information about the meaning of 'false positive', and the value of the true positive rate, had been supplied, a majority of respondents still gave much too high a figure for their answer. This seems to be symptomatic of a widespread phenomenon, first identified by Kahneman and Tversky, of so-called 'neglect of the base rate'—i.e. of the tendency to ignore the incidence in the population of the disease *D*. However, readers of Chapter 13 will be aware that the identification of 'single-case' probabilities (which is presumably what is meant by asking for the chance that someone has *D*) with frequency probabilities (which is presumably what the percentages are intended to represent) has long been and still is a subject of controversy. Add in the continued puzzlement about the relation between conditional statements involving probabilities and conditional probabilities (Lewis 1976), and the fact that informed opinion on the correct probability model(s) to apply to experience is still far from unanimous, and it is hardly surprising that untrained people have difficulty in obtaining the 'correct' answers to problems like the one above (for a similarly sceptical discussion see also Levi, 1985).

Cosmides and Tooby themselves claim that results like the ones they record do not at all show that people are not Bayesian reasoners (though by 'Bayesian' they mean only 'capable of using Bayes's Theorem'; they certainly do not mean 'Bayesian' in the sense which it is given in this book). Pursuing a line of inquiry stemming from Gigerenzer (1991), they vary the test format to show that if the questions asked, and the information supplied, are put in terms of frequencies in populations of a specified size—e.g. "out of 1000 people who are perfectly healthy, 50 of them test positive for the disease" (p. 59)—then the vast majority of respondents do obtain the 'correct' answer 1 in 51. But if caution is advised over Kahneman and Tversky's interpretation of their results, then it is positively urged for Cosmides and Tooby's interpretation of theirs, which is no less than that people do possess a faculty for probabilistic, even Bayesian, reasoning, which is manifested however for a purely frequentist concept of probability.

A more accurate conclusion is that their respondents are competent at whole number arithmetic, which is anyway hardly surprising in view of the fact that they are often university students. But with *probabilistic* reasoning, and especially with reasoning about frequency probabilities, Cosmides and Tooby's results have very little to do at all, despite their dramatic claims. As Chapter 13 also demonstrates, the relation between sample frequencies and frequency probabilities is far from direct, and can only be forged successfully within a theory of non-frequency probabilities. To claim that people are statistical reasoners because they can get the right answer to the disease problem when it is posed in terms of numbers in an actual finite reference population is fallacious, and the data tells us little except, as we have said, about the subjects' ability to do arithmetic.

It does seem, however, that people do not always follow Bayesian canons of reasoning. In fact, we should be extremely surprised if subjects were invariably to employ impeccable Bayesian reasoning, even where the calculations were mathematically tractable. It is instructive to compare Kahneman and Tversky's, and others', apparently negative results with a rather striking and very uniform result (one of the present authors has tested it himself on a group of American freshman and sophomore students) of an experiment, devised by P. C. Wason (1966) to test subjects' performance of a simple deductive task. Four cards are placed flat on a table. Each card has an integer printed on one face and a letter on the other. The uppermost faces of the cards are

| E | K | 4 | 7 |
|---|---|---|---|

and the subjects are asked to name those cards, and only those cards, which need to be turned over in order to determine whether the statement, 'if a card has a vowel on one side, then it has an even number on the other', is false or not. Wason discovered that the vast majority of his subjects indicated either the pair of cards E and 4, or only the card 4. The correct answer is, of course, the pair E and 7.

This empirical result has proved to be remarkably persistent:

> Time after time our subjects fall into error. Even some professional logicians have been known to err in an embarrassing fashion, and only the rare individual takes us by surprise and gets it right. It is impossible to predict who he will be. This is all very puzzling. . . . (Wason and Johnson-Laird, 1972, p. 173)

Puzzled Wason and Johnson-Laird may be, but about one thing they are clear: these subjects did get the answer wrong. Moreover, even the subjects themselves eventually agreed on that. Now this observation has an obvious relevance to Kahneman's and Tversky's dramatic claim, made in the light of evidence analogous to Wason's, that we are not Bayesians. Wason has shown, by this and other empirical studies, that we are not consistently deductive logicians in practice—but he has not shown, nor did he claim to have shown, that we are not deductive logicians in some other important sense. For we ourselves nevertheless constructed those deductive standards and consciously attempt to meet them, even though we sometimes fail, and in some cases nearly always fail. By the same token, it is not prejudicial to the conjecture that *what we ourselves take to be correct inductive reasoning* is Bayesian in character that there should be observable and sometimes systematic deviations from Bayesian precepts.

## ■ k THE DEMPSTER-SHAFER THEORY

### k.1 Belief Functions

The personalist Bayesian theory is a mathematical theory of uncertain reasoning. It is not the only one, however, and the last twenty years have seen the emergence of several others. With the rapid development of artificial intelligence, and in particular of rule-based expert systems, it has become a matter of some practical urgency to find an acceptable mathematical model of uncertainty, and the supply of candidates has been at least commensurate with the demand.

One of the most influential rivals to the personalist Bayesian theory is that of *belief functions* put forward by Glenn Shafer (1976), and it is usually referred to as the Dempster-Shafer theory, because one of its central principles is Dempster's rule for combining evidence. Because the Dempster-Shafer theory derives much of its current support from Shafer's influential objections to the personalist Bayesian theory as a theory of evidence, we shall outline the Dempster-Shafer theory, argue that it itself is inadequate, and then proceed to consider—and rebut—Shafer's criticisms of the Bayesian theory.

The fundamental idea of the Dempster-Shafer (DS) theory is the equation of degree of belief with evidential support. Formally, it takes expression in a class of functions, called *basic probability assignments,* defined on the subsets of what is called a *frame of discernment,* denoted by $\Theta$. The latter corresponds to a class of exclusive and exhaustive hypotheses. The basic probability assignment to a subset $A$ of $\Theta$ is constrained to lie in the closed unit interval, and is intended to be that amount of belief you commit to $A$ (which represents the *disjunction* of all its constituent hypotheses; the subsets correspond to the $M(a)$ in Chapter 2, section **f**) but not to any proper subset of $A$, i.e., not to any stronger proposition. The empty set, representing a contradiction, is assigned the basic probability number 0 by any assignment, and the sum of all the basic probability numbers assigned to the subsets of $\Theta$ is 1.

Basic probability assignments, despite the name, are not formally probability functions, and only in the exceptional case that the singletons of $\Theta$ and nothing else are assigned non-zero basic probability numbers will the associated belief function Bel be. Bel$(A)$ is supposed to represent the total belief assigned directly and indirectly to $A$ (by every $B$ such that $B$ implies $A$) and is the sum of the basic probability numbers assigned to $A$ itself and the proper subsets of $A$. It follows that Bel$(A)$ also lies between 0 and 1 inclusive, that Bel$(\emptyset) = 0$ and Bel$(\Theta) = 1$ and that the basic probability assignment is uniquely determined by Bel. Those subsets of $\Theta$ assigned positive basic probability numbers are called the *focal elements* of Bel.

In this theory, the body of data causing your belief to be as it is, is not represented explicitly but only implicitly in the Bel function: Bel$(A)$ is allegedly the total support given by the evidence to $A$. Since Bel is not required to obey the probability axioms, there is no systematic relation between Bel$(A)$ and Bel$(A^*)$, where $A^*$ is the complement of $A$ with respect to $\Theta$ (note that it is possible for $A$ and its complement both to be assigned belief 0). But there is nevertheless a relation between $A$ and $A^*$ in the Dempster-Shafer theory, expressed in the function $\mathrm{Pl}(A) = 1 - \mathrm{Bel}(A^*)$. $\mathrm{Pl}(A)$ is the *plausibility* of $A$, and it is intended to represent the degree to which the evidence *fails* to support $A^*$. Since Bel determines and is determined by the basic probability assignment, it follows that so also is Pl, and expressing Pl in terms of that basic probability assignment, it is easy to see that $\mathrm{Bel}(A) \leq \mathrm{Pl}(A)$.

The pair $[\mathrm{Bel}(A),\mathrm{Pl}(A)]$ is called a *belief interval* for $A$, and when (and only when) Bel = Pl, then it is easy to show that Bel is a probability function. The analogy with lower and upper probabilities seems very powerful, and indeed formally Bel and Pl are lower and upper probabilities determined by a closed convex set $\Pi$ of probability functions; i.e., $\mathrm{Bel}(A)$ is the minimum of the values of $P(A)$ for every $P$ in $\Pi$, while $\mathrm{Pl}(A)$ is the maximum of $P(A)$ for every $P$ in $\Pi$ (Kyburg, 1987). The converse is not true, however: it is not the case that every such family of probability functions determines a belief function.

The analogy breaks down more seriously when it comes to the method for updating on additional evidence employed in the respective theories. In the DS theory, this is achieved by Dempster's rule for combining the belief functions representing the distinct bodies of evidence; this may give a different overall belief function than that obtained by updating Bel by conditionalisation in a frame in which some subset $B$ (representing the conditioning statement) becomes established with certainty. Dempster's rule states that if $m_1$ and $m_2$ are the basic probability assignments associated with two belief functions $\mathrm{Bel}_1$ and $\mathrm{Bel}_2$ over the same frame and generated by independent bodies of evidence, then the combined $m(A)$ is the orthogonal sum $\Sigma m_1(C)m_2(D)$ of products of basic probability numbers $m_1(C),m_2(D)$ of all pairs $C,D$ of subsets of $\Theta$ whose intersection is equal to $A$. However, if there are two sets $X$ and $Y$ such that $X \cap Y = \emptyset$ and $m_1(X),m_2(Y) > 0$, then $m$ would not be well-defined by this scheme, since $m(\emptyset)$ would be non-zero. These pairs must be omitted from the sum, which is then normalised by dividing by $1 - \Sigma\, m_1(X)m_2(Y)$ for every pair $X,Y$ whose intersection is empty. The combined belief function $\mathrm{Bel}_1 \oplus \mathrm{Bel}_2$ is defined in the usual way from $m$. Note that it will not be defined when there is some $A$ such that $\mathrm{Bel}_1(A) = 1$ and $\mathrm{Bel}_2(A^*) = 1$; this would mean that $\mathrm{Bel}_1$ contradicts $\mathrm{Bel}_2$, and intuitively two such functions cannot be consistently welded into a new one.

$\text{Bel}_1 \oplus \text{Bel}_2$ is associative and commutative. When for some $B$, $m_2(B) = 1$, so that $B$ and every proposition implied by it is rendered certain by the new evidence, then $\text{Bel}_1$ and $\text{Bel}_2$ are combinable if and only if $\text{Bel}_1(B^*) < 1$, and the combined function $\text{Bel}(A \mid B) = \text{Bel}_1 \oplus \text{Bel}_2(A)$ is such that

$$\textbf{(1)}\ \text{Bel}(A \mid B) = \frac{\text{Bel}_1(A \cup B^*) - \text{Bel}_1(B^*)}{1 - \text{Bel}_1(B^*)}$$

(1) is called *Dempster's Rule of Conditioning*. It easily follows that $\text{Pl}(A \mid B) = \dfrac{\text{Pl}_1(A \cap B)}{\text{Pl}_1(B).}$ Also, if $\text{Bel}_1$ is a probability function, then so is $\text{Bel}(. \mid B)$ and (1) reduces to ordinary Bayesian conditionalisation: $\text{Bel}(A \mid B) = \dfrac{\text{Bel}_1(A \cap B)}{\text{Bel}_1(B).}$ The contrast with lower and upper probabilities becomes apparent in the inequality

$$\min_{P \in \Pi} P(A \mid B) \leq \text{Bel}(A \mid B) \leq \text{Pl}(A \mid B) \leq \max_{P \in \Pi} P(A \mid B),$$

where $\Pi$ is the family of probability functions determined by $\text{Bel}(. \mid B)$. Kyburg (1987) constructs a simple example where the outer inequalities are strict and $\text{Bel}(A \mid B) = \text{Pl}(A \mid B)$, though $\text{Bel}_1(A) \neq \text{Pl}_1(A)$.

### k.2 What Are Belief Functions?

That is enough by way of exposition. Bel and Pl cannot, as the results of the previous paragraph demonstrate, easily be interpreted as lower and upper probabilities. How, then, is the formal apparatus of the DS theory to be interpreted? Shafer has, of course, already provided an answer of sorts to this question: $\text{Bel}(A)$ is a measure of the degree to which a contextually implied body of data is evidence in favour of $A$, and Dempster's Rule of Combination then allegedly tells us how evidence from distinct sources combines to support each proposition in the frame of discernment.

The trouble with this answer is that it does not, as it stands, provide us with any means of assessing how adequately Bel and Dempster's rule fulfil these roles. Why should an adequate measure of evidence, and the belief it evokes, satisfy the conditions imposed by Shafer? In logicians' terminology, we have a syntax but as yet very little in the way of a semantics for it. The contrast with the Bayesian theory in this respect could not be more marked, for in the latter, support is measured explicitly in terms of changes in consistent degrees of belief induced by clearly specified evidence, and it is *demonstrable* that consistent measures of belief have the formal structure of probabilities. You may not want support to be defined in this way, but at least you know why, if it is, it must have the formal structure it has; for that structure is then unambiguously determined.

Shafer has, however, attempted to provide a more definite key to understanding his theory in terms of what he calls a set of *canonical examples*. These are randomly coded messages. You receive a coded message, and you know that the codes are chosen at random from a specified set, where the chance of the $i$th code being chosen is $p_i$ (Shafer, 1981a). The decoded messages are all of the form 'The true hypothesis is in $A$' for some $A$, where $A$ is a subset of some frame of discernment, so with probability $p_i$ the true message is 'The truth is in $A_i$'. Let $m(A)$ be the total chance that the message was 'The truth is in $A$', i.e., $m(A)$ is the sum of all $p_i$ such that $A_i = A$. Bel$(A)$ is therefore the chance that the message *implies* that the truth is in $A$. We are also asked to assume that the true message is itself true and that the coded message is our only evidence, so that Bel$(A)$ measures our degree of belief that the truth is in $A$. According to Shafer, "Our task, when we assess evidence using belief functions, is to choose values of $m(A)$ that make the canonical 'coded-message' example most like that evidence" (1981a, p. 6).

Now suppose that two messages are transmitted in such a way that the code selected for the first is probabilistically independent of that selected for the second. If $\text{Bel}_1(C)$ and $\text{Bel}_2(D)$ correspond to the chances that the first message implies that the truth is in $C$ and that the second implies that the truth is in $D$, then $\text{Bel}_1 \oplus \text{Bel}_2(A)$ turns out to give the chance that the two messages jointly imply that the truth is in $A$ (the multiplication of the $m$-values in the orthogonal sum results from the codes being chosen independently).

It is no doubt a picturesque idea that bodies of observational data are randomly encrypted messages sent by God or

Nature, but in the late twentieth century it hardly provides a convincing justification for the syntax of belief functions. If this is all that can be mustered to support its adoption, then it is hardly enough to warrant the considerable degree of acceptance that the theory has received by workers in artificial intelligence. In a variant semantics recently advanced by Pearl (1990), the randomly encoded messages are replaced by randomly selected theories, and Bel$(A)$ is equated with the total chance that $A$ is a consequence of the theory selected. No direct significance is given to the basic probability assignments. But this account is really no advance in plausibility on Shafer's own (which theories are in the population of theories? who put them there? who or what arranged the selection mechanism?).

These scenarios, we submit, are not to be taken seriously. Nor do we believe they are by those who nevertheless take the DS syntax seriously. They do so because they accept at their face value the objections brought by Shafer against the only worked out and plausible alternative, the Bayesian theory, and which his own theory is expressly designed to avoid. We shall show that these objections are either based on desiderata for a theory of evidence that in fact no adequate theory should embody, or else they arise from a misunderstanding of what the Bayesian theory actually says.

Let us deal with the second point first. Just as he adduces 'canonical examples' designed to make his own theory intelligible, Shafer adduces them also for the Bayesian theory, and then points out that they are inappropriate for a theory of belief based on evidence. The 'canonical examples' for the Bayesian theory, according to Shafer, are states of affairs chosen by a chance mechanism (1981). It is these chance mechanisms that allegedly support the probability distributions over the hypothesis-evidence spaces. Now while it is true that any probability distribution *can* be modelled by a suitable arrangement of random drawings from urns (for example), or limiting cases based on these, it does not follow that they must be, and in the personalist Bayesian theory they certainly are not. At the risk of stating the truth too often, we repeat that the probability distributions over hypotheses are degree-of-belief distributions and nothing else.

Shafer does nevertheless recognise this fact but maintains that it represents a retreat "from the idea that the prior

Bayesian belief function is based on any particular evidence" (1976, pp. 26–27). It is not entirely clear what this is intended to mean. If it means that Bayesian priors may reflect very diverse bodies of accumulated evidence, then it is true also of Shafer's own belief functions. The only retreat that we can think of which may be what Shafer is referring to, is the abandonment of any attempt to represent current epistemic probability distributions as posterior distributions obtained by conditioning successively on each piece of evidence thrown up during the agent's lifetime, relative to an original prior distribution purporting to represent total ignorance.

That such a programme existed is true: it was Carnap's, as we know. It was abandoned, and rightly, because the choice of any distribution prior to all experience would of necessity be quite arbitrary, as also would be the choice of language within which to represent all the diverse 'deliverances of experience'. But the personalist theory has never professed such grandiose and unattainable ambitions. It is merely a theory of inference from prior to posterior distributions, as we keep stressing, and it is unfair, to put it mildly, to charge it with retreating from a position it never occupied and never wanted to occupy. What Shafer fails to point out is how close his own theory actually is in this respect to the personalist theory. Both incorporate prior distributions which each theory represents simply as given, and both incorporate the effect of new evidence *given* those priors. They just do it in different ways.

### k.3 Representing Ignorance

The mention of *ignorance* brings us to the most influential of Shafer's objections, and it is an objection to *any* probabilistic attempt to represent uncertain reasoning. The objection is that no such theory can adequately represent ignorance between alternatives. This charge appears to carry weight, in view of the well-known difficulties with the Principle of Indifference, difficulties on which Shafer himself dwells. He points out that the natural way to represent total ignorance is to assign null belief to each of the alternatives stated, and that in his theory, this is possible because there are belief functions that assign the value 0 not only to each atom of the frame but also even to *every* proper subset. In a probabilistic theory ignorance can, it seems, be expressed only by a uniform distribution of probabilities over atoms. Shafer sees two problems with this. (i) It might be that an atom in one frame is composite in another; e.g., a counter may be classified as blue or non-blue, or it may be classified as blue, green, red, etc. Yet a uniform distribution over one classification leads to a skewed distribution over the other, as we know. (ii) You are not allowed to be ignorant about $\sim h$ whatever your state of belief about *h;* for the probability of $\sim h$ is fixed once you have determined that of *h*.

Our answer to these objections is very simple: it is that the notion of pure ignorance *is not well-defined,* and it is a virtue, not a vice, of a probabilistic theory that it brings this fact out very clearly. Such a theory shows you quite explicitly that you cannot be *uniformly unopinionated* between all possible sets of alternatives, as we saw in Chapter 4, and this is surely intuitively correct. If you have equal degrees of belief in each of the numbers from 0 to 10 being called, then obviously you cannot, or at any rate certainly should not, have equal degrees of belief in the propositions '0 will be called' and 'A non-zero number will be called'. But in Shafer's theory you can, and that is why Shafer's theory is *false* as a theory of reasonable belief.

That an increase in belief for *h* is automatically a decrease in belief for its negation is also, we believe, a very basic intuition; it is, of course, an immediate consequence of the probability axioms. That Shafer's denial of it, and the embodiment of that denial in his theory, should be taken as a reason for rejecting the Bayesian theory rather than his own we find perplexing. In our opinion, the difficulties in giving any coherent interpretation of the syntax of the Dempster-Shafer theory merely reflect the fact that no coherent interpretation exists. The theory is radically at odds with sound reasoning, and all of Shafer's attempts to convict the Bayesian theory of that offence merely have the unintended consequence of convicting his own.

## ■ I EVALUATING PROBABILITIES WITH IMPRECISE INFORMATION

It is often said that one of the factors standing in the way of a widespread use of the Bayesian theory is that the component probabilities—the likelihoods and especially the priors—of a Bayes's Theorem calculation are often not readily computable, because the data are too vague, or too numerous and diverse, or all these things together. This is a common complaint of workers in artificial intelligence, among others, who want a neatly packaged set of algorithms to apply, and who consequently tend to favour classical techniques (which are now frequently produced in the form of easily usable software packages), or even to invent their own (like the MYCIN and EMYCIN calculi of certainty factors).

Much in fact has been done to render Bayesian techniques more algorithmic. Bayesian software is now produced, there are Bayesian expert systems, and the recent and very promising representation of probabilistic relationships in so-called Bayesian networks is a specifically computer-oriented development of the Bayesian formalism (Pearl, 1988, contains a comprehensive account). But citing this work should not divert attention away from the fact that the Bayesian theory is not set up primarily as a source of algorithms but as a general logic of consistent belief. It is no criticism of this function that there is a vast array of inferential problems for which no algorithmic solution exists and where fallible personal judgment will consequently play a major role.

Indeed, most problems of uncertain inference—whether a witness's testimony is reliable, what the weather will do tomorrow, how the disintegration of the Soviet Union will affect world politics in the next ten years, and so on—are of a characteristically diffuse kind. Even those which social scientists frequently have to deal with often involve no precisely characterised statistical or other mathematical model. Shafer tells us that applied statisticians who have tried to apply Bayesian method to problems in which no such model can be assumed have even found more difficulty in evaluating likelihoods than the notorious priors (1981b). In the circumstances stated, this is not at all surprising, but we do not infer from this that the Bayesian methodology is inapplicable. *On the contrary: the Bayesian methodology, being a general theory of uncertainty, is always applicable.* Indeed, it is indispensable.

It is surprising how readily a complaint about the world—the data it supplies are often complex and difficult to interpret—is turned into a complaint against a theory which acknowledges its vagaries and into praise for one which does not. For example, that classical techniques appear to offer easily computable solutions where the Bayesian theory does not (because of its dependence on priors) is frequently cited as not merely a practical but a *theoretical* disadvantage of the latter. The fact is, however, as we have often emphasised, that classical procedures can do this only because they systematically ignore relevant information that the Bayesian conscientiously attempts to represent.

And the way such information is represented will be a matter of personal judgment (though expert systems have been developed whose prior probabilities are based on the pooled opinions of many specialists). To want some universal 'inference engine' into which you can feed data and which will duly output probabilities is a natural enough desire. Carnap, among others, tried to build one. His attempt failed, and if in this uncertain world there is one thing we can be certain of—and we can be certain precisely because the world is uncertain—it is that all attempts must fail. The complaint that the Bayesian leaves too much to individual judgment and not enough to formula is therefore fundamentally misconceived.

## ■ m ARE WE CALIBRATED?

We have already mentioned more than once the concept of a calibrated probability measure. The basic idea is simple. Suppose that $S$ is an indefinitely continued sequence of propositions each describing the occurrence of a single event. For example, $S$ might be a sequence of weather predictions for successive days. Let $P$ be a subjective probability measure defined on a domain including the members of $S$. Select the subsequence $S_r$ of those members of $S$ assigned probability $r$ conditional on all information gathered by the owner of $P$ up to that point. $P$ is *calibrated for* $S$ if, for all $r$, the relative frequency of truths in $S_r$ converges to $r$.

The calibration criterion is generalised in Dawid's classic paper (1982) to the condition that relative frequencies in any rule-selected subsequence of $S$ converge to the average conditional probability of the propositions selected. *S* is now explicitly an infinite sequence, and Dawid, using results from the theory of martingales (a martingale is a mathematical representation of a fair game), derives the remarkable and surprising theorem that with *P*-probability one, *P* is calibrated for all sequences *S*.

Calibration is frequently held up as a criterion of how well your probability function matches or reflects empirical reality, and a deep-seated worry of many subjectivists is that the subjectivist theory clearly provides no guarantee of any such matching. Scoring-rules, of the type we discussed in Chapter 5, section **c.2,** are often suggested as a method of inducing better calibration by imposing penalties for miscalibration; indeed, they have been used with considerable success.

But Dawid's theorem tells us that we should be certain *a priori* that we are calibrated—when we know perfectly well that we may be very badly miscalibrated indeed. In other words, it seems to be true both (i) that we should recognise the very real possibility, if not likelihood, of miscalibration, and (ii) that we should nevertheless be certain that we shall be calibrated. Here there seems to be at best a paradox, at worst an outright contradiction, as Dawid himself pointed out in his paper. Can it be resolved?

The answer is—fortunately—that it can. As Joseph B. Kadane pointed out in the discussion of Dawid's paper (1982, p. 610), Dawid's theorem is a theorem about the properties of a conditional probability function *in the limit where complete information exists* (this is a condition of the theorem's validity). Thus it is really hardly surprising to learn that you ought to believe that your probabilities, conditional on the evidence to hand, will *eventually* converge to the empirical relative frequencies—although it remains of course logically possible that they will not, a fact allowed for in the convergence only occurring 'with *P*-probability one'.

Where the members of *S* are of the form '$X_i = 1$', where the $X_i$ are a sequence of independent, identically distributed random variables, we already know that, given suitable regularity conditions, with probability one the relative frequencies will converge to the posterior distribution; i.e., the probability function is calibrated for such sequences. This is just a consequence of the Strong Law of Large Numbers, and Dawid's theorem can therefore be regarded as an interesting extension of that result.

## n RELIABLE INDUCTIVE METHODS

Hume has a famous argument (1739, Book I, Part III) which appears to show that inductive arguments will inevitably at some point presuppose what they set out to establish; they will be circular, in other words. As we saw in Chapter 4, section **c,** Carnap's $\lambda$ calculus affords a striking vindication of Hume's thesis.

Hume's thesis has now achieved the status of conventional wisdom, though attempts are periodically made to circumvent his arguments. One of the more recent, due to Reichenbach (1949), exploits the possibility that an inductive method may generate hypotheses which demonstrably converge to the truth in the limit as data accumulate, *whatever those data might be.* The notion of an inductive method, or 'truth detecting paradigm', as it has been called, which robustly identifies the truth in the limit as the evidence accumulates, was taken up by Putnam, and more recently developed by Osherson, Glymour, Weinstein, Kelly, and others, and made the foundation of a new discipline, called *Formal Learning Theory.* One of the conclusions they draw from their researches is that the Bayesian theory can be formally demonstrated to be suboptimal in the class of all inductive methods. We shall examine this claim, but to do so means that we shall have to examine briefly first the basic ideas and theses of Formal Learning Theory itself.

The precise characterisation of an inductive method within formal learning theory varies from author to author, but the differences seem to be inessential. We can define (roughly following Kelly, 1990) such a method to be any function $F$ which associates with a given hypothesis and a finite data sequence a truth-value. $F$ is reliable, or successful, for a specified class $K$ (usually required to be countable) of structures and hypothesis $h$ if, for each $w$ in $K$, $F$ stabilises on the truth-value of $h$ in $w$ when given enough data about $w$.

Clearly, for any given $K$ and $h$ one of the first questions to be asked is whether such a function exists. The question can be extended by considering additional constraints. A natural one is that the hypothesis conjectured true at any stage be consistent with the data. Another it might seem natural to impose, and which we shall consider later at some length, is that the method be effective, i.e. performable by a suitably idealised computer equivalent in power to a Turing machine.

Among the many interesting and deep results which have been obtained (see Glymour's survey article, 1991, for some of them), are those which appear to suggest that the Bayesian theory determines an inductive method which is demonstrably suboptimal. Whether this is so or not turns on two issues: (i) whether the Bayesian theory determines an inductive method in this sense at all; and (ii) whether, even if it could be made to do so in some acceptable manner, the condition of its being effective is a desirable one.

As to (i), Bayesianism does not, on the face of it, prescribe a procedure for outputting the truth-values of hypotheses in response to data inputs. It is a great advance, as we see it, that the Bayesian theory does not have to resort to these crude and misleading qualitative categories. However, formally speaking we could take the range of the function $F$ to be the set $\{0,1\}$ rather than the set {true, false}, and say that $F$ successfully identifies the truth just in case any hypothesis to which $F$ eventually assigns only the value 1 is in fact true. The Bayesian theory can then be taken to determine an inductive method by setting $F(h) = 1$ if a suitable probabilistic condition is satisfied; e.g. if $P(h) > \frac{1}{2}$, or even if $h$ is just the mode of the posterior distribution.

Granted some such definition, it can be shown that if any method at all reliably identifies the truth over all possible states of affairs left open by some specified background information, then there is a Bayesian method which does (Juhl, 1993, Theorem 2.1). But this happy situation appears to change when it is required that the Bayesian method be computable. Putnam had earlier shown by a diagonalisation argument that, given that the probability function $P$ was Turing-computable, then it is possible to construct a hypothesis $h$ such that *if h is true* then its so-called instance confirmation (the probability that the next observation will yield an instance of $h$, conditional on the evidence up to that point) will never rise above one half (Putnam, 1975). So it looks as if the answer to the question whether there is always a reliable *computable* Bayesian learner is likely to be negative. Osherson, Stob, and Weinstein (1988), and, using a simpler argument Juhl (op. cit.), argue that it is, and the latter shows that there is always a reliable computable non-Bayesian learner.

How damaging to the Bayesian theory are results like these? In our opinion, not at all. In the first place, it is not clear exactly how a computable Bayesian should be characterised. Juhl's, and Osherson, Stob, and Weinstein's, negative results require that a computable Bayesian be one which has a computable procedure for selecting the mode of the posterior distribution. However, it is difficult to see how this is possible unless at the very least the probability function itself be computable. Yet this is an exceedingly, and unrealistically, strong condition to impose. We can easily see why by noting that no computable probability function can be strictly positive, in the sense of assigning the unconditional probability 0 only to contradictions. For a computable, strictly positive function would immediately provide a decision procedure for logical truth (truth in all interpretations of the extralogical items in the sentence), and a celebrated theorem of Church (1936) shows this to be impossible.

Computable probability functions are a very artificial and restricted class (In Gaifman and Snir's definability ordering for probability functions on a first order language (1980, §3), the computable functions are, in terms of what they ascribe probability 0 to, the most dogmatic). To restrict the Bayesians to these is rather like restricting mathematicians to arithmetical methods only in their proofs. Nor is this the only reason for disputing the inference to the suboptimality of Bayesian procedures from claims about what 'computable Bayesians' can't do. For it turns out that computable non-Bayesian procedures succeed where 'computable Bayesian' ones do not because the former allow hypotheses to be proposed at certain stages of data acquisition which are actually inconsistent with those data. This sort of liberalism is hardly a feature on which to base a claim to superiority.

Nothing in the preceding discussion, however, should be taken to depreciate the new and powerful discipline of formal learning theory, or the intrinsic interest and depth of its results. One in particular, due to Kelly, provides additional information about a well-known Bayesian convergence-of-opinion theorem. This theorem (Halmos, 1950, p. 215, Theorem B) says that if the hypothesis $h$ is that the infinite sequence $w$, whose initial segment $s(w)$ is being observed, is in a set $H$ of sequences, then $P(h|s(w))$ tends to 1 if $w$ is in $H$ and 0 if not, except on at most a class $C$ of sequences of a priori probability 0. Kelly's result (1990, Theorem 2.2) implies that $H$ occurs at a certain rather low level in the Borel hierarchy relativised to the complement of $C$ (the level of a set in this hierarchy indicates the degree of quantificational complexity involved in defining it in terms of a basis consisting of those subsets of $C$ determined by specifying only finitely many sequence coordinates).

## ■ o FINALE

One of the reasons why one expects deductive reasoning to exercise a more-or-less widely felt and obeyed constraint on the way people reason is because it is truth-preserving. Probabilistic reasoning also possesses a characteristic which authorises it to exercise no less a regulatory function: its rules, as we observed in Chapter 5, are broken on pain of committing inconsistency. It is, we suggest, for this reason that divergence from the norm set by the probability calculus is so widely regarded as deviant. Certainly, ever since people chose to express their uncertainty in terms of the odds they thought fair, they have felt themselves explicitly constrained by the axioms of the probability calculus, and while it was not until this century that it was explicitly proved that obedience to the calculus is a necessary condition for fairness, there can be little doubt that that result was taken for granted.

The discovery of the probability calculus, together with the usual formula connecting (fair) odds and probabilities, can now be seen to be part of the great scientific renaissance of the seventeenth century. The probability calculus became the foundation of a mathematical theory of uncertainty, of enormous potential scope and power, which simultaneously generated a quantitative logic of inductive inference and bound together the new mathematical concept of probability with another developed at about the same time, utility, to produce a theory of rational action. The mathematicians of the eighteenth century, and to a lesser extent the nineteenth, divided their time between developing the new physics and extending the probability calculus and the theory of inductive inference and rational decision based on it: among these pioneers, Huyghens, James and Daniel Bernoulli, Laplace, and Poisson stand out as pre-eminent.

On the way, however, paradoxes began to appear in the programme, mostly connected with the Principle of Indifference but also—as a criterion of rational action—with the principle of expected utility. These problems, especially those within the theory of probability itself, seemed at one time, in the early years of this century, so intractable that many people, like Fisher and Popper (as we have seen), wrote off the account of probability on which the programme was based. But they were wrong: in the middle years of this century, shortly after Fisher and Popper penned their obituaries, secure foundations were finally laid. Ramsey first, and then von Neumann and Morgenstern, put utility theory on a consistent basis, and Ramsey and de Finetti realised that an adequate theory of epistemic probability can dispense with pseudo-objective principles like that of Indifference without giving up its claim to impose quite objective standards of consistency in reasoning involving such probabilities. The probabilities might be personal, but the constraints imposed on them by the condition of consistency are certainly not—a distinction still not widely grasped even today, and whose failure to be appreciated continues to vitiate so much contemporary discussion.

We have written this book in an attempt to convince believers in 'objective' standards in science that there is nothing subjective in the Bayesian theory *as a theory of inference:* its canons of inductive reasoning are quite impartial and objective. We want this simple truth to be more widely appreciated. We maintain also that this is the *only* theory which is adequate to the task of placing inductive inference on a sound foundation, and we believe that we have demonstrated this fact in this and the previous chapters.

Why, though, we have been asked, is there nothing in this book about *rational decisions?* If we now have in the personalist Bayesian theory a satisfactory foundation of epistemic probability, as we claim, and if the theory of utility is also now set on a satisfactory basis, why not include an exposition of what, admittedly, is a highly successful and fertile discipline, that of Bayesian decision theory—especially, it is often argued, since any adequate account of theory-choice must include reference to some theory or other of rational decision-making?

We choose to stop at this point because we are far from convinced that decision theory does have any useful, let alone crucial, role to play in a theory of *inference,* which is what we have been expounding in this book. In fact, we believe that it does not. The intellectual tools one needs to assess such things as empirical support, weight of evidence, and probability itself, have nothing to do with making decisions, in any but a purely trivial sense. In our view, decision theory presupposes these tools, not they it. And while Bayesian decision theory is a worthwhile and rewarding study, it is not one which we are concerned with here. If we have presented an acceptable account of scientific inference, we are satisfied enough.

## EXERCISES

1. Show that $P(e^n) = P(e_n \mid e^{n-1})\text{x} \ldots xP(e_2 \mid e_1)xP(e_1)$.
2. Show that if $h \vdash e_i$ for every $i$ and $P(h) > 0$, then $P(e_n \mid e^{n-1})$ tends to 1.
3. A sequence of functions $f_n(x)$ is said to converge uniformly to $g(x)$ over a subset $X$ of its domain if for any positive real number $r$, however small, there is a number $m$ such that for all $x$ in $X$ and all $n > m$ the absolute value of the difference between $f_n(x)$ and $g(x)$ is less than $r$. Define $c_{ki}$ as in section **c** and suppose $h$ entails $e_i$ for all $i$. Show that if $f_n(k) = P(c_{kn} \mid c_{k(n-1)})$ converges uniformly to the constant function $g(k) = 1$ over the set $\{k : k \geq 1\}$, then there is a number $m$ such that $P(e_m \mid e_m - 1) > \frac{1}{2}$ and $P(\sim e_m \mid e_{m-1}) > \frac{1}{2}$.
4. Show that $P(\sim e \vee h \mid e) - P(\sim e \vee h) < 0$ if $P(h \mid e) < 1$.

5. Show that all the consequences of $\sim e \lor h$ are consequences of $\sim e$.

6. Suppose $h \vdash e$. Show

   (i) that $\sim e \lor h$ shares no non-tautologous consequences with $e$
   (ii) that $\sim e \lor h$ is the strongest statement which conjoined with $e$ is equivalent to $h$.

7. Let $Ct(a) = 1 - P(a)$. Show that $ct(e \lor h) + ct(\sim e \lor h) = ct(h)$. Hence show that when $h$ entails $e$, $ct(e) + ct(\sim e \lor h) = ct(h)$.

8. Define $S(h,e) = P(h \mid e) - P(h)$. Show

   (i) that $S(h,e) = S(\sim e \lor h,e) + S(e \lor h,e)$
   (ii) that $S(h,e) = S(\sim e \;\&\; h,e) + S(e \;\&\; h,e)$.

9. Let $Cn(a)$ be the set of consequences of $a$, and define $Cn(a).Cn(b) = Cn(a \lor b)$ and $Cn(a) + Cn(b) = Cn(a \;\&\; b)$ and $Cn(a)' = Cn(\sim a)$. Define a mapping $Cn(a)^* = M(a)^c$, where $M(a)^c$ is the set-theoretic complement of $M(a)$ with respect to $M(t)$, $t$ a tautology. Show that * is an isomorphism, i.e., that $[Cn(a)']^* = Cn(a)^{*c}$, $[Cn(a).Cn(b)]^* = Cn(a)^* \cap Cn(b)^*$, and $[Cn(a) + Cn(b)]^* = Cn(a)^* \cup Cn(b)^*$.

10. Where $h$, $X$, and $P_m(h)$ and $E_m(X)$ are defined as in section **f**, show that $P_m(h) = E_m(X)$, $m \leq 1000$.

11. Show from the stated properties of basic probability assignments that $0 \leq \text{Bel} \leq 1$.

12. A *simple support function* f assigns the value $s$ to every set in a frame of discernment $\Theta$ which includes some specified set $A$, 1 to $\Theta$, and 0 to every other set in the frame, where $0 \leq s \leq 1$. Show that $f$ is a belief function, with basic probability numbers $m(A) = s$, $m(\Theta) = 1 - s$, and $m(B) = 0$ for every other subset $B$ of $\Theta$.

13. Show how Bel determines a unique basic probability assignment.

14. Where Pl(A) is defined as in section **k.1**, show that $\text{Pl}(A|B) = \dfrac{\text{Pl}(A \cap B)}{\text{Pl}(B)}$.

**15.** Show that every probability function defined on the set of all subsets of $\Theta$ is a belief function. Exhibit a frame $\Theta$ and a basic probability assignment to the subsets of $\Theta$ whose associated belief function is not a probability function.

**16.** Show that a necessary and sufficient condition for a belief function to be a probability function is that $\text{Bel}(A) = 1 - \text{Bel}(A^*)$ for every $A$.

# BIBLIOGRAPHY


Altman, D. G.; Gore, S. M.; Gardner, M. J.; and Pocock, S. J. 1983. 'Statistical Guidelines for Contributors to Medical Journals', *British Medical Journal,* vol. 286, 1489–493.

Andersen, B. 1990. *Methodological Errors in Medical Research.* Oxford: Blackwell.

Anscombe, F. J. 1963. 'Sequential Medical Trials', *Journal of the American Statistical Association,* vol. 58, 365–83.

Armendt, B. 1980. 'Is There a Dutch Book Argument for Probability Kinematics?', *Philosophy of Science,* vol. 47, 583–89.

Armitage, P. 1975. *Sequential Medical Trials.* 2nd edition. Oxford: Blackwell.

Atkinson, A. C. 1985. *Plots, Transformations and Regression.* Oxford: Clarendon Press.

Atkinson, A. C. 1986. 'Comment: Aspects of Diagnostic Regression Analysis', *Statistical Science,* vol. 1, 397–402.

Babbage, C. 1827. 'Notice respecting some Errors common to many Tables of Logarithms', *Memoirs of the Astronomical Society,* vol. 3, 65–67.

Bacon, F. 1620. *Novum Organum.* In *The Works of Francis Bacon,* vol. 4, edited by J. Spedding, R. L. Ellis, and D. D. Heath, 1857–58. London: Longman and Company.

Barnett, V. 1973. *Comparative Statistical Inference.* Chichester: Wiley.

Bayes, T. 1763. 'An essay towards solving a problem in the doctrine of chances', *Philosophical Transactions of the Royal Society,* vol. 53, 370–418. Reprinted with a biographical note by G. A. Barnard in *Biometrika,* 1958, vol. 45, 293–315.

Belsey, D. A.; Kuh, E.; and Welsch, R. E. 1980. *Regression Diagnostics: Identifying Influential Data and Sources of Collinearity.* New York: John Wiley and Sons, Inc.

Benveniste, J. et al. 1988. *Nature,* vol. 333, 816.

Bernoulli, J. 1713. *Ars Conjectandi.* Basiliae.

Berry, D. A. 1989. 'Ethics and ECMO', *Statistical Science,* vol. 4, 306–10.

Berry, D., and Berger, J. O. 1988. 'Statistical Analysis and the Illusion of Objectivity', *American Scientist,* vol. 76, 159–165.

Blackwell, D., and Dublns, L. 1962. 'Merging of Opinions with Increasing Information', *Annals of Mathematical Statistics,* vol. 33, 882–87.

Bland, M. 1987. *An Introduction to Medical Statistics*. Oxford: Oxford University Press.

Bolzano, B. 1837. *Wissenschaftstheorie* (English translation by Rolf George published under the title *Theory of Science*, Blackwell, 1972).

Bourke, G. J.; Daly, L. E.; and McGilvray, J. 1985. *Interpretation and Uses of Medical Statistics*. 3rd edition. St. Louis: Mosby.

Bowden, B. V. 1953. 'A Brief History of Computation', in *Faster than Thought*, edited by B. V. Bowden. London: Pitman Publishing.

Brandt, R. 1986. 'Comment' on Chatterjee and Hadi (1986), *Statistical Science*, vol. 1, 405–7.

British Medical Association Panel on Collagen Diseases and Hypersensitivity. 1957. *British Medical Journal*, 611.

Broemeling, L. D. 1985. *Bayesian Analysis of Linear Models*. New York: Marcel Dekker, Inc.

Brook, R. J. and Arnold, G. C. 1985. *Applied Regression Analysis and Experimental Design*. New York: Marcel Dekker, Inc.

Byar, D. P.; Simon, R. M.; Friedewald, W. T.; Schlesselman, J. J.; DeMets, D. L.; Ellenberg, J. H.; Gail, M. H.; and Ware, J. H. 1976. 'Randomized Clinical Trials', *The New England Journal of Medicine*, 74–80.

Byar, D. P. et al. (22 co-authors). 1990. 'Design Considerations for AIDS Trials', *New England Journal of Medicine*, vol. 323, 1343–48.

Carnap, R. 1947. 'On the Applications of Inductive Logic', *Philosophy and Phenomenological Research*, vol. 8, 133–48.

______. 1950. *Logical Foundations of Probability*. Chicago: University of Chicago Press.

______. 1952. *The Continuum of Inductive Methods*. Chicago: University of Chicago Press.

______. 1968. 'Inductive Logic and Inductive Intuition', *The Problem of Inductive Logic*. ed. I. Lakatos. Amsterdam: North Holland.

Carnap, R., and Jeffrey, R., eds. 1971. *Studies in Inductive Logic and Probability*. Berkeley: University of California Press.

Chatterjee, S., and Hadi, A. S. 1986. 'Influential Observations, High Leverage Points, and Outliers in Linear Regression', *Statistical Science*, vol. 1, 379–416.

Chatterjee, S., and Price, B. 1977. *Regression Analysis by Example*. New York: John Wiley and Sons.

Christensen, D. 1991. 'Clever Bookies and Coherent Beliefs', *The Philosophical Review*, vol. C, no. 2, 229–47.

Church, A. 1936. 'A Note on the Entscheidungsproblem', *Journal of Symbolic Logic*, vol. 1, 40–41, 101–2.

______. 1940. 'On the Concept of a Random Sequence', *Bulletin of the American Mathematical Society*, vol. 46, 130–35.

Cochran, W. G. 1952. 'The $\chi^2$ Test of Goodness of Fit', *Annals of Mathematical Statistics*, vol. 23, 315–45.

______. 1954. 'Some Methods for Strengthening the Common $\chi^2$ Tests', *Biometrics*, vol. 10, 417–51.

Colton, T. 1974. *Statistics in Medicine*. Boston: Little, Brown.

Cook, R. D. 1986. 'Comment' on Chatterjee and Hadi (1986), *Statistical Science*, vol. 1, 393–97.

Cosmides, L., and Tooby, J. 1992. 'Are Humans Good Intuitive Statisticians After All? Rethinking Some Conclusions from the Literature on Judgment under Uncertainty', *ZIF Research Group Biological Foundations of Human Culture,* Report 3/92. Bielefeld.

Cournot, A. A. 1843. *Exposition de la Théorie des Chances et des Probabilités*. Paris.

Cox, D. R. 1985. 'Notes on Some Aspects of Regression Analysis', *Journal of the Royal Statistical Society,* vol. 131A, 265–79.

Cox, R. T. 1961. *The Algebra of Probable Inference*. Baltimore: The Johns Hopkins Press.

Cramér, H. 1946. *Mathematical Methods of Statistics*. Princeton: Princeton University Press.

Cranberg, L. 1979. 'Do Retrospective Controls Make Clinical Trials "inherently fallacious"?', *British Medical Journal,* 1265–66.

Daniel, C., and Wood, F. S. 1980. *Fitting Equations to Data*. New York: John Wiley and Sons, Inc.

Daniel, W. W. 1991. *Biostatistics: A Foundation for Analysis in the Health Sciences*. 5th edition. New York: John Wiley and Sons.

Darwin, C. 1868. *The Variation of Animals and Plants under Domestication*. 2 vols. London: John Murray.

Dawid, A. P. 1982. 'The Well-Calibrated Bayesian', *Journal of the American Statistical Association,* vol. 77, 605–13.

Dawid, A. P.; Stone, M.; and Zidek, J. V. 1973. 'Marginalization Paradoxes in Bayesian and Structural Inference', *Journal of the Royal Statistical Society,* B, 189–223.

de Finetti, B. 1974. *Theory of Probability,* vol. 1. New York: John Wiley and Sons, Inc.

Dempster, A. P. 1968. 'Upper and Lower Probabilities Induced by a Multivalued Mapping', *Annals of Mathematical Statistics,* vol. 38, 325–39.

Descartes, R. 1637. *Discourse on Method, Optics, Geometry and Methodology,* edited by P. J. Olscamp, 1965. Indianapolis: Bobbs-Merrill, Inc.

Diaconis, P., and Zabell, S. L. 1982. 'Updating Subjective Probability', *Journal of the American Statistical Association,* vol. 77, 822–30.

Dobzhansky, T. 1967. 'Looking Back at Mendel's Discovery', *Science,* vol. 156, 1588–89.

Dorling, J. 1979. 'Bayesian Personalism, the Methodology of Research Programmes, and Duhem's Problem', *Studies in History and Philosophy of Science,* vol. 10, 177–87.

———. 1982. 'Further Illustrations of the Bayesian Solution of Duhem's Problem'. Unpublished.

Downham, V., ed. 1973. *Issues in Political Opinion Polling*. Chichester: John Wiley and Sons, Inc.

Duhem, P. 1905. *The Aim and Structure of Physical Theory* (translated by P. P. Wiener, 1954). Princeton: Princeton University Press.

Dunn, J. M., and Hellman, G. 1986. 'Dualling: A Critique of an Argument of Popper and Miller', *British Journal for the Philosophy of Science,* vol. 37, 220–23.

Earman, J. 1986. *A Primer on Determinism*. Dordrecht: Reidel.

———. 1992, *Bayes or Bust? A Critical Examination of Bayesian Confirmation Theory*. Cambridge, Mass.: MIT Press.
Edgeworth, F. Y. 1895. 'On some Recent Contributions to the Theory of Statistics', *Journal of the Royal Statistical Society,* vol. 58, 505–15.
Edwards, A. L. 1984. *An Introduction to Linear Regression and Correlation*. 2nd edition. New York: W. H. Freeman and Company.
Edwards, A. W. F. 1972. *Likelihood*. Cambridge: Cambridge University Press.
———. 1986. 'Are Mendel's Results Really Too Close?', *Biological Reviews of the Cambridge Philosophical Society,* vol. 61, 295–312.
Edwards, W. 1968. 'Conservatism in Human Information Processing', in *Formal Representation of Human Judgment,* B. Kleinmuntz, ed., 17–52.
Edwards, W., Lindman, H., and Savage, L. J. 1963. 'Bayesian Statistical Inference for Psychological Research', *Psychological Review,* vol. 70, 193–242.
Ehrenberg, A. S. C. 1975. *Data Reduction: Analysing and Interpreting Statistical Data*. London: John Wiley and Sons, Inc.
Ellenberg, S. 1981. 'Studies to Compare Treatment Regimens', *Journal of the American Medical Association,* vol. 246, 2481–482.
Elston, R. C., and Johnson, W. D., 1987. *Essentials of Biostatistics*. Philadelphia: F.A. Davis Company.
FDA. 1988. *Guideline for the Format and Content of the Clinical and Statistical Sections of New Drug Applications*. Center for Drug Evaluation and Research, Food and Drug Administration, Rockville, Md.
Feiblman, J. K. 1972. *Scientific Method*. The Hague: Martinus Nijhoff.
Feller, W. 1950. *Introduction to Probability Theory and its Applications,* vol. 1. New York: John Wiley and Sons, Inc.
Feyerabend, P. 1975. *Against Method*. London: New Left Books.
Fine, T. L. 1973. *Theories of Probability*. New York: Academic Press.
Finetti, B. de. 1937. 'La prévision; ses lois logiques, ses sources subjectives', *Annales de l'Institut Henri Poincaré,* vol. 7, 1–68. (Reprinted in 1964 in English translation as 'Foresight: its Logical Laws, its Subjective Sources', in *Studies in Subjective Probability,* edited by H. E. Kyburg, Jr., and H. E. Smokler.) New York: John Wiley and Sons, Inc.
———. 1979. *Theory of Probability*. New York: John Wiley and Sons, Inc.
Fisher, R. A. 1922. 'On the Mathematical Foundations of Theoretical Statistics', *Philosophical Transactions of the Royal Society of London,* vol. A222, 309–368.
———. 1935. 'Statistical Tests', *Nature,* vol. 136, 474.
———. 1936. 'Has Mendel's Work Been Rediscovered?', *Annals of Science,* vol. 1, 115–137.
———. 1947. *The Design of Experiments*. 4th edition. Edinburgh: Oliver and Boyd. (First published 1926.)
———. 1956. *Statistical Methods and Statistical Inference*. Edinburgh: Oliver and Boyd.
———. 1970. *Statistical Methods for Research Workers*. 14th edition. Edinburgh: Oliver and Boyd. (First published 1925.)
Franklin, A., and Howson, C. 1984. 'Why do scientists prefer to vary their

experiments?', *Studies in History and Philosophy of Science,* vol. 15, 51–62.
Gaifman, H., and Snir, M. 1980. 'Probabilities over Rich languages, Testing and Randomness', *Journal of Symbolic Logic,* 47, 495–548.
Garber, D. 1983. 'Old Evidence and Logical Omniscience in Bayesian Confirmation Theory', in *Testing Scientific Theories,* edited by J. Earman. Minneapolis: University of Minnesota Press, 99–131.
Gärdenfors, P. 1988. *Knowledge in Flux.* Cambridge, Mass.: MIT Press.
Gärdenfors, P., and Sahlin, N. E. 1988, *Decision, Probability and Utility.* Cambridge: Cambridge University Press.
Gardner, M. J., and Altman, D. G., eds. 1989. *Statistics With Confidence.* London: British Medical Journal.
Giere, R. N. 1973. 'Objective Single Case Probabilities and the Foundations of Statistics', in P. Suppes et al., eds. *Logic, Methodology and Philosophy of Science IV.* North Holland: Amsterdam, 467–483.
———. 1976. 'A Laplacean Formal Semantics for Single Case Propensities', *Journal of Philosophical Logic,* vol. 5, 321–53.
———. 1984. *Understanding Scientific Reasoning.* 2nd edition. New York, London: Holt, Rinehart and Winston.
Gigerenzer, G. 1991. 'How to Make Cognitive Illusions Disappear: Beyond Heuristics and Biases.' *European Review of Social Psychology,* vol. 3, 83–115.
Gillies, D. A. 1973. *An Objective Theory of Probability.* London: Methuen.
———. 1989. 'Non-Bayesian Confirmation Theory and the Principle of Explanatory Surplus', *Philosophy of Science Association 1988,* edited by A. Fine and J. Leplin, vol 2, Pittsburgh: Pittsburgh University Press, 373–81.
———. 1990. 'Bayesianism versus Falsificationism', *Ratio,* vol. 3, 82–98.
Glymour, C. 1980. *Theory and Evidence.* Princeton, New Jersey: Princeton University Press.
———. 1991. 'The Hierarchies of Knowledge and the Mathematics of Discovery', *Minds and Machines,* 1, 75–95.
Good, I. J. 1950. *Probability and the Weighing of Evidence.* London: Griffin.
———. 1961. 'The Paradox of Confirmation', *British Journal for the Philosophy of Science,* vol. 11, 63–64.
———. 1962. 'Subjective Probability as the Measure of a Nonmeasurable Set', in *Logic, Methodology and Philosophy of Science, Proceedings of the 1960 International Congress,* edited by E. Nagel, P. Suppes, and A. Tarski. Stanford: Stanford University Press.
———. 1965. *The Estimation of Probabilities.* Cambridge, Mass: MIT Press.
———. 1969. 'Discussion of Bruno de Finetti's Paper "Initial Probabilities: A Prerequisite for any Valid Induction"', *Synthese,* vol. 20, 17–24.
———. 1977. 'Dynamic Probability, Computer Chess, and the Measurement of Knowledge', *Machine Intelligence 8,* edited by E. W. Elcock and D. Michie, 139–50. New York: John Wiley and Sons, Inc.
———. 1981. 'Some Logic and History of Hypothesis Testing', in *Philo-*

*sophical Foundations of Economics,* edited by J. C. Pitt. Dordrecht: Reidel.
———. 1983. 'Some History of the Hierarchical Bayes Methodology', *Good Thinking.* Minneapolis: University of Minnesota Press, 95–105.
Goodman, N. 1954. *Fact, Fiction and Forecast.* London: The Athlone Press.
Gore, S. M. 1981. 'Assessing Clinical Trials—Why Randomize?', *British Medical Journal,* vol. 282, 1958–960.
Grünbaum, A. 1976. 'Is the Method of Bold Conjectures and Attempted Refutations *Justifiably* the Method of Science', *British Journal for the Philosophy of Science,* vol. 27, 105–36.
Gumbel, E. J. 1952. 'On the Reliability of the Classical Chi-Square Test', *Annals of Mathematical Statistics,* vol. 23, 253–63.
Gunst, R. F., and Mason, R. C. 1980. *Regression Analysis and its Application.* New York: Marcel Dekker, Inc.
Hacking, I. 1965. *Logic of Statistical Inference.* Cambridge: Cambridge University Press.
———. 1967. 'Slightly More Realistic Personal Probability', *Philosophy of Science,* vol. 34, 311–25.
Haldane, J. B. S. 1945. 'On a Method of Estimating Frequencies', *Biometrika,* vol. 33, 222–25.
Halmos, P. 1950. *Measure Theory.* New York: Van Nostrand Reinhold.
Harnett, D. L. 1970. *Introduction to Statistical Methods.* Reading, Mass.: Addison-Wesley Publishing Company.
Hays, W. L. 1969. *Statistics.* London: Holt, Rinehart and Winston. (First published 1963.)
Hays, W. L., and Winkler, R. L. 1970. *Statistics: Probability, Inference and Decision,* vol. 1. New York: Holt, Rinehart and Winston, Inc.
Hempel, C. G. 1945. 'Studies in the Logic of Confirmation', *Mind,* vol. 54, 1–26 and 97–121. Reprinted in Hempel, 1965.
———. 1965. *Aspects of Scientific Explanation.* New York: The Free Press.
———. 1966. *Philosophy of Natural Science.* Englewood Cliffs, N.J.: Prentice-Hall.
Henkel, R. E. 1976. *Tests of Significance.* Beverley Hills, California: Sage Publications.
Hesse, M. 1974. *The Structure of Scientific Inference.* Berkeley: University of California Press.
Hintikka, J. 1965. 'Towards a Theory of Inductive Generalisation', in *Logic, Methodology and Philosophy of Science,* edited by Y. Bar Hillel. Amsterdam: North Holland, 274–88.
———. 1966. 'A Two-Dimensional Continuum of Inductive Methods' in *Aspects of Inductive Logic,* edited by J. Hintikka and P. Suppes. Amsterdam: North Holland, 113–32.
———. 1968. 'Induction by Enumeration and Induction by Elimination', in *Problems in the Philosophy of Science,* edited by I. Lakatos. Amsterdam: North Holland, 191–216.
Hobson, A. 1971. *Concepts in Statistical Mechanics.* New York: Gordon and Breach.
Hodges, J. L., Jr., and Lehmann, E. L. 1970. *Basic Concepts of Probability and Statistics.* 2nd edition. San Francisco: Holden-Day.

Horn, A., and Tarski, A. 1948, 'Measures in Boolean Algebras', *Transactions of the American Mathematical Society,* vol. 64, 467–97.

Horwich, P. 1982. *Probability and Evidence.* Cambridge: Cambridge University Press.

Howson, C. 1973. 'Must the Logical Probability of Laws be Zero?', *British Journal for the Philosophy of Science,* vol. 24, 153–63.

______. 1984. 'Bayesianism and Support by Novel Facts', *British Journal for the Philosophy of Science,* vol. 35, 245–51.

______. 1987. 'Popper, Prior Probabilities and Inductive Inference', *British Journal for the Philosophy of Science,* vol. 38, 207–24.

______. 1988. 'On the Consistency of Jeffreys's Simplicity Postulate, and its Role in Bayesian Inference', *The Philosophical Quarterly,* vol. 38, 68–83.

______. ed. 1976. *Method and Appraisal in the Physical Sciences.* Cambridge: Cambridge University Press.

Howson, C., and Franklin, A. 1986. 'A Bayesian Analysis of Excess Content and the Localisation of Support', *British Journal for the Philosophy of Science,* vol. 36, 425–31.

______. 1991. 'Maher, Mendeleev and Bayesianism', *Philosophy of Science,* vol. 58, 574–85.

Howson, C., and Oddie, G. 1979. 'Miller's so-called Paradox of Information', *British Journal for the Philosophy of Science,* vol. 30, 253–61.

Hume, D. 1739. *A Treatise of Human Nature,* Books 1 and 2. Fontana Library.

______. 1777. *An Inquiry Concerning Human Understanding,* edited by L. A. Selby-Bigge. Oxford: The Clarendon Press.

Jaynes, E. T. 1968. 'Prior Probabilities', *Institute of Electrical and Electronic Engineers Transactions on Systems Science and Cybernetics,* SSC-4, 227–241.

______. 1983. *Papers on Probability, Statistics and Statistical Physics,* edited by R. Rosenkrantz. Dordrecht: Reidel.

______. 1985. 'Some Random Observations', *Synthese,* vol. 63, 115–38.

Jeffrey, R. C. 1970. 'Review of Eight Discussion Notes', *Journal of Symbolic Logic,* vol. 35, 124–27.

______. 1983. *The Logic of Decision.* 2nd edition. Chicago and London: University of Chicago Press.

Jeffreys, H. 1961. *Theory of Probability.* 3rd edition. Oxford: The Clarendon Press.

Jeffreys, H., and Wrinch, D. 1921. 'On Certain Fundamental Principles of Scientific Enquiry', *Philosophical Magazine,* vol. 42, 269–98.

Jennison, C., and Turnbull, B. W. 1990. 'Statistical Approaches to Interim Monitoring: A Review and Commentary', *Statistical Science,* vol. 5, 299–317.

Jevons, W. S. 1874. *The Principles of Science.* London: Macmillan and Co.

Juhl, C. 1993. 'Bayesianism and Reliable Scientific Inquiry'.

Kadane, J. B. 1986. 'Progress Toward a More Ethical Method for Clinical Trials', *Journal of Medicine and Philosophy,* vol. 2, 385–404.

Kadane, J. et al. 1980. 'Interactive Elicitation of Opinion for a Normal

Linear Model', *Journal of the American Statistical Association,* vol. 75, 845–54.
Kahneman, D., and Tversky, A. 1972. 'Subjective Probability: a Judgment of Representativeness', in *Cognitive Psychology,* vol. 3, 430–54.
Kant, I. 1783. *Prolegomena to any Future Metaphysics,* edited by L. W. Beck, 1950. Indianapolis: The Bobbs-Merrill Company, Inc.
Kelly, K. 1990. *General Characteristics of Inductive Inference over Arbitrary Data Sets.* Report CMU-PHIL-13, Carnegie Mellon University.
Kempthorne, O. 1966. 'Some Aspects of Experimental Inference', *Journal of the American Statistical Association,* vol. 61, 11–34.
———. 1971. 'Probability, Statistics and the Knowledge Business', in *Foundations of Statistical Inference,* edited by V. P. Godambe and D. A. Sprott. Toronto: Holt, Rinehart and Winston of Canada.
———. 1979. *The Design and Analysis of Experiments.* Huntington, N.Y.: Robert E. Krieger.
Kendall, M. G., and Stuart, A. 1979. *The Advanced Theory of Statistics,* vol. 2. 4th edition. London: Charles Griffin and Company Limited.
———. 1983. *The Advanced Theory of Statistics,* vol. 3. 4th edition. London: Charles Griffin and Company Limited.
Keynes, J. M. 1921. *A Treatise on Probability.* London: Macmillan.
Kitcher, P. 1985. *Vaulting Ambition.* Cambridge, Mass: MIT Press.
Kolmogorov, A. N. 1950. *Foundations of the Theory of Probability* (Translated from the German of 1933 by N. Morrison.) New York: Chelsea Publishing Company. Page references are to the 1950 edition.
———. 1965. 'Three Approaches to the Quantitative Definition of Information', *Problemy Peredacii Informacii,* vol. 1, 4–7.
Koopman, B. O. 1940. 'The Bases of Probability', *Bulletin of the American Mathematical Society,* vol. 46, 763–74.
Kries, J. von 1886. *Die Principien der Wahrscheinlichkeitsrechnung. Eine logische Untersuchung.* Freiburg.
Kuhn, T. S. 1970. *The Structure of Scientific Revolutions.* 2nd edition. Chicago: University of Chicago Press. (First published 1962.)
Kuipers, T. A. F. 1978. *Studies in Inductive Probability and Rational Expectation.* Dordrecht: Reidel.
———. 1980. 'A Survey of Inductive Systems', *Studies in Inductive Logic and Probability,* vol. II, edited by R. C. Jeffrey. Berkeley: University of California Press.
Kyburg, H. E., Jr. 1974. *The Logical Foundations of Statistical Inference.* Dordrecht and Boston: D. Reidel Publishing Company.
———. 1983. *Epistemology and Inference.* Minneapolis: University of Minnesota Press.
———. 1987. 'Bayesian versus non-Bayesian Evidential Updating', *Artificial Intelligence,* 31, 271–93.
Kyburg, H. E., Jr., and Smokler, E., eds. 1980. *Studies in Subjective Probability.* Huntington, N.Y.: Robert E. Krieger.
Lakatos, I. 1968. 'Criticism and the Methodology of Scientific Research Programmes', *Proceedings of the Aristotelian Society,* vol. 69, 149–86.

______. 1970. 'Falsification and the Methodology of Scientific Research Programmes', in *Criticism and the Growth of Knowledge,* edited by I. Lakatos and A. Musgrave. Cambridge: Cambridge University Press.
______. 1974. 'Popper on Demarcation and Induction', in *The Philosophy of Karl Popper,* vol. 2, ch. 5, edited by P. A. Schilpp. La Salle, Illinois: Open Court.
______. 1978. *Philosophical Papers.* 2 vols, edited by J. Worrall and G. Currie. Cambridge: Cambridge University Press.
Laplace, P. S. de. 1820. *Essai Philosophique sur les Probabilités.* Page references are to *Philosophical Essay on Probabilities,* 1951. New York: Dover Publications.
Levi, I. 1967. *Gambling with Truth.* New York: Knopf.
______. 1980. *The Enterprise of Knowledge.* Cambridge, Mass.: MIT Press.
Lewis, C. I. 1946. *An Analysis of Knowledge and Valuation.* La Salle, Illinois: Open Court.
Lewis, D. 1973. *Counterfactuals.* Oxford: Basil Blackwell.
______. 1976. 'Probabilities of Conditionals and Conditional Probabilities', *The Philosophical Review,* vol. LXXXV, 297–315.
______. 1981. 'A Subjectivist's Guide to Objective Chance', in *Studies in Inductive Logic and Probability,* edited by R. C. Jeffrey, 263–93. Berkeley and Los Angeles: University of California Press.
Lewis-Beck, M. S. 1980. *Applied Regression.* Beverley Hills, California: Sage Publications.
Lindgren, B. W. 1976. *Statistical Theory.* 3rd edition. New York: Macmillan.
Lindley, D. V. 1956. 'On a Measure of the Information Provided by an Experiment', *Annals of Mathematical Statistics,* vol. 27, 986–1005.
______. 1957. 'A Statistical Paradox', *Biometrika,* vol. 44, 187–92.
______. 1965. *Introduction to Probability and Statistics, from a Bayesian Viewpoint.* 2 vols. Cambridge: Cambridge University Press.
______. 1970. 'Bayesian Analysis in Regression Problems', in *Bayesian Statistics,* edited by D. L. Meyer and R. O. Collier. Itasca, Illinois: F. E. Peacock.
______. 1971. *Bayesian Statistics, a Review.* Philadelphia: Society for Industrial and Applied Mathematics.
______. 1982. 'Scoring Rules and the Inevitability of Probability', *International Statistical Review,* vol. 50, 1–26.
______. 1982a. 'The Role of Randomization in Inference', *Philosophy of Science Association,* vol. 2, 431–46.
______. 1985, *Making Decisions.* 2nd edition. London: John Wiley and Sons, Inc.
Lindley, D. V., and El-Sayyad, G. M. 1968. 'The Bayesian Estimation of a Linear Functional Relationship', *Journal of the Royal Statistical Society,* vol. 30B, 190–202.
Lindley, D. V., and Phillips, L. D. 1976. 'Inference for a Bernoulli Process (a Bayesian View)', *The American Statistician,* vol. 30, 112–19.
Lucas, J. R. 1970. *The Concept of Probability.* Oxford: Oxford University Press.
McDonald, C. 1991. Letter. *British Medical Journal,* vol. 302, 1271.

McIntyre, I. M. C. 1991. 'Tribulations for Clinical Trials', *British Medical Journal,* vol. 302, 1099–1100.

Mackie, J. L. 1963. 'The paradox of confirmation', *British Journal for the Philosophy of Science,* vol. 38, 265–277.

Maddox, J.; Randi, J.; and Stewart, W. W. 1988. '"High-Dilution" Experiments a Delusion', *Nature,* vol. 334, 278–90.

Maher, P. 1988, 'Prediction, Accommodation and the Logic of Discovery', *Philosophy of Science Association 1988,* edited by A. Fine and J. Leplin, vol. 1, 273–86.

———. 1990. 'Why Scientists Gather Evidence', *British Journal for the Philosophy of Science,* vol. 41, 103–19.

Makinson, D. 1985. 'How to Give It Up: a Survey of Some Formal Aspects of the Logic of Theory-Change', *Synthese,* vol. 62, 347–63.

Mallet, J. W. 1880. 'Revision of the Atomic Weight of Aluminum', *Philosophical Transactions,* vol. 171, 1003–1035.

Mann, H. B., and Wald, A. 1942. 'On the Choice of the Number of Intervals in the Application of the Chi-Square Test', *Annals of Mathematical Statistics,* vol. 13, 306–17.

Medawar, P. 1974. 'More Unequal than Others', *New Statesman,* vol. 87, 50–51.

Meier, P. 1975. 'Statistics and Medical Experimentation', *Biometrics,* vol. 31, 511–29.

Mellor, D. H. 1971. *The Matter of Chance.* Cambridge: Cambridge University Press.

Mendelson, E. 1964. *Introduction to Mathematical Logic.* New York: Van Nostrand Reinhold Company.

Miller, D. 1991. 'On the Maximization of Expected Futility', PPE Lectures, Lecture 8. Department of Economics: University of Vienna.

Miller, D. W. 1966. 'A paradox of information', *British Journal for the Philosophy of Science,* vol. 17, 59–61.

Miller, R. 1987. *Bare-faced Messiah.* London: Michael Joseph.

Milne, P. M. 1983. 'A Note on Scale Invariance', *British Journal for the Philosophy of Science,* 34, 49–55.

———. 1987. 'Physical Probabilities', *Synthese,* vol. 73, 329–59.

———. 1989. 'Coherence, Cancellation and Conditionalisation' (unpublished ms.).

———. 1993. 'The Foundations of Probability and Quantum Mechanics', *Journal of Philosophical Logic,* vol. 22, 129–68.

Mises, R. von. 1939. *Probability, Statistics and Truth.* Originally published in German in 1928. First English edition, prepared by H. Geiringer. London: George Allen & Unwin.

———. 1957. Second English edition, revised, of *Probability, Statistics and Truth.*

———. 1964. *Mathematical Theory of Probability and Statistics.* New York: Academic Press.

———. 1963–64. *Selected Papers,* edited by Ph. Frank, S. Goldstein, M. Kac, 2 vols. Providence, Rhode Island: American Mathematical Society.

Mood, A. M. 1950. *Introduction to the Theory of Statistics*. New York: McGraw-Hill Book Company, Inc.
Mood, A. M., and Graybill, F. A. 1963. *Introduction to the Theory of Statistics*. New York: McGraw-Hill Book Company, Inc.
Musgrave, A. 1975. 'Popper and "Diminishing Returns from Repeated Tests"', *Australasian Journal of Philosophy,* vol. 53, 248–53.
Neyman, J. 1935. 'On the Two Different Aspects of the Representative Method: the Method of Stratified Sampling and the Method of Purposive Selection', reprinted in Neyman, 1967, 98–141.
______. 1937. 'Outline of a Theory of Statistical Estimation Based on the Classical Theory of Probability', *Philosophical Transactions of the Royal Society,* vol. 236A, 333–80.
______. 1941. 'Fiducial Argument and the Theory of Confidence Intervals', *Biometrika,* vol. 32, 128–150. (Page references are to the reprint in Neyman, 1967.)
______. 1952. *Lectures and Conferences on Mathematical Statistics and Probability*. 2nd edition. Washington: U.S. Department of Agriculture.
______. 1967. *A Selection of Early Statistical Papers of J. Neyman*. Cambridge: Cambridge University Press.
Neyman, J., and Pearson, E. S. 1928. 'On the Use and the Interpretation of Certain Test Criteria for Purposes of Statistical Inference', *Biometrika,* vol. 20, 175–240, part I; 263–94, part II.
______. 1933. 'On the Problem of the Most Efficient Tests of Statistical Hypotheses', *Philosophical Transactions of the Royal Society,* vol. 231A, 289–337. (Page references are to the reprint in Neyman and Pearson's *Joint Statistical Papers,* 1967. Cambridge: Cambridge University Press.)
Niiniluoto, I. 1983. 'Novel Facts and Bayesianism', *British Journal for the Philosophy of Science,* vol. 34, 375–79.
Pearl, J. 1988, *Probabilistic Reasoning in Intelligent Systems*. San Mateo: Morgan Kaufmann Publishers.
______. 1990. 'Reasoning with Belief-Functions: an Analysis of Compatibility', *International Journal of Approximate Reasoning,* vol. 4, 363–89.
Osherson, D. N.; Stob, M.; and Weinstein, S. 1988. 'Mechanical Learners Pay a Price for Bayesianism', *Journal of Symbolic Logic,* 1245–51.
Pearson, E. S. 1966. 'Some Thoughts on Statistical Inference', in *The Selected Papers of E. S. Pearson,* 276–83. Cambridge: Cambridge University Press.
Pearson, K. 1892. *The Grammar of Science*. (Page references are to the edition of 1937, London: J.M. Dent & Sons Ltd.)
Peto, R. 1978. 'Clinical Trial Methodology', in *Proceedings of the International Meeting on Comparative Therapeutic Trials*. Paris: Springer International.
Peto, R. et al. 1988. 'Randomised Trial of Prophylactic Daily Aspirin in British Male Doctors', *British Medical Journal,* vol. 296, 313–31.
Phillips, L. D. 1973. *Bayesian Statistics for Social Scientists*. London: Nelson.
______. 1983. 'A Theoretical Perspective on Heuristics and Biases in Prob-

abilistic Thinking', in *Analysing and Aiding Decision,* edited by P. C. Humphreys, O. Svenson, and A. Vari. Amsterdam: North Holland.

Pocock, S. 1983. *Clinical Trials: A Practical Approach.* Chichester: Wiley.

Poincaré, H. 1905, *Science and Hypothesis,* translated by W. J. G. (Page references are to the edition of 1952, New York: Dover Publications.)

Polanyi, M. 1962. *Personal Knowledge.* 2nd edition. London: Routledge and Kegan Paul.

Pollard, W. 1985, *Bayesian Statistics for Evaluation Research: An Introduction.* Beverly Hills: Sage Publications.

Polya, G. 1954. *Mathematics and Plausible Reasoning,* vols. 1 and 2. Princeton: Princeton University Press.

Popper, K. R. 1959. 'The Propensity Interpretation of Probability', *British Journal for the Philosophy of Science,* vol. 10, 25–42.

———. 1959a. *The Logic of Scientific Discovery.* London: Hutchinson.

———. 1960. *The Poverty of Historicism.* London: Routledge and Kegan Paul.

———. 1963. *Conjectures and Refutations.* London: Routledge and Kegan Paul.

———. 1972. *Objective Knowledge.* Oxford: Oxford University Press.

———. 1983. 'A proof of the impossibility of inductive probability', *Nature,* vol. 302, 687–88.

Popper, K. R., and Miller, D. W. 1987. 'Why Probabilistic Support is not Inductive', *Philosophical Transactions of the Royal Society of London,* A 321, 569–91.

Pratt, J. W. 1962. In Birnbaum, A. 'On the Foundations of Statistical Inference', *Journal of the American Statistical Association,* vol. 57, 269–326.

———. 1965. 'Bayesian Interpretation of Standard Inference Statements', *Journal of the Royal Statistical Society,* 27B, 169–203.

Pratt, J. W.; Raiffa, H.; and Schlaifer, R. 1965. *Introduction to Statistical Decision Theory.*

Prout, W. 1815. 'On the Relation Between the Specific Gravities of Bodies in Their Gaseous State and the Weights of Their Atoms', *Annals of Philosophy,* vol. 6, 321–30. (Reprinted in *Alembic Club Reprints,* No. 20, 1932, 25–37. Edinburgh: Oliver and Boyd.)

Putnam, H. 1975. 'Probability and Confirmation', reprinted in *Mathematics, Matter and Method (Philosophical Papers,* vol. 1). Cambridge: Cambridge University Press.

Ramsey, F. P. 1931. 'Truth and Probability', *The Foundations of Mathematics and Other Logical Essays.* London: Routledge and Kegan Paul.

Redhead, M. 1974. 'On Neyman's Paradox and the Theory of Statistical Tests', *British Journal for the Philosophy of Science,* vol. 25, 265–71.

———. 1985. 'On the Impossibility of Inductive Probability', *British Journal for the Philosophy of Science,* vol. 36, 185–91.

———. 1986. 'Novelty and Confirmation', *British Journal for the Philosophy of Science,* vol. 37, 115–18.

Reichenbach, H. 1935. *Wahrscheinlichkeitslehre. Eine Untersuchung uber die logischen und mathematischen Grundlagen der Wahrscheinlich-*

*keitsrechnung*. Leiden. English translation: *The Theory of Probability*. Berkeley: University of California Press, 1949.

———. 1949. *The Theory of Probability*. Berkeley: University of California Press.

Rhoades, D. A. 1986. 'Predicting Earthquakes', in *The Fascination of Statistics,* edited by R. J. Brook, C. A. Arnold, T. H. Hassard and R. M. Pringle, 307–19. New York: Marcel Dekker, Inc.

Rivadulla, A. 1991. 'Probabilistic Support, Probabilistic Induction, and Bayesian Confirmation Theory', *British Journal for the Philosophy of Science*. Forthcoming.

Rosenkrantz, R. D. 1977. *Inference, Method, and Decision: Towards a Bayesian Philosophy of Science*. Dordrecht: Reidel.

Royall, M. R. 1991. 'Ethics and Statistics in Randomized Clinical Trials', *Statistical Science,* vol. 6, 52–88.

Russell, B. 1912. 'On Induction', *The Problems of Philosophy*. (Reprinted, 1973) Oxford: Oxford University Press.

Salmon, W. C. 1981. 'Rational Prediction', *British Journal for the Philosophy of Science,* vol. 32, 115–25.

Savage, L. J. 1954. *The Foundations of Statistics*. New York: John Wiley and Sons, Inc.

———. 1962. 'Subjective Probability and Statistical Practice', in *The Foundations of Statistical Inference,* edited by G. A. Barnard and D. R. Cox 9–35. New York: John Wiley and Sons, Inc.

———. 1962a. A prepared contribution to the discussion of Savage, 1962, 88–89, in the same volume.

Schlaifer, R. 1959. *Probability and Statistics for Business Decisions*. New York: McGraw-Hill.

Schlesinger, G. N. 1991. *The Sweep of Probability*. Notre Dame: University of Notre Dame Press.

Schroeder, L. D.; Sjoquist, D. L.; and Stephan, P. E. 1986. *Understanding Regression Analysis*. California: Sage Publications, Inc.

Schwartz, D.; Flamant, R.; and Lellouch, J. 1980. *Clinical Trials [L'essai therapeutique chez l'homme]*. New York: Academic Press. Translated by M. J. R. Healy.

Seal, H. L. 1967. 'The Historical Development of the Gauss Linear Model', *Biometrika,* vol. 57, 1–24.

Seber, G. A. F. 1977. *Linear Regression Analysis*. New York: John Wiley and Sons, Inc.

Seidenfeld, T. 1979. *Philosophical Problems of Statistical Inference*. Dordrecht: Reidel.

Senn, S. 1991. Review of *Scientific Reasoning: The Bayesian Approach*. *Statistics in Medicine,* vol. 10, 1161–62.

Shafer, G. 1976. *A Mathematical Theory of Evidence*. Princeton: Princeton University Press.

———. 1981a. 'Constructive Probability', *Synthese,* vol. 48, 1–60.

———. 1981b. 'Jeffrey's Rule of Conditioning', *Philosophy of Science,* vol. 48, 337–62.

Shafer, G., and Tversky, A. 1985. 'Languages and Designs for Probability Judgements', *Cognitive Science,* vol. 9.

Shimony, A. 1970. 'Scientific Inference,' in *Pittsburgh Studies in the Philosophy of Science,* vol. 4, edited by R. G. Colodny. Pittsburgh: Pittsburgh University Press.

———. 1985. 'The Status of the Principle of Maximum Entropy', *Synthese,* vol. 63, 35–53.

Shore, J. E. and Johnson, R. W. 1980. 'Axiomatic Derivation of the Principle of Maximum Entropy and the Principle of Minimum Cross-Entropy', *IEEE Transactions on Information Theory,* IT-26, 26–37.

Silvey, S. D. 1970. *Statistical Inference.* Baltimore, Maryland: Penguin Books.

Skyrms, B. 1977. *Choice and Chance.* Belmont: Wadsworth Publishing Company.

———. 1984. *Pragmatics and Empiricism.* New Haven: Yale University Press.

———. 1987. 'Dynamic Coherence and Probability Kinematics', *Philosophy of Science,* vol. 54, 1–20.

Smart, W. M. 1947. 'John Couch Adams and the Discovery of Neptune', *Occasional Notes of the Royal Astronomical Society,* no. 11.

Smith, C. A. B. 1961. 'Consistency in Statistical Inference and Decision', *Journal of the Royal Statistical Society,* Series B, vol. 23, 1–25.

Smith, T. M. 1983. 'On the Validity of Inferences from Non-random Samples', *Journal of the Royal Statistical Society,* vol. 146A, 394–403.

Snedecor, G. W., and Cochran, W. G. 1967. *Statistical Methods.* 6th edition. Ames, Iowa: The Iowa State University Press.

Spielman, S. 1976. 'Exchangeability and the Certainty of Objective Randomness', *Journal of Philosophical Logic,* vol. 5, 399–406.

———. 1977. 'Physical Probability and Bayesian Statistics', *Synthese,* vol. 36, 235–69.

Sprent, P. 1969. *Models in Regression.* London: Methuen and Co. Ltd.

Sprott, W. J. H. 1936. Review of K. Lewin's: *A Dynamical Theory of Personality, Mind,* vol. 45, 246–51.

Stainaker, R. 1975. 'A Theory of Conditionals', *Causation and Conditionals,* ed. E. Sosa. Oxford: Oxford University Press, 165–80.

Stas, J. S. 1860. 'Researches on the Mutual Relations of Atomic Weights', *Bulletin de l'Académie Royale de Belgigue,* 208–336. (Reprinted in part in *Alembic Club Reprints,* No. 20, 1932, 41–47. Edinburgh: Oliver and Boyd.)

Strike, P. W. 1991. *Statistical Methods in Laboratory Medicine.* Oxford: Butterworth Heinemann.

Stuart, A. 1954. 'Too good to be true', *Applied Statistics,* vol. 3, 29–32.

———. 1962. *Basic Ideas of Scientific Sampling.* London: Charles Griffin.

Suppes, P. 1981. *Logique du Probable.* Paris: Flammarion.

Swinburne, R. G. 1971. 'The Paradoxes of Confirmation—a Survey', *American Philosophical Quarterly,* vol. 8, 318–29.

Tarski, A. 1956. *Logic, Semantics, Metamathematics.* Oxford: Clarendon Press.

Teller, P. 1973. 'Conditionalisation and Observation', *Synthese,* vol. 26, 218–58.

Thomson, T. 1818. 'Some Additional Observations on the Weights of the Atoms of Chemical Bodies', *Annals of Philosophy,* vol. 12, 338–50.

Urbach, P. 1981. 'On the Utility of Repeating the "Same" Experiment', *Australasian Journal of Philosophy,* vol. 59, 151–62.

———. 1985. 'Randomization and the Design of Experiments', *Philosophy of Science,* vol. 52, 256–73.

———. 1987. *Francis Bacon's Philosophy of Science.* La Salle, Illinois: Open Court.

———. 1987a. 'Clinical Trial and Random Error', *New Scientist,* vol. 116, 52–55.

———. 1987b. 'The Scientific Standing of Evolutionary Theories of Society', *The LSE Quarterly,* vol. 1, 23–42.

———. 1989. 'Random Sampling and the Principles of Estimation', *Proceedings of the Aristotelian Society,* vol. 89, 143–64.

———. 1991, 'Bayesian Methodology: Some Criticisms Answered', *Ratio (New Series),* vol. 4, 170–84.

———. 1992a. Review of Schlesinger, *The Sweep of Probability,* 1991. *Mind,* vol. 101, 395–97.

———. 1992b. 'Regression Analysis: Classical and Bayesian', *British Journal for the Philosophy of Science,* vol. 43, 311–42.

———. 1993. 'The Value of Randomization and Control in Clinical Trials', *Statistics in Medicine,* vol. 12, 1421–31.

Van Fraassen, B. C. 1980. *The Scientific Image.* Oxford: The Clarendon Press.

———. 1984. 'Belief and the Will', *Journal of Philosophy,* vol. LXXXI, 235–56.

———. 1989. *Laws and Symmetry.* Oxford: Clarendon Press.

Velikovsky, I. 1950. *Worlds in Collision.* London: Victor Gollancz, Ltd. (Page references are to the 1972 edition, published by Sphere Books, Ltd.)

Velleman, P. F. 1986. 'Comment' on Chatterjee, S., and Hadi, A. S. (1986), *Statistical Science,* vol. 1, 412–15.

Velleman, P. F., and Welsch, R. E. 1981. 'Efficient Computing of Regression Diagnostics', *American Statistician,* vol. 35, 234–42.

Venn, J. 1866. *The Logic of Chance.* London: John MacMillan and Co.

Walley, P. 1991. *Statistical Reasoning with Imprecise Probabilities.* London: Chapman and Hall.

Wason, P. C. 1966. 'Reasoning', in *New Horizons in Psychology,* edited by B. M. Foss. Harmondsworth: Penguin.

Wason, P. C. and Johnson-Laird, P. N. 1972. *Psychology of Reasoning.* London: Batsford.

Watkins, J. W. N. 1985. *Science and Scepticism.* London: Hutchinson and Princeton: Princeton University Press.

———. 1987. 'A New View of Scientific Rationality', in *Rational Change in Science,* edited by J. Pitt and M. Pera. Dordrecht: Reidel.

Weisberg, S. 1980. *Applied Linear Regression.* New York: John Wiley and Sons, Inc.

Welsch, R. E. 1986. 'Comment' on Chatterjee, S., and Hadi, A. S. (1986), *Statistical Science,* vol. 1, 403–5.

Whitehead, J. 1983. *The Design and Analysis of Sequential Clinical Trials.* Chichester: Ellis Horwood.

Whitehead, J. 1993. 'The Case for Frequentism in Clinical Trials', *Statistics in Medicine,* vol. 12, 1405–13.

Williams, P. M. 1976. 'Indeterminate Probabilities', in *Formal Methods in the Methodology of Empirical Sciences,* edited by M. Przelecki, K. Szaniawski, and R. Wojcicki. Ossolineum and Reidel, 229–46.

———. 1980. 'Bayesian Conditionalisation and the Principle of Minimum Information', *British Journal for the Philosophy of Science,* vol. 31, 131–44.

Wonnacott, T. H., and Wonnacott, R. J. 1980. *Regression: A Second Course in Statistics.* New York: John Wiley and Sons, Inc.

Worrall, J. 1978. 'The Ways in which the Methodology of Scientific Research Programmes improves on Popper's Methodology', *Progress and Rationality in Science,* eds. Radnitzky, G., and Anderson, G., Dordrecht: Reidel.

———. 1989. 'Fresnel, Poisson, and the White Spot: the Role of Successful Predictions in the Acceptance of Scientific Theories', *The Uses of Argument,* edited by D. Gooding et al., Cambridge: Cambridge University Press.

Yates, F. 1981. *Sampling Methods for Censuses and Surveys.* 4th edition. London: Charles Griffin.

# INDEX

## A



## B

C

D



E

Q



R

S